Flexible Crossroads

Roger Hayter

Flexible Crossroads: The Restructuring of British Columbia's Forest Economy

UBC Press · Vancouver · Toronto

Printed in Canada on acid-free paper ∞

ISBN 0-7748-0775-X (hardcover)
ISBN 0-7748-0776-8 (paperback)

Canadian Cataloguing in Publication Data

Hayter, Roger, 1947-
Flexible crossroads

Includes bibliographical references and index.
ISBN 0-7748-0775-X (bound)
ISBN 0-7748-0776-8 (pbk.)

1. Forests and forestry – Economic aspects – British Columbia. 2. Forest policy – British Columbia. I. Title.
SD146.B7H39 2000 338.1'749'09711 C00-910103-9

This book has been published with the help of a grant from the Humanities and Social Sciences Federation of Canada, using funds provided by the Social Sciences and Humanities Research Council of Canada.

UBC Press acknowledges the financial support of the Government of Canada through the Book Publishing Industry Development Program (BPIDP) for our publishing activities.

Canadä

We also gratefully acknowledge the support of the Canada Council for the Arts for our publishing program, as well as the support of the British Columbia Arts Council.

This book has also been published with the help of a grant from Simon Fraser University's Publication Committee.

Set in Stone by Artegraphica Design Co. Ltd.
Printed and bound in Canada by Friesens
Copy-editor: Maureen Nicholson
Proofreader: Deborah Kerr

UBC Press
University of British Columbia
2029 West Mall, Vancouver, BC V6T 1Z2
(604) 822-5959
Fax: (604) 822-6083
E-mail: info@ubcpress.ubc.ca
www.ubcpress.ubc.ca

For Bucquie

Contents

Preface

For two decades, perhaps longer, British Columbia's forest economy has experienced fundamental restructuring of its competitive foundations and global role. Yet the direction of change remains uncertain and volatile, underlined by the severity of the crisis facing the forest industries in the late 1990s. The metaphor of the crossroads eloquently captures the dilemma of restructuring (Hayter 1987; Hammond 1991; Drushka et al. 1993; Marchak 1991; Binkley 1997a). As many observers emphasize, the crossroads represents a shift in resource dynamics "between forests provided by providence and those created through human husbandry and stewardship" (Binkley 1997a: 39). But BC's forest economy is at a crossroads in another sense, one rooted in industrial dynamics that are global in scope (Hayter and Barnes 1997). In this regard, Bell's (1974) famous study of the postindustrial society constitutes a clarion call for attempts to explain deep-seated ("paradigmatic") changes in patterns of production and consumption across modern societies (Harvey 1990; Castells 1996). In this crossroads, a significant theme in the restructuring of production systems centres on the "imperatives of flexibility" (Freeman 1987). From this perspective, the prosperity of BC's forest economy depends on its flexibility in responding not only to resource dynamics, but also to technological, market, and political changes that are increasingly rapid, intrusive, and global.

The BC forest economy, in other words, finds itself at a "double crossroads," ultimately rooted in resource and industrial dynamics respectively. Inevitably, these dynamics interact (Graham and Martin 1990). If resource conditions define physical limitations and challenges to social action, the latter have modified, with growing impact, the resource for at least 100 years. Moreover, industrial dynamics are themselves fundamentally shaped by their own inheritance. Future directions simply cannot be divorced from past choices, choices reflecting the actions and policies of individuals and organizations in the public and private sectors that have created massive rounds of investment in productive facilities and infrastructure while simultaneously shaping attitudes and values. At the same time, as the crossroads metaphor highlights, new industrial and resource dynamics are demanding change. Indeed, transformation of the BC forest economy is well under way.

As much rancorous debate attests, the crossroads at which the BC forest economy finds itself is not clearly signposted. Rather, the way ahead is highly contested. Within the context of resource dynamics, the meaning of sustained yield, whether old-growth forests are "renewable" following industrial forestry, the "balance" between conservation and use for industrial purposes, and how forest use should be regulated are problematical issues (Kimmins 1992; M'Gonigle and Parfitt 1994; Drushka et al. 1993). Issues related to Aboriginal rights and justice are rapidly taking provincial forest policy into uncharted waters. From the perspective of industrial dynamics, the imperatives of flexibility are profoundly ambiguous, or to put the matter in another way, the meaning of flexibility is itself flexible (Gertler 1988). Two polar views on "flexible production" illustrate without exhausting this ambiguity (Hayter and Barnes 1992). To some observers, flexibility implies leaner, cost-conscious production systems based on worker exploitation and short-term economic considerations. In this view, "leaner" means "meaner" and no private-sector responsibility exists for community and environmental priorities. In the opposite view; flexibility extols innovative firms networking among one another, each organized by participatory decision making structures and multi-skilled high-wage workers who manufacture high-value products in a way that reduces demands on the environment. Moreover, these two models of flexibility are often blurred in the real world and, in practice, the crossroads at which the BC forest economy finds itself offers multiple directions.

This book analyzes the transformation of the BC forest economy in the last quarter of the twentieth century. The orientation of the analysis favours explanations rooted in industrial dynamics so that special attention is paid to the imperatives of flexibility; through this lens, questions related to resource dynamics are explored. More specifically, the book adopts a political economy perspective that recognizes the power of institutions to shape the economic landscape – the location of economic activity, the structure of production and communities, the nature of trade relations, and the pattern of land use. In this view, the chief agent of economic change is the modern corporation, and the implications of corporate structure and strategy for the organization of BC's forest economy is a central theme in this book. But other institutions are powerful and cannot be ignored. In particular, governments determine the regulatory framework in which corporations operate, and in BC where forest land is overwhelmingly under public control, the provincial government has been historically significant in shaping how forests have been exploited. Labour unions and, more recently, Aboriginal and environmental groups are other institutions actively responsible for shaping BC's forest landscape.

It needs to be emphasized that this analysis does not discount the significance of market forces – after all, the BC forest economy is a quintessential capitalist enterprise that has been buffeted by globalization since its inception. But "market forces" themselves are institutional arrangements that are altered by policy as well as by consumer preference. Contemporary export markets for BC lumber, for example, cannot be understood without reference to American protectionist policies or environmental boycotts. Within BC, the struggles among industry, governments, unions, Aboriginal groups, and environmentalists implicitly assume markets can be regulated, albeit in different ways, and that there are fundamental choices to be made regarding the ownership and use of BC's forest resource.

Flexible Crossroads is in two parts. In Part 1, three chapters provide conceptual and historical perspective to understanding the flexible crossroads facing the BC forest economy. Conceptually, the contemporary evolution of BC's forest industries is placed within a wider context of a shift from so-called Fordist principles of production to more "flexible" principles of production, a shift that is occurring, albeit in disputed forms, across key industries throughout North America and beyond (Chapter 1). This discussion is linked to BC and the forest industries through models of resource development, particularly the prism of Innis's staple thesis, which provides a "Canadian" theory of export-led development. Chapter 2 sketches the evolution of BC's forest economy from its origins to the 1970s. Large-scale industrialization of BC's forests began in the late nineteenth century, centring on logging and lumber and organized primarily as an entrepreneurial model. Following the Great Depression of the 1930s and the Second World War, the next thirty years witnessed a remarkable expansion of the forest industry to all parts of the province. During these years, often referred to as the "Fordist long boom," pulp and paper became the key industry, and the entrepreneurial organization of development was replaced by large corporations that form the basis of the model of production known as Fordism.

Indeed, Fordism provides the template for assessing the current phase of transformation towards a more flexible system of production (Chapter 3). In general terms, the shift to flexibility in the BC forest economy has been extremely volatile, featuring strong booms and busts and what also seems to be an endless rethinking of forest policy. In practice, forest policy is caught in a maelstrom. A simple model of the BC forest economy as a production system in transition from Fordism to flexibility concludes Chapter 3.

In Part 2, the book progressively explores the transformation of the BC forest economy in more detail, especially in relation to the strategies and structure of MacMillan Bloedel, foreign ownership, the role of small firms, trade relations, employment and labour relations, forest community development, environmentalism, and technology. The trends are highly diverse, and, inevitably, the shift towards flexibility implies a more differentiated and unruly landscape than that associated with Fordism. In policy terms, from the point of industrial forestry, the central challenge is to establish innovation as the priority that guides forest policy in terms of overall strategy and specific execution.

This book examines the BC forest economy from an industrial perspective, specifically that associated with the subdiscipline of economic geography. Thus, the concept of the "forest economy" centres on the "forest-product industries." As such, the book complements recent investigations of BC's forest economy that have given analytical priority to resource dynamics or at least to resource policy (Kimmins 1992; Nemetz 1992; Drushka et al. 1993; M'Gonigle and Parfitt 1994; Hammond 1993; Binkley 1997a,b; Burda et al. 1998; Wilson 1998), as well as to Barnes and Hayter's (1997) study that provides a spectrum of views in both the resource and the social traditions. Marchak's (1983) landmark study of the political economy of BC's forest industries, written before the end of the recession of the early 1980s, before flexibility had become such an important theme in the theoretical literature and the practical world, and before the onset of controversies over trade, environment, and Aboriginal issues, is an important antecedent of this discussion. Special mention should also be made of Schwindt and Heaps's (1996) parsimonious account of the economics of the BC forest sector. Clearly, much has changed in the BC forest economy since the mid-1980s, and understanding these changes is the objective of this book.

From a disciplinary point of view, *Flexible Crossroads* is an exercise in economic (more specifically, industrial) geography whose mandate is to explain "how place makes a difference to the economic process" (Barnes 1987). In recent years, economic geographers have characterized the global economy as a "mosaic of regions" (Scott and Storper 1986; Storper and Salois 1997), as a "kaleidoscope" of changing patterns (Patchell 1996), or as an integration of "local models" (Barnes 1987). From this perspective, the BC forest economy offers a distinct "local model" in which global industrial transformations, such as those captured by the shift from Fordism to flexibility, are uniquely interpreted by BC's situation on the geographic margin of production and its distinct history and culture that collectively combine to create local attitudes, behaviour, and policy. This study contributes towards an understanding of the industrial anatomy of BC, notably with respect to the influence of the forest economy in shaping BC as a distinct region and its role in the global economy.

On a personal note, this book allows me to integrate the research I have conducted on the BC forest economy over the past twenty-five years on a variety of topics, notably the long-run strategies of corporations (Hayter 1976), locational decision making (Hayter 1978c), labour supply (Hayter 1979; Ofori-Amoah and Hayter 1989), foreign investment (Hayter 1981, 1982a, 1985), technology and innovation (Hayter 1987, 1988), employment restructuring (Hayter 1997b; Grass and Hayter 1989; Barnes and Hayter 1990;

Hayter and Barnes 1992; Hayter, Grass, and Barnes 1994), trade relations, conflicts, and organization (Hayter 1992; Hayter and Barnes 1992; Edgington and Hayter 1997; Hayter and Edgington 1997), forest community development (Barnes and Hayter 1992, 1994; Hayter and Barnes 1997), remanufacturing (Rees and Hayter 1996), and environmentalism (Hayter and Soyez 1996). While much material is drawn from these studies, the book develops new arguments and lines of integration, and contains more recent data.

Acknowledgments

Over the years, many people in the forest industry throughout British Columbia – Vancouver, New Westminster, Burnaby, Chemainus, Port Alberni, Kitimat, Quesnel, Prince George, Mackenzie, Kelowna, Powell River, and elsewhere – have shown unusual warmth, patience, and help in response to my inquiries. I can't possibly thank everybody. However, I would like to mention some of the individuals in Chemainus and Port Alberni who for over twelve years have willingly given their time and knowledge to my Simon Fraser University Geography 426 class on its annual three-day field trip. Neil Malbon, Bill Elwyn, Janet Schlackl, Bill Cafferata, Murray Hess, Ed Melnyck, Tony Sudar, Jim Ritchie, Ray Dyer, Norm Koffler, Karl Schutz, Cecile McKinley, Paul Beltgens, Ray Carrol, Keith Wyton, Darrell Frank, John MacFarlane, and others, have all contributed greatly to the success of this course which, not incidentally, provides me with my best teacher evaluations. In Vancouver, Mike Apsey and Don Lidstone have also greatly contributed to this class. Needless to say, consistent with my Yorkshire upbringing, I have never paid them a cent or offered even a token gift. I hope that they are not too offended by the following pages. My students in Geography 426, whom I have shamelessly used in developing this book, have always been enthusiastic contributors to my thinking.

Flexible Crossroads has been further enriched by my collaborations of over a decade with Trevor Barnes and more recently with David Edgington. I have also greatly benefited as supervisor of an outstanding group of graduate students, several of whom, notably Ben Ofori-Amoah, Eric Grass, Kevin Rees, Tanya Behrisch, Elizabeth Hay, and Tim Reiffenstein have written theses on BC's forest economy; and Jerry Patchell helped in research related to various forest products. At UBC Press, Randy Schmidt, Holly Keller-Brohman, Maureen Nicholson, and Ann Macklem have been most helpful and efficient. Financially, the SSHRC has been an invaluable source of support for the research on which this book is based, and I am most grateful. Geoff Ironside and two referees provided many constructive comments and editorial help that I should have already done. In my university department,

Paul DeGrace has done a superlative job drawing the maps; Ray Squirrel continues to provide much welcome advice, as well as vital supplies of jujubes; and Diane Sherry and friends offer all kinds of office support.

Finally, I would like to dedicate this book to my wife, Bucquie, a continuing source of love, inspiration, and that unique combination of western Canadian tolerance, openness, and frankness. In these respects, Alison, Lynn, and Megan have played their part, too.

Acronyms

ACDS	Alberni-Clayoquot Development Society
AEP	Alberni Enterprise Project
APD	Alberni Pacific Division
AVCIS	Alberni Valley Cottage Industry Society
BCFP	British Columbia Forest Products
BCTMP	bleached chemi-thermo-mechanical pulp mill
BCWSG	British Columbia Wood Specialties Group
BOD	biochemical oxygen demand
CBD	central business district
CEPU	Communications, Energy and Papermakers' Union
CFCLI	US Coalition for Fair Canadian Lumber Imports
CFIC	Canadian Forest Industries Council
CIT	Court of International Trade
Cocel	Columbia Cellulose
CORE	Commission on Resources and Environment
CP&P	Cariboo Pulp and Paper
CPR	Canadian Pacific Railway
CPU	Canadian Papermakers Union
CVAWP	Council of Value-Added Wood Processors
CWC	Communication Workers of Canada
CZC	Crown Zellerbach Canada
DOC	US Department of Commerce
ECW	Energy and Chemical Workers
ETP	Entrepreneur Training Project
FDI	foreign direct investment
FEN	forest ecosystem network
FL	forest licence
FTO	foreign testing organization
GATT	General Agreement on Tariffs and Trade
HBC	Hudson's Bay Company
ICT	information and communication techno-economic paradigm

ILAP	Industry and Labour Adjustment Program
ILRA	Independent Lumber Remanufacturers' Association
ITA	International Trade Administration
ITC	International Trade Commission
IWA	International Woodworkers of America
JAS	Japan Agricultural Standard
JBV	joint business ventures
LEAC	Local Employment Assistance Development Program
LIA	low intensity area
LRMP	land and resource management plans
LRSY	long run-sustained yield
MB	MacMillan Bloedel
MDF	medium-density fibreboard
MNC	multinational corporation
MOU	memorandum of understanding
NAWLA	North American Wholesalers Lumber Association
NDP	New Democratic Party
NGO	nongovernmental organization
NIFM	Northwest Independent Forest Manufacturers
NTB	nontariff barriers
OSB	oriented strandboard
OTT	old temporary tenure
OUW	Organization of Unemployed Workers
PAAC	Port Alberni Adjustment Committee
PFC	platform frame construction
PHA	pulpwood harvesting area
PSYU	public sustained yield unit
R&D	research and development
RAN	Rainforest Action Network
SBFEP	Small Business Forest Enterprise Program
SLA	Softwood Lumber Agreement
SME	small and medium-sized enterprises
TFL	tree farm licence
VEG	visually effected greenup
WCRIC	West Coast Research and Information Cooperative
WTO	World Trade Organization

Part 1
Global and Historical Perspective

1
Global Industrial Transformation, Resource Peripheries, and the Canadian Model

The contemporary restructuring of British Columbia's forest economy is part of a wider, global transformation of economy, society, and industry (Bell 1974; Castells 1996). Beginning in the 1970s, in the industrial heartlands of North America and Europe, the winds of deep-seated economic change were heralded by rapid and unanticipated increases in energy prices, stagflation, and deep recessions. The severity and compounding nature of these shocks to the core industrial regions of the industrially most powerful countries in the world provide reminders that, historically, capitalism and industrialization are crisis-ridden processes, or in Schumpeter's (1943) famous phrase, "creatively destructive."

To help understand these processes of creative destruction, which vary so much in place and time, there has been in recent years considerable contemplation of the idea that industrialization comprises long-term impulses (or waves or cycles) of sustained economic growth that ultimately end in crisis, only for some kind of transformation to then pave the way for a new (and different) phase of long-term growth (Mensch 1979; Mandel 1980; Freeman and Perez 1988). In this context, the global industrial dynamics since the 1980s or so are presented, with much debate and qualification, as a transformation from Fordism to something more "flexible," variously labelled "flexible specialization," "flexible accumulation," "post-Fordism," and the "information and communication techno-economic paradigm" (Best 1990; Freeman and Perez 1988; Harvey 1988; Lipietz 1986; Hirst and Zeitlin 1991; Piore and Sabel 1984). BC is inevitably caught in the maelstrom of global economic change, and the ideas underlying Fordist and flexible production systems provide appropriate templates to examine contemporary restructuring in BC's forest economy.

This chapter establishes the global context for understanding the restructuring of BC's forest industries. This global context, in reality complex and many layered, is summarized in terms of (sets of) models of industrialization that capture significant processes in BC's forest industries. First, the chapter summarizes two related models of (long-run) industrialization, considered broadly as processes of creative destruction, with particular attention to the key features associated with Fordism and flexible systems of production. Second, the chapter reviews models of the development trajectories available to resource-based regions within the industrialization process, especially in relation to forestry-based regions. The last part of the chapter examines resource-based development from a Canadian perspective, specifically the so-called staple thesis.

These three contexts – of global industrial transformation, resource dynamics, and Canadian development – define the broad conceptual framework for examining the role of the forest staple in BC. Each of these contexts provides important, progressively more focused insights and yardsticks of evaluation for understanding the dynamics of BC's forest industries. Moreover, the complex intersection of these contexts counters tendencies towards teleological and deterministic explanations. Indeed, global contexts are themselves not simply implemented, but also shaped by local conditions and actors. In the next chapter, which focuses on the transformation of BC's forest resource, these local contributions are made more explicit.

Industrialization as a Process of Creative Destruction

In the late eighteenth century, beginning in England, the Industrial Revolution, heralded by the factory system and complementary revolutions in agriculture and transportation, witnessed increases in the pace, scale, and nature of economic and population growth. These changes have scarcely abated since, and industrialization has become truly global within the past 100 years. Industrialization's imprints, however, are highly uneven in temporal, sectoral, and geographical terms. As Schumpeter (1943) emphasizes, industrialization since the late eighteenth century has been a process of "creative destruction." On the one hand, industrialization is creative through the development of product and process innovations that enhance the division of labour and productivity, offer new opportunities for investment, consumption, and employment, and generally provide the material basis for population growth and the prospect of improvements in standards of living. Innovation also promises hope for the resolution of environmental pollution. On the other hand, industrialization exploits workers, undermines existing operations and communities, and harms the environment while excessive competition encourages overcapacity and cost cutting. If the wealth of (some) nations has massively increased over the past 200 years, then depressions, regional problems, and disadvantaged groups all testify to the unevenness of the industrialization process.

Historically, the industrialization process is both virtuous and exploitative. A constant stream of innovation ensures that new industries and new industrial spaces are continually created, while old industries and places constantly seek to adjust and diversify. Moreover, industrial evolution is highly competitive, and the life cycles of factories, firms, industries, and communities take on very different forms. Attempts to explain this increasingly complex evolution – and there are many – have frequently identified "stages of development." Two recent and widely discussed models are outlined by Piore and Sabel (1984) and Freeman (1987). Both models are consistent with the idea of industrialization as a process of creative destruction, both stress the importance of institutional as well as technological innovation for long-term economic growth, and both emphasize contemporary fundamental changes in the global economy towards flexibility. For Piore and Sabel, the transition is a "second industrial divide" between Fordist mass production and flexible specialization, and for Freeman a shift from the Fordist to the information and communication techno-economic (ICT) paradigm. They differ in that the latter emphasizes the secular nature of industrialization, whereas the former interprets industrialization as a constant interplay between the force of mass production and the force of flexible specialization.

Shifts in Techno-Economic Paradigms

According to Freeman (1987), industrialization is structured through fifty-year cycles or so-called Kondratiev waves that he labels "techno-economic paradigms" (Table 1.1). Emerging techno-economic paradigms centre on major new technologies that have pervasive effects throughout the economy, creating new forms of production and engineering principles, new organizational arrangements, and government economic and social policies. In technological terms, each paradigm is associated with distinctive key factor industries, main carrier or dominant industries, and supporting infrastructure, plus some newer, rapidly growing industries that typically become the dominant industries of the next paradigm. In this framework, key factor industries represent "new" inputs to production and related activities that are abundant and used extensively throughout the industrial system, cheaply, and without fear of supply bottlenecks. In this model, infrastructure refers primarily to economic overhead capital, such as canals, roads, railways, pipelines, and ports, although new techno-economic paradigms have also involved innovations in social overhead capital related to housing, public services, hospitals, and schools.

The rationale for a shift in the techno-economic paradigm occurs when new forms of production, technology, and engineering principles (and institutional arrangements) offer substantial improvements in productivity over prevailing systems and ways of thinking, especially if the benefits of the latter have more or less "played themselves out." In the early years of the Industrial Revolution (the early mechanization paradigm), for example, especially in the English textile industry, the factory system and the adoption of machinery, powered by water wheels and increasingly by steam engines, created economies of scale (declines in average cost per unit of output

Table 1.1

Selected characteristics of techno-economic paradigms

Techno-economic paradigm (key factor industries)	Dominant industries; infrastructure (plus new industries)	Source of productivity improvement	BC forest economy: product developments
(1) Early mechanization, 1770s-1830s (cotton, pig iron)	Textiles and related activities, iron working and castings, water power, potteries; canals, turnpike roads (steam engines, machinery)	Mechanization and factory-level scale economies	Captain Cook takes some masts for ships
(2) Steam power and railway, 1830s-80s (coal, transport)	Steam engines, steamships, machine tools, iron, railway equipment; railways, shipping (heavy engineering)	Steam engine and new transportation systems enhance mechanization possibilities	First sawmills established in 1850s; logging
(3) Electrical and heavy engineering, 1890s-1930s (steel)	Electrical engineering, electrical machinery, cable and wire, heavy engineering, armaments, steel strips, heavy chemicals, synthetic dyes; electricity supply and distribution (autos, aircraft, telecommunications, and others)	Cheap steel is superior engineering material to iron. More flexible drive systems for electrical machinery, overhead cranes, and power tools improved layout. Standardization facilitates worldwide operations	Lumber is lead industry, generating many backward linkages in machinery and logging. Pulp and paper (newsprint), shingles and shakes, plywood, box plants established

(4) Fordist mass production, 1930s-1980s (energy-oil)	Autos, trucks, tractors, tanks, armaments, aircraft, consumer durables, process plant, synthetic materials, petrochemicals; highways, airports, airlines (computers, NC-controlled machine tools, microelectronics, missiles)	Flow processes and assembly line extend scale economies. Standardization of components	Kraft pulp is lead industry; plywood also grows rapidly; other commodities grow substantially. Particleboard and paperboard established
(5) Information and communication, 1980s ("chips")	Computers, electronic capital goods, software, optical fibres, telecommunications, robotics, flexible manufacturing systems, ceramics, information services, digital telecommunications network, satellites (advanced biotech products, space activities, fine chemicals)	Flexible manufacturing systems, networking and economies of scope. Electronic control systems. Networking of design, production and marketing	Commodity growth levels off (lumber, pulp) or declines (plywood); backward linkages decline. Engineered wood products and silviculture established

Source: Based on Freeman 1987: 68-75; see also Hayter 1997a: 27-35.

with increasing size of factory) not possible in traditional industry. Cotton and pig iron constituted the key factor industries while investments in canals and turnpike roads provided new infrastructure required to access larger markets and sources of materials.

The technological changes associated with each techno-economic paradigm are paralleled by new regimes of regulation and institutional innovation involving new forms of business organization, research and development (R&D), and labour relations, as well as macroeconomic government policy initiatives and even new ways of organizing the international economy (see Freeman 1987; Freeman and Perez 1988). Thus, in the early mechanization paradigm, the "industrial entrepreneur" emerged as the leading economic agent, ideas about free markets were mooted, restrictions on trade and worker mobility were attacked, and factory-based waged labour signalled changes in the organization of production. Indeed, Freeman (see also Hirst and Zeitlin 1991) emphasizes the intertwining or matching of technological and institutional innovation in long-run processes of industrialization. In addition, industrialization is a strongly geographical phenomenon that consistently favours some places over others (Pollard 1981). The dominant industries of new paradigms often occur in "new industrial spaces," to coin Scott's (1988) term, both because these spaces offer appropriate location conditions and because they are less constrained by inertial forces. Established industries and regions face much more painful prospects in adjusting to the demands of new phases of industrialization.

Shifts in the techno-economic paradigm inevitably involve societies in crises of structural adjustment. There is tremendous inertia in long-established technological and institutional arrangements. These inertias, represented by immobile investments in productive and infrastructural capital and ingrained human attitudes, capabilities, and relationships encased in law, tradition, and prevailing notions of engineering common sense, are powerful forces resisting change. Moreover, it is not easy to implement new paradigms because of the expense of new capital projects (including infrastructure), uncertainty over appropriate choices in the public and private sector, and the effort required to develop new skills and relationships. In the techno-economic paradigm framework, historical turning points, when the balance shifts from one paradigm to another, occur during periods of economic depression and/or war.

In Freeman's long-cycle view of industrialization, the growth potentials of established paradigms eventually peter out, and overcapacity and cost cutting contribute towards increasingly severe recessions. In turn, severe recessions illuminate the productivity problems of existing industries while widespread business failure and job loss undermine inertial forces. Indeed, crisis stimulates fundamental rethinking of economic structures, revealing the decisive advantages of new paradigms for enhanced productivity and profitability. In the absence of crisis, the motivation is insufficient for established economic agents to change patterns of behaviour, and insufficient for investment in radical new systems of production that are uncertain and expensive.

The Fordist Techno-Economic Paradigm

Fordism, with its roots in Henry Ford's assembly line and Taylor's scientific management (also commonly labelled Taylorization) at the beginning of the twentieth century, emerged as the dominant techno-economic paradigm of the Depression of the 1930s and the Second World War (Table 1.1). For western capitalism, Fordism defined a booming, but stable, international system anchored by US hegemony and fixed exchange rates, and a general balance between the demand and supply of goods that supported full employment. Fordism also implied a distinct "industrial state" dominated by the related interests of Big Business and Big Government, fully abetted by Big Labour (Galbraith 1966). Mass production underpinned these interests.

In technological terms, the dominant industries were in transportation-related activities, none more so than auto production, as well as various petrochemical industries and other types of consumer durables, while several new industries such as computers and microelectronics grew extremely rapidly. The key factor industry was oil – abundant, cheap, and widely applicable – and heavy expenditures in highways and airports took advantage of this energy source and reinforced and promoted the dominant industries. In technological terms, the advantage of Fordism had been anticipated by Ford: the innovation of the assembly line and the interchangeability of parts in the manufacture of autos provided the exemplar for massive standardization and the realization of substantial economies of scale in a wide variety of industries. Moreover, Ford's assembly line principles – an internal just-in-time system that brought work to the workers without pause – were complemented by Taylorism, which recommended highly refined task and job demarcation to increase labour productivity through worker specialization on routine, frequently repeated tasks that in turn facilitated further mechanization.

Fordism's technological innovations were interwoven and underpinned by various institutional innovations, typically pioneered in the US. Thus, in the US, the M (multidivisional) form of corporate organization, led by General Motors, allowed effective, hierarchical control of huge, integrated organizations, the most important expression of which were multinational in scope. Indeed, the multinational corporation (MNC) became the dominant organization of the Fordist paradigm. With respect to labour relations, rising labour militancy, which threatened scientific management and industrial stability in general, was offset by the emergence of industrial unions and collective bargaining sanctioned by government legislation. While collective bargaining during Fordism accepted Taylorization as a principle of work organization, it also ensured social stability because workers received substantial wage and nonwage benefits in return for their discipline and productivity. Social stability was further supported by macroeconomic policies based on Keynesianism, while the hegemony of the US in the capitalist part of the world provided structure and order to the growth and liberalization of trade and investment within a regime of fixed exchange rates, established by the Bretton Woods Agreement of 1944, which lasted until the early 1970s. For western countries, the thirty golden years after the Second World War were literally a long boom.

Fordism, in other words, implies a particular international regime of regulation and production-system organization. In the latter context, a Fordist template may be summarized by the following attributes:

- Factories feature the mass production of standardized goods on assembly lines or in continuous-flow processes that emphasize cost minimization and the full exploitation of economies of scale and size.
- Vertically and horizontally integrated MNCs dominate production so that substantial flows of goods and services, including exports and imports, are among affiliated plants within the firm.
- Purchases from outside suppliers are typically arm's length and made mainly to minimize costs and uncertainty, for example, by the use of subcontractors to temporarily provide excess capacity.
- Decision-making structures of MNCs are strongly hierarchical and organized according to M-form principles.
- Labour is organized according to the strictures of Taylorism; that is, limited as much as possible to highly specialized, operating tasks.
- Unions provide labour with considerable power to gain high wages and good benefits and in return provide discipline and agree to Taylorism.
- Technological change is internalized in specialized R&D departments separate from manufacturing units. Innovation itself is planned as a linear process naturally progressing through R&D stages and into manufacturing systems.

With the US as the undisputed economic and military power of capitalism, Fordism soon spread to other parts of the world, led by a rapid expansion of US-based MNCs, first in Canada and BC. Just how extensively Fordist production systems developed in the US and elsewhere is a matter of debate. Other forms of organization existed and Fordism itself has several variants (Lipietz 1986). Even in the US, the models of business segmentation developed in the 1950s and 1960s recognized both a core Fordist segment and a peripheral small and medium-sized firm ("market system") segment (Galbraith 1966; Averitt 1968). Similar segmentation models of labour markets distinguished the relatively stable, well-paid jobs of the Fordist sector from the much more variable employment conditions that prevailed in the rest of the economy (Doeringer and Piore 1971; Peck 1996). At the same time, these segmentation models reinforce that Fordism pertained primarily to the dominant or leading carrier industries that served to define organizational and technological "best-practice" characteristics exemplified in the US. For Servan Schreiber (1968), "the American challenge" meant the European duplication of the technological and marketing know-how of American MNCs.

The Information and Communication Techno-Economic Paradigm

In the US and elsewhere after the mid-1970s, Fordism was increasingly threatened as a system of regulation and as a set of productive arrangements. According to Freeman, recessionary crises, especially the economic slump of the early 1980s that ravaged industrial structures across North America and Europe, marked the turning point from Fordism to the ICT paradigm. As a system of international regulation, Fordist arrangements were undermined by free exchange rates, financial deregulation, neoconservative economic policies, and the formation of trade blocs that uneasily evolved alongside multilateralism, creating new difficulties for tried and trusted Keynesian macroeconomic policies as national economies became more open and vulnerable. In the context of globally significant MNCs and declining national autonomy, powerful nongovernment organizations (NGOs), most notably environmental NGOs, emerged rapidly to challenge the forces of industrialization.

As a system of production, Fordist arrangements were undermined by technological change, market dynamics, and recession. New (main carrier) industries, notably computers, microelectronics, telecommunications, optical fibres, and robotics, based on the key factor input of silicon chips, have expanded significantly and exerted powerful, pervasive effects on all other activities, including forest products (Table 1.1). These industries have transformed how information is communicated, how materials are processed, and the nature of the linkages between markets and suppliers. CAD-CAM systems, for example, have redefined relationships among R&D, engineering, and manufacturing, typically enhancing precision, reliability, quality, and speed. Machines have become much smarter and efficient, in some cases replacing and deskilling labour, in other cases preserving and creating jobs while requiring entirely different skills. Moreover, the new technologies have allowed producers seemingly infinite variety in differentiating outputs, consumers infinite variety in their demands, and new ways of rapidly communicating supply and demand. In the new systems of production of the ICT paradigm, economies of scope – how firms use their assets and capabilities to serve higher-value, more varied markets – rival economies of scale as engineeering common sense; flexibility has become a key imperative for firms, factories, and workers as markets have become more volatile and differentiated.

Globally, once overwhelmingly dominant American industries and MNCs have been challenged by the German and especially Japanese industries and MNCs that most fully pursued flexibility within their productive arrangements. Established Fordist production structures, exposed as too rigid and inefficient, have been forced to adapt, typically painfully. In Freeman's (1987) account, production systems of the ICT paradigm are organized according to imperatives of flexibility, summarized by the following attributes:

- Flexible production is highly efficient in the use of materials, space, and workers, but it gives priority to value maximization and product differentiation. Based on economies of scope, flexible production may occur as batches in small factories or as flexible mass production in large factories.
- MNCs remain important in flexible production, but small and medium-sized firms play significant roles in performing highly specialized activities.
- Networking among firms, for example, in terms of subcontracting, design collaboration, and strategic alliances is an important way of realizing economies of scope; interfirm relationships are stable and mutually reinforcing as well as competitive and arm's length.
- Decision-making structures of MNCs are relatively flat, implying greater personal responsibility and self-supervision.
- Labour is based on principles of flexibility, implying that work supply is closely adjusted to demand, for example, through greater use of part-time workers, workers who are readily hired and fired, or more multi-skilled workers.
- Unions are less powerful in general, and worker benefits are closely related to skill as well as productivity.
- All employees are expected to contribute to innovation, and R&D activities are closely integrated with manufacturing activities. Innovation is structured as a "loopy" process that "naturally" progresses by constant interaction among R&D, manufacturing, and marketing departments.

Admittedly, flexibility is an ambiguous, contested term, and ideas about flexible factories, firms, production systems, and workers evince multiple attributes with different implications for local development. Nevertheless, in one form or another, flexible production systems are widely regarded as a global juggernaut necessary for competitive advantage. According to the theory of flexible specialization, in which flexible and Fordist production systems are opposites, this juggernaut is a socially virtuous one.

Industrialization as Flexible Specialization

The contemporary significance of the idea of flexible specialization originated with Piore and Sabel (1984). In their view, industrialization has occurred historically, not as a series of techno-economic paradigms, but as a kind of ongoing debate, or confrontation, between the forces of flexible specialization (also labelled craft production) and mass production (Figure 1.1). On the one hand, flexible specialization is most strongly revealed by networks of interacting small and medium-sized enterprises (SMEs) manufacturing a highly differentiated and changing array of products by skilled workers using multipurpose (flexible) machinery, including CAD-CAM systems. Each firm specializes in a particular component, product, or function, while collectively the population of firms offers considerable flexibility in responding quickly to demands. The advantages of flexible specialization are particularly well developed in industrial districts where the proximity of firms facilitates entrepreneurial and synergistic interactions, the exploitation of market niches, the development and sharing of information, the development of common labour pools, the creation of specialized infrastructure and institutions, and other forms of external economies. In contrast, mass production is represented by the Fordist assembly line in which high volumes of standardized outputs are manufactured by relatively unskilled workers using expensive special-purpose machinery (Hirst and Zeitlin 1991: 2).

In this framework, different forms of flexible specialization have historically provided alternatives to the mass production model. The chosen options, however, do not depend solely on efficiency considerations but also on the interest and visions of powerful individuals and organizations within society (Sabel and Zeitlin 1985: 161-4). Once chosen, a particular technology

Figure 1.1

Industrialization since the Industrial Revolution: Flexible specialization versus mass production

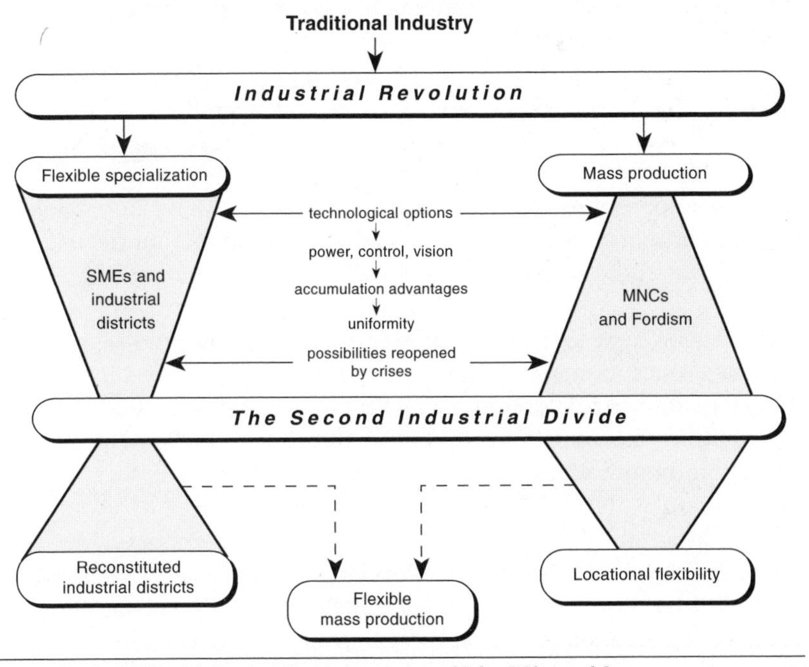

Source: Hayter 1997a: 36. Reprinted with permission of John Wiley and Sons.

strategy develops, supported by capital investments and infrastructure, accumulating advantages that limit other possibilities. For Piore and Sabel (1984), the mass production option, ultimately led by the US, became the dominating vision during the nineteenth century and most of the twentieth century, confirmed by Ford's innovation of the assembly line and related ideas. Moreover, as more production came under corporate control, MNCs geographically dispersed their activities, in search of markets, resources, cheap labour, and subsidies. SMEs were not eliminated during Fordism, even within the main Fordist industries, such as auto production. Yet, under Fordism, as mass production organized by integrated MNCs defined the core of the economy, SMEs were relegated to peripheral roles, either as vulnerable and dependent subcontractors to giant MNCs or as constituents of industrial districts in decline, relics of previous rounds of industrialization (Hiebert 1990; Hayter and Patchell 1993).

Piore and Sabel's (1984) main thesis is that in the 1970s a second industrial divide occurred that marked a renaissance of flexible specialization in Europe and elsewhere, based on agglomerations of SMEs manufacturing a wide range of high-quality consumer goods. In their view, market uncertainties, differentiated consumer demands, and technological change are combining to increase the competitive advantages of SMEs, especially when SMEs cooperate as well as compete with one another. The emergence of Japan as an industrial superpower offers further support to this view. Thus, in Japan, SMEs are more important than in western economies, forming the basis for myriad small-scale community-based industries (Ide and Takeuchi 1980; Patchell and Hayter 1992) and playing dynamic roles in flexible mass production organized around core companies such as Toyota (Fruin 1992; Patchell 1993; Hayter 1997a: 349-72). In this view, in the case of flexible specialized mass production, SMEs have developed in collaborative, stable relationships that act as incentives for continuing innovation and rapid, nuanced responses to consumer demands. Indeed, the enhanced role of SMEs as collaborators and innovators underlies proposals for flexible specialization as a policy prescription for local development (Cooke and Morgan 1993).

The flexible specialization thesis is criticized by those who suggest that corporate concentration is increasing and that MNCs remain committed to geographic dispersal ("location flexibility"), constantly threatening locally polarized development (Amin 1993; see also Harvey 1989b). Flexible specialization, however, does reveal the real, ongoing tensions that exist between small scale and large scale, between the forces of geographic concentration and geographic dispersal, and at least offers hope, along with not inconsequential evidence, that flexible production can contribute positively to local development. The flexible specialization perspective also emphasizes that economic history varies considerably from region to region, more so than anticipated by the techno-economic paradigm model.

At the same time, significant points of overlap are evident between the flexible specialization model and techno-economic paradigm. Both interweave institutional and technological innovations in accounts of industrialization, and both consider the historical and geography contexts of economic behaviour. Finally, both models emphasize that industrial countries since the 1970s have experienced a fundamental change from models dominated by Fordist mass production to ones dominated by flexibility.

Both models also share a neglect: the experience of resource peripheries. Although the techno-economic paradigm recognizes the importance of resources as inputs (coal and oil are identified as key factor industries in different paradigms) and iron and steel and petrochemicals as main carrier branches, explanatory emphasis is on the technological and institutional innovations of leading (that is, core) countries and leading (that is, manufacturing) industries. Meanwhile, the flexible specialization model, which interprets small firm networks, has agglomerative force as a trend and as a policy prescription, and has been primarily conceived in the context of urban-based manufacturing, not easily reconciled with the dispersal (and restructuring) of resource-based towns.

Industrialization in Resource Peripheries

Although tangential to theories of economic growth, industrialization fundamentally depends on resources. Both the short- and long-term rhythms of capitalist development closely implicate resource regions, new and old. Industrialization is as much a search for resources as for markets, reaching into the outermost parts of the world simply because resources are present. The economic structures, and indeed the very settlement, of remote regions around the globe have been fundamentally shaped by the demands of industrial heartlands for their resources. The demands of the UK, the US, and Japan, for example, have underpinned the evolution of BC's forest industries over the last 100 years (see Table 1.1, last column).

Two Models of the Organization of Resource-Based Development

The global pattern, implications, and organization of resource exploitation are highly varied (Baldwin 1956; Auty 1995). Thus, Baldwin's (1956) classic, theoretical study contrasts two polar forms of export-based, resource-based development in remote regions supplying industrial heartlands. His scenarios are presented in the context of two agricultural regions, one based on a plantation economy established within a poor country dominated by a traditional, largely self-sufficient rural sector, and the other by an entrepreneur-farmer economy created by settlers occupying an otherwise empty land. Both regions export agricultural goods, but the development implications are very different.

Baldwin contrasts the plantation and entrepreneurial economy in terms of their respective production functions. On the one hand, the plantation model is part of a foreign-owned MNC and output is exported by means of the MNC's marketing network. Financial capital, management, fertilizer, and equipment are all imported, and the only local inputs required are land, climate, and unskilled labour. In contrast to indigenous farming, however, the plantation is relatively capital intensive and labour inputs are limited. To be viable, the plantation requires transportation and other types of infrastructure dedicated to the special needs of the plantation, whether financed from local or international sources. All decisions affecting the operation of the plantation, and any associated R&D inputs, are controlled from a head office in a distant heartland.

Baldwin's conclusion is that the local development impacts of this type of plantation investment are limited. The plantation offers some employment at higher wages than the local average and generates exports. However, the provision of relatively high-wage income to a few serves to underline income disparities and the sense of a dual economy, whereas export income is offset by imports and the repatriation of profits plus any tax concessions the plantation may receive. The linkages or multiplier effects the plantation generates for the surrounding economy are likely to be meagre. The relatively low level of education and technical training of people in the surrounding population limits their role to menial positions. It is hard to see how plantation-style management can provide a meaningful demonstration effect for local firms in terms of best-practice technology or management. Moreover, the plantation receives key service and material inputs from affiliated operations overseas that undermine the possibility of their provision locally. Indeed, any change in the structure of the plantation depends on priorities dictated by a distant head office.

The situation of the entrepreneur-farmer economy is different. Immigrants to an empty land face substantial difficulties in homesteading and overcoming the barriers in serving distant markets. From the beginning, however, land use is determined according to principles of supply and demand throughout the territory, while supporting investments in infrastructure are created as public goods – available to everyone. In an empty land, no problems are inherited from an established culture; that is, there are no "inhibiting traditions" to worry about. Moreover, immigrant settlers by definition establish an indigenous capitalist class of entrepreneurs. Over time, export income is funnelled by farmer-entrepreneurs to promote local diversification by increasing broadly based demands for local goods and services. Local businesses are more inclined than plantations to replace imports with local production, and no institutional restraints operate to prevent them from doing so. A farmer-entrepreneurial society is also more egalitarian than a plantation economy, and the population is better educated and capable of innovation. Unlike its plantation counterpart, despite problems of remoteness, a farmer-entrepreneurial society is capable of self-directed change and of coping with the dynamics of capitalism on its own terms. Assuming the resource is maintained, the farmer-entrepreneurial model offers a potential path for sustained economic development diffused throughout the region and shaped by local abilities.

The plantation (or branch-plant) and farmer-entrepreneurial (or simply entrepreneurial) models readily apply to the resource sector as a whole, including forest products. Clearly, not all possibilities are defined by these two radically different forms of resource-based export development. In practice, development is affected by specific historical and geographical circumstances; other starting points can be proposed as well as those of an empty land or of a dense population using traditional technology. In addition, the two polar cases may not be mutually exclusive in any particular territory. The Canadian case is instructive in this regard.

In Canada, including BC, different resources have evolved along different lines, variously reflecting the influences of the plantation/branch-plant and entrepreneurial models. After all, two historically important planks of Canadian economic strategy have been to promote immigration and to welcome foreign investment by powerful MNCs. Consequently, resources have been primarily developed along the lines of the farmer-entrepreneur model, notably agriculture, logging, and the fur trade, and to some extent fishing and sawmilling. On the other hand, some resources were developed more along the lines of the plantation model, such as various kinds of mineral extraction and pulp and paper, which operate as branch-plant operations of foreign firms. In some cases, as in fishing and sawmilling, entrepreneurial forms of development have been challenged over time by plantation forms of development. In Canada, however, plantation-type development is not solely the prerogative of foreign-based MNCs, because large Canadian-based MNCs also exist. In the Canadian case, the plantation model was also not implanted within the context of the inhibiting traditions of an already dense population of Aboriginal peoples. Rather, if the organization of the fur trade was interwoven with an indigenous culture, subsequent resource development largely, if not entirely, assumed an empty land because an already dispersed Native population was effectively ignored and marginalized (Willems-Braun 1997b).

Moreover, given Baldwin's (1956) plea for the superiority of the entrepreneurial model from the perspective of local development, it is interesting to note that in Canada and elsewhere, virtually across the resource spectrum, including forest products, the plantation model has been the much more prevalent form of organizing resource exploitation. Part of the reason is technological, including the rapidly increasing importance of economies of scale based on processing innovations and massive fixed costs in plant and infrastructure. Part of the reason is geographical, because increasingly dispersed patterns of resource exploitation are based on large-volume production to keep transportation and production costs down. A third reason is that resource exploitation for industrial purposes is not primarily motivated by local development. Rather, the opening of resource peripheries is motivated to serve the needs of industrial cores.

The Export Imperative

Resource-industry spaces represent "pure" export-based forms of industrialization. In core industrial countries, such as the US, exports are typically to other core regions within the country. In peripheral countries, rich or poor, resource peripheries are dependent on true export markets over which they have no influence, and the pressure to remain cost-competitive is especially severe. Thus, resource peripheries, from their inception, are unusually exposed and vulnerable to exogenous (global) forces. Given extreme export dependency, resource peripheries are vulnerable to the full impact of globally based industrial and social dynamics – technological change, changes in demand, new supply sources, and policies ranging from trade to labour relations.

Moreover, the viability of resource peripheries is also intricately affected by resource dynamics comprising fundamental forces of change that do not affect secondary manufacturing centres. Specifically, with industrial use, resources deplete or at least change, typically declining in both quantity and quality.

The Resource Cycle and Forests

A basic categorization of natural resources is renewable and nonrenewable resources, or flow and stock resources. Nonrenewable or stock resources are permanently depleted by exploitation, and the duration of exploitation is clearly limited by the size of the resource and the rate of exploitation. For nonrenewable resources, a finite resource-based cycle makes sense, even if technological change periodically redefines the size of the economically available resource. Renewable resources enjoy permanently or potentially indefinite resource cycles. Forests are an example of a potentially renewable resource that can either be "mined" to levels beyond which renewal is compromised, perhaps absolutely (what Rees [1985] labels the "critical zone"), or be used in a way that "sustains" the forest.

History shows that civilization has typically resulted in forest depletion and that societies over the long term have chosen to treat forests as a stock resource (Clawson and Sedjo 1984). One well-known estimate suggests that before discernible impacts by human societies, the world's land area included over 13,000 billion hectares of forests around 10,000 B.C., comprising more or less equal areas of closed forest and open woodland (Haden-Guest et al. 1956). A respected recent estimate places the world forest area at around 4,100 billion hectares, a figure that includes closed and open forests as well as regenerated forests (Persson 1974). Persson himself regarded his estimate as optimistic although other estimates are both higher (up to 6,050 billion hectares) and lower (down to 3,031 billion hectares) (Mather 1990: 59). Clearly, the world's forest area is much less than it used to be, and forests that may be considered as natural "climax communities" are of limited extent (Kimmins 1992: 42).

Clapp (1998b: 138-9) cogently argues that resource cycles, comprising three main phases of exploration and initial boom, large-scale exploitation, and ultimate collapse, apply to renewable resources (such as forests) as well as nonrenewable resources. In his view, the basic cause of resource depletion is the application of a narrow economic calculus that discounts future benefits in assessing the value of forests while ignoring the many environmental benefits of forest ecosystems. Given this philosophy, resource exploitation is relentlessly driven by a cost-price squeeze that occurs as the best-quality, most accessible timber is harvested first; competitive tendencies among rival firms seeking timber supplies; regulations that favour large-scale, export-oriented commodity production that cannot be easily adapted to small-scale (and flexible) operations; short-term political horizons that evolve in close alliance with the dominant resource industry; and community attitudes unable to extricate themselves from the expectations of boom-and-bust cycles (see Freudenburg 1992). Indeed, it seems that the forest industry itself fails to recognize that logging changes the nature of the forest resource (Graham and Martin 1990). Eventually, the "falldown effect," the decline in timber harvests from overharvesting and the replacement of old-growth forests by second-growth stands comprising smaller trees and lower wood volumes, signals the beginning of resource exhaustion and social crisis (Clapp 1998b: 138).

Historically, societies have made many attempts to regulate the forest cycle and to "sustain" forest values (Clawson and Sedjo 1984; Steen 1984). An important thesis in this context is that such regulations are more likely to occur after the onset of crisis, that is when forest depletion is already well advanced (Lee 1984). Indeed, as forest depletion accelerated with industrialization in the nineteenth century, more countries grappled seriously with sustained yield principles, as pioneered in Germany, and with forest conservation. Admittedly, regulations did not always work and, as in the key US case, effective response to the initial recognition of the forest crisis took several decades (Mather 1990: 45-9). Nevertheless, Mather (1990; 1992) believes forests "turn-rounds" or "transitions," measured most readily by an increase in the forest area of a country, have become important. Among forest regions such as BC, the forest transition implies innovation of effective sustained yield forestry, as measured by the nature of reforestation and afforestation, and conservation efforts.

The Forest Transition Model

According to Mather (1990: 31), a sequential pattern of land use is evident in many parts of the world (Table 1.2). In the first stage, the forest is, and is seen to be, an unlimited resource in no danger of exhaustion. Forests may be "open access resources" freely available to all or the "common property resources" of particular groups whose use is shaped by tradition. Although demands on forests remain light, and natural regeneration effective, the open-access system may be the efficient and democratic method of allocation (Clawson and Sedjo 1984). With economic development, and as populations grow, however, demands on forestry resources escalate. Since ancient times, wood provided the basic material of construction for many purposes (ships, castles, temples, houses, and buildings of all kinds), a basic source of fuel (including for mining and smelting), and land under forest was needed for agriculture. In addition to the well-known case of the Mediterranean in classical times (Thirgood 1981; Meiggs 1982), forest depletion (the onset of Stage 2 in Table 1.2) became a significant problem in many parts of Europe and Japan before the Industrial Revolution. The onset of the Industrial Revolution signalled the global spread of forest depletion. Indeed, the full incorporation of areas, including BC, into the global economy in the nineteenth century rapidly increased demands on the forest and stimulated a rapid transition from hunter-gathering economies in the early phases of Stage 1 to the full-blown exploitation of Stage 2.

As demands for and on forest resources increased, "institutions were required to insure that [the] resource was not destructively exploited [and] to provide reasonable assurance that investments in maintaining and improving the resource were allowed to reach fruition" (Clawson and Sedjo 1984; see Bromley 1992). In classical times, responses to the problems of deforestation were largely, if not completely, a failure (Thirgood 1981; Mather 1990: 35). In recent centuries, successful forest transitions are claimed to have occurred (Stages 3 and 4 in Table 1.2), most notably among advanced countries (Lee 1984; Totman 1989; Mather 1992; LeHeron and Roche 1985; Mather and Needle 1998), but not exclusively so (Clapp 1995). Indeed, Mather and Needle (1998) argue that the forest transition has considerable global potential because increases in agricultural productivity allow for the return of land to forest use.

Table 1.2

Sequential model of forest resource trends

Stage	Trend in resource area	Perception of trend
(1) "Unlimited" resource	contraction	positive or neutral
(2) Depleting resource	contraction	negative
	- - - - forest transition - - - -	
(3) Expanding resource	re-creation/expansion	neutral/negative
(4) Equilibrium?	(stability)	n/a

Source: Mather 1990: 31. Reprinted with permission of John Wiley and Sons.

The forest transition has occurred at varying times in different places. The most significant early cases are Germany and Japan where the forest transition occurred before large-scale industrialization. In the German context, Lee (1984) rejects the conventional explanation of the adoption of sustained-yield forestry in the eighteenth century as an outcome of the stratified and stable nature of German society, with its long history of forest management creating a known forest situation of more or less equal areas of land in each age class. Rather, Lee emphasizes the volatility of Germany, as a result of wars, unrest, and increasing economic change. In these circumstances, he suggests sustained-yield and related investments in forestry were responses to crisis and attempts to achieve stability. Another early, if less well-known European example of the forest transition is provided by Finland in the nineteenth century (Raumolin 1984, 1992). For Japan, Totman (1989) argues that while natural and social pressures were seemingly leading to an inexorable trend towards forest devastation in what was then a closed society, for a variety of reasons, effective measures were introduced to implement a forest transition beginning in the late eighteenth century.

Following large-scale industrialization, an increasing number of countries, especially developed countries, have chosen to engage in forest transition, the most important being the US (Mather 1990: 38-49; Westoby 1989). Indeed, in the US, massive exploitation of forests was a primary stimulus to the conservation movement and to the introduction of policies, beginning at the end of the nineteenth century, by which time all major forest areas of the US had been exploited, which led to a substantial expansion of the forest area, especially in areas formerly cleared for agriculture. Indeed, net annual timber growth in the US between 1920 and 1970 increased from 6 to 18 billion cubic feet (Clawson 1979). The rapid growth of younger forests, increased regulation, and the role of science and technology in breeding rapid-growth, drought-resistant species have all been important factors underlying this trend (Humphrey 1990: 44). The basis of the contemporary position of the American South as the North American, possibly global, low-cost region for forest-product manufacture stems from this turn-round.

Forest exploitation is clearly deepened by industrialization. Mather et al. (1999) further suggest that there is a basis for linking long-term trends in forest use with economic development, including the emergence of forest transitions as society's income rises. Yet, the forestry transition is not an automatic response to the crisis of forest depletion. As Mather (1990) notes, the evolution of forest use around the globe is extremely complex, and his simple unilinear model (Table 1.2) provides a template for comparison and point of departure to raise questions about macrolevel trends. Globally, there may be "no biological shortage of wood" (Westoby 1989). But controversy over forests remains, especially over the few surviving remnants of old-growth forest that environmental arguments seek to protect from further depletion and from the need for a "transition." Even among successful

Figure 1.2

Stages of forest use in terms of objectives and ownership

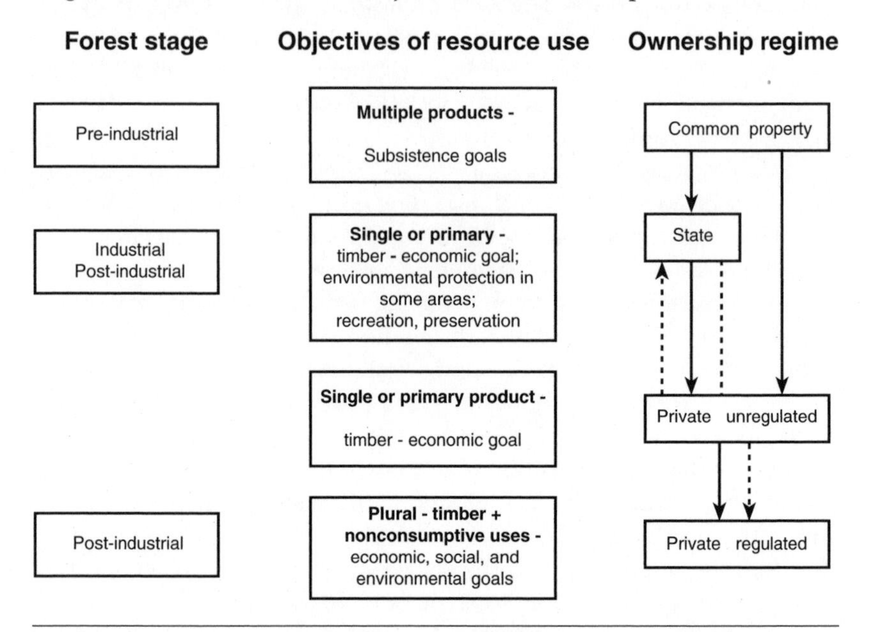

Note: Solid lines indicate main sequential trend of ownership; dash lines show subsidiary trends.
Source: Based on Mather 1990: 90.

examples of the forest transition, questions arise about forest productivity and health, and controversy remains about the nature of forests and how they should be used. Questions can also be raised about the purposes and organization of the transition (Clapp 1995, 1998b). The idea of the forest transition, however, at least suggests that public policy can have possible effects, even if the purposes of that policy are difficult to sort out.

Evolution of Forest Resource Objectives and Regulation

To help summarize the complex evolution of forestry objectives and ownership, Mather (1990) distinguishes preindustrial, industrial, and postindustrial societies (Figure 1.2). In the preindustrial stage, society's goals relate to subsistence, and forest use is likely regulated by an implicit policy of open access or some form of common property arrangement in which specific groups have rights based on tradition. As societies develop and eventually engage in full-blown industrialization, forest use becomes dominated by economic values associated with the forest industries. According to this model, as demands on the forest resource increase, the state takes control, and then during industrialization, forests are often, but not invariably, privatized without regulatory constraints. In the postindustrial stage, economic goals other than those related to timber are important, and environmental values emerge as preeminent concerns. Meanwhile, privatized forests become subject to increased regulation.

The pattern of evolution of the objectives and management of the forest resource outlined by this global model is highly generalized and by no means unvarying or inevitable. Yet, the model does capture important features of forest use and provides at least a yardstick by which to compare trends in specific regions including BC. For example, in much of Canada, including BC, forest ownership has remained under state control, to a greater degree than is typical for developed economies (Marchak 1995). The shift towards environmental values has been as strong in BC as elsewhere, however. Indeed, with successful economic growth and development, the model recognizes that societies place greater value on environmental amenities and the needs of future generations, and they have the funds to commit to meeting environmental goals. It is also commonly supposed that environmental regulation is crisis motivated; that is, resource management occurs after, rather than in anticipation of, depletion and degradation. Among developed economies throughout the world, the escalation of environmental legislation regarding forest use since the 1970s (Mather 1990: 120) has been associated with substantial widespread public support in favour of environmental values (Mather 1990: 268-75).

In the 1960s, the idea of "multiple use" gained much currency as a framework to reconcile conflicting environmental and industrial goals. This consensus, if indeed there was such an agreement, has been severely tested in recent decades, however, most notably over the related issues of old growth and biodiversity. Thus, international environmentalism, led by Greenpeace, rejects virtually any form of industrial harvesting of old-growth forests on the basis that the complex and interdependent diversity of such forests cannot be replaced once harvested, even if reforestation is practised. They stress that biodiversity is a global issue and responsibility, and it should not be compromised by localized needs for job creation. In those regions within the temperate forest zone, such as BC where the falldown effect is occurring (that is, a decline in the volume of timber harvested in the switch from old-growth to managed forests), "intrusion" of such environmental demands adds to economic dilemmas.

Traditionally, in many parts of the world, including Canada, the forest industry has argued that it is environmentally as well as economically sound to cut old-growth forests containing large volumes of timber that are no longer growing and to replace such old but "no-growth" forests with new, fast-growing forests. This argument is reinforced by reference to concerns about increases in "overmature" or "decadent" timber (that is, timber in old-growth forests that has lost value for industrial purposes) (Percy 1986). The environmental counterargument is that old growth has unique ecological characteristics that cannot be replaced by managed forests (Maser

1990; Kimmins 1992). This concern inevitably increases as falldown becomes a reality, because the extinction of old growth is now a real possibility. From industry's point of view, the imposition of conservation measures at a time of falldown constitutes a double whammy. In the Pacific Northwest, especially BC, the forest industry faces such a situation.

The Staple Thesis of Canadian Resource Development

The techno-economic paradigm and flexible specialization models suggest that all regions, from industrial heartlands to resource peripheries, are affected by global forces of industrialization, which in the past quarter century are represented by shifts from Fordist to flexibility paradigms. In addition, resource peripheries are directly affected by resource cycles that reflect natural as well as industrial dynamics. At the same time, these global forces of industrialization and resource development are grounded in particular places where they are shaped by particular histories and inheritances. From this perspective, the staple thesis, especially that version associated with Harold Innis (1933) and the "Innisians" (Barnes 1993; Drache 1995; Wilkinson 1997), provides a specifically Canadian model of resource-based development relevant to BC's resource industries.

The contribution of staples to Canadian economic development, and most particularly in understanding Canada's global role, has long been recognized (Aitken 1961; Innis 1967; Mackintosh 1967; Marshall et al. 1936; Clement and Williams 1989). In the conventional or orthodox view, staple development is simply a response to the dictates of comparative advantage. In this view, staples have provided Canada with an inevitable and appropriate trajectory of development and basis for diversification. The Innisian view, on the other hand, interprets staple exploitation within the framework of creative destruction, of instability and crisis, in which the manner of exploitation is itself subject to policy.

Innis was a relentless advocate of theories developed out of, and that reflect, Canadian experience (Barnes 1993). Innis argued that ideas developed in one context cannot be readily applied to another one, and in particular, a theory of development articulated within and for the industrial powers of the world is not relevant to resource peripheries. As Innis wrote: "Canadians are obliged to treat the economic theory of old countries and to attempt to fit their analysis of new economic facts into an old economic background. The handicaps of this process are obvious, and there is evidence to show that the application of the economic theories of old countries to the problems of new countries is a new form of exploitation with dangerous consequences" (1956: 3). Innis recognized that the key point of departure for an understanding of Canadian economic history was the observation that Canada was developed as a source of staple products for metropolitan countries. For Innis, staple exports shaped Canada's institutions and internal structures. Given the central role of staple exports, Innis sought to develop a distinctively Canadian theory by wedding three types of concerns: geographical, institutional, and technological (see Neill 1972: 42; Rotstein 1977: 6). Indeed, Watkins (1963: 141) long ago concluded that Innis's "staples approach [is] Canada's most distinctive contribution to political economy." However, not all staples approaches are Innisian (Watkins 1981).

Staple Development as a Diversification Process

Conventionally, staple production in Canada is defined as comprising primary resource activities and primary manufacturing activities, such as lumber, pulp-and-paper mills, and fish-processing plants, in which resources are major inputs to the production process. As such, the staple sector as a whole offers Canada substantial comparative advantages for exporting. These advantages are so great that collectively staples become "the leading sector of the economy and sets the pace for economic growth" (Watkins 1963: 144). General economic development is then a process of diversification around this export base. As Watkins (1963: 144) writes: "The central concept of a staple theory, therefore, is the spread effects of the export sector, that is, the impact of export activity on domestic economy and society."

The diversification process is realized through three types of interindustry linkages, namely backward, forward, and final demand linkages. These linkages can be hypothetically presented as part of a simple chronology of development in the context of an "empty land" (Figure 1.3). Thus, staple investment initially occurs to provide raw materials to a distant heartland (assuming complementarity, transferability, and no intervening opportunities); indeed, there are no other reasons for investment in an empty, distant resource periphery. Entrepreneurs or management, labour, capital, machinery, and other nonresource inputs are imported. As the staple industry grows, demands for crucial inputs may reach a sufficient scale to justify backward-linked local production, defined as investments in activities that supply inputs to the staple sector, including investments in machinery, equipment, and other inputs such as chemicals and fertilizer. Producer services, such as engineering consultants, accountants, lawyers, transportation companies, or other kinds of service or information to the staple sector, also constitute backward linkages. More broadly, backward linkages extend to public-sector investments in roads, railways, ports, and other types of economic infrastructure, plus investments in social infrastructure such as housing, schools, and hospitals. If resource exploitation faces unusual conditions in terms of climate, geology, topography, and vegetation, then local innovation in manufacturing special-performance characteristics in equipment is likely to be encouraged.

Forward linkages refer to investments in activities that use the staple output. As noted, activities such as sawmills, smelters, and fish-processing factories are conventionally considered as part of the staple sector. The many activities defined as secondary wood-processing exemplify forward linkages

Figure 1.3

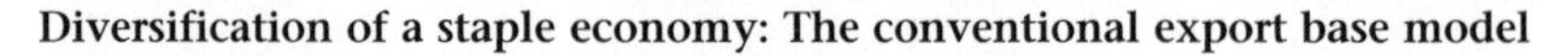

Diversification of a staple economy: The conventional export base model

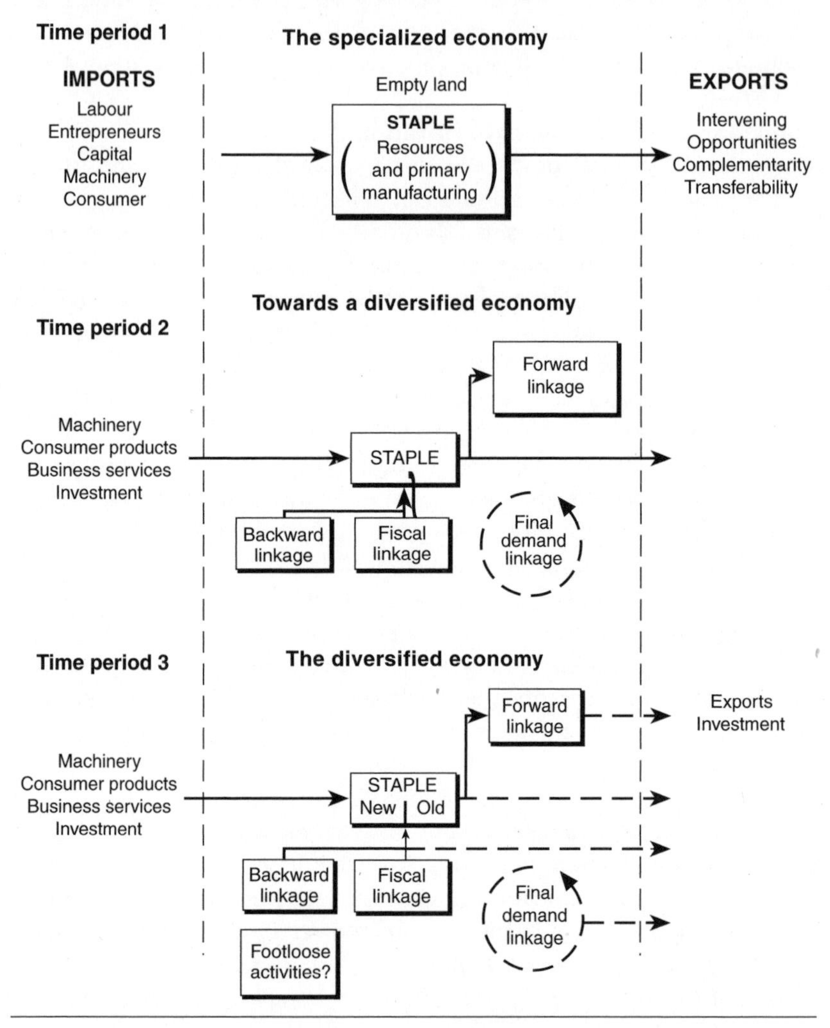

around the forest staple. Typically, forward and backward linkages concentrate in urban areas to access external economies of scale related to specialized labour pools, infrastructure, and information flows. While opportunities for forward linkages, in contrast to backward linkages, are often thought to be sharply curtailed by small local markets (Watkins 1963), their development is now crucial to the current emphasis on "value added" (Chapter 6).

Final demand linkages are investments in activities that supply consumption goods to workers in staple-good industries. Thus, local consumer goods, such as retail and entertainment services of all kinds, are introduced as population size and income warrants. Ultimately, population size and income depend on the size, range, and quality of resources that are exploited, as well as the nature and extent to which backward and forward linkages are developed. Over time, firms that developed as backward and forward linkages may develop export potentials of their own, and some remote, resource regions offering high levels of amenity may attract footloose activities (that is, activities whose location is scarcely influenced by transportation cost considerations) (North 1955).

The resource industries also make payments to governments, or "fiscal linkages" (Auty 1990). These fiscal linkages in turn contribute to service provision and some governments in resource regions have created rainy day funds built up during booms for use during busts and as a source of funds for diversification. The weight of opinion, however, based on the experience of resource-rich developing countries (Auty 1993, 1995) and resource-rich peripheries within advanced countries, including Canada (Watkins 1963; Gunton and Richards 1987; Freudenburg 1992), is that diversification around staples sectors is highly problematical. Theories vary, however, in interpreting the limits to diversification.

Staple Diversification as a Response to Comparative Advantage

Diversification around the staple is not automatic. In the comparative advantage model, it is market driven. On the one hand, diversification is encouraged by a diversity and richness of resources, by investment in related activities, and by serving the consumption needs of a growing population. For some regions, diversification may be further enhanced by footloose activities and by the achievement of a sufficient population size, wealth, and degree of urbanization that facilitate various kinds of agglomeration of scale, allowing both for a deepening of the division of labour and for greater propensities to innovate. On the other hand, resources may run out, cheaper resources may be found in other regions, substitute products may be developed with different location requirements, the need for backward and forward linkages may lessen, and the size of local consumer markets may remain relatively small and even stagnant.

Comparative advantages, in conventional theory, are created and lost by competition. From this perspective, there is nothing intrinsically wrong or inappropriate about resource-based comparative advantages. Indeed, the role of resource exports in patterns of regional economic growth has been judged a distinctive historical feature of North American, not just Canadian, regional growth patterns, in contrast to European models (North 1955). In BC, at the end of the long postwar boom, Shearer et al.'s (1973) well-known study interpreted the province's resources as comparative advantages, the export of which had provided the main engine of economic growth of BC since its incorporation into the world economy, beginning with the arrival of the Hudson's Bay Company and its search for beaver pelts. Following the decline of the fur trade in the 1860s, they noted, the most influential staple sectors in the province's development have been minerals, such as gold, coal, and copper, and especially forest products, specifically lumber, plywood, pulp, newsprint, and paperboard. They also noted the existence of a relatively small-scale secondary manufacturing sector, especially backward linkages into machinery, and final demand linkages. Ultimately, Shearer et al. saw the global role of the BC economy as narrowly based but stable, and as an economy that had, more or less, evolved according to free trade (market) principles. Interestingly, they concluded that any further commitment by Canada towards free trade would have few implications for BC, good or bad.

The idea of comparative advantage runs deep in the psyche of the Canadian and British Columbian staple economy (Marchak 1983). Ultimately, it defines an attitude that the structure and health of the staple industries are governed by competition and foreign markets over which local producers have no control. If markets are volatile, they are so because that is how laws of supply and demand work. Meanwhile, in the face of consumer sovereignty and competition, the responsibility of local producers is to strive to keep costs lower than revenues. In turn, this striving ensures efficiency and innovation. Comparative advantage is an idea that appeals to common sense. From a policy perspective, its main recommendations are invariably to promote competition; thus, free trade and open door policies to foreign investment are celebrated, whereas unions that want to restrict competition for work are treated with greater suspicion.

Innis offered another interpretation of the nature of staple development in Canada. Indeed, he saw the comparative advantage model as a theory developed by and for the interests of the major industrial countries ("metropoles"). In short, the theory justified the large-scale export of raw materials and relatively unprocessed commodities in support of industrial machines elsewhere.

Staple Diversification as a Distinctly Canadian Process

In the Innisian view, the diversification process and the limits to diversification are not simply a matter of the fair play of market forces that apply universally. Rather, in the Canadian case, institutions interweave with geography and technology in shaping how staples are developed (Figure 1.4). In this view, institutions, geography, and technology, which combine to make Canada a distinctive place, are omitted or understated in comparative advantage accounts of staples development (Parker 1988: 65; Hayter and Barnes 1990).

Geography is vital to an Innisian perspective in two senses. First, geography refers to the natural resources and, more generally, the physical geography of a region. As Innis notes, "geography provides the grooves which determine the course and to a large extent the character of economic life" (Innis 1946: 87). Second, Innis interpreted geography in terms of spatial

Figure 1.4

The Innisian triad

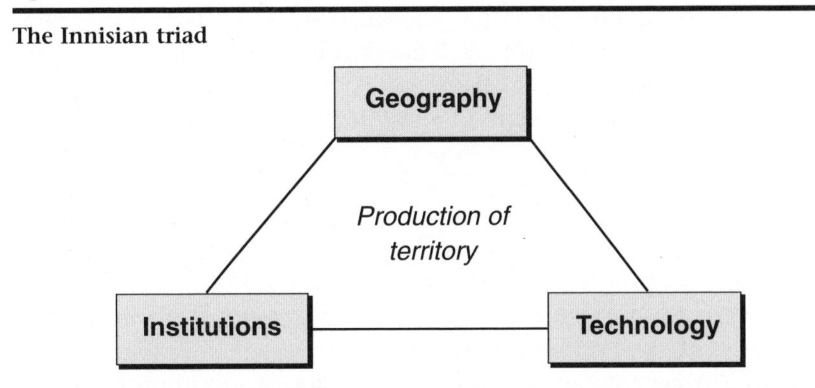

Source: Willems-Braun 1997a: 110. Reprinted with permission of Western Geographical Press, University of Victoria.

relationships, notably in the tension between forces creating centralization and forces resulting in decentralization (Neill 1972: 84). In this context, centralization meant the creation of industrial heartlands or cores, first in the UK and later in the US. But the very nature of industrial centralization in the cores creates its antithesis: forces of decentralization. Thus, growth and stability in the cores require and shape resources, or staples production occurs in the periphery in such countries as Canada. As Neill writes: "The mass production techniques of mechanized, integrated economies require the mass production of raw materials. Consequently, the Canadian economy was biased in favour of large-scale production of primary commodities required for the continual advance of mechanized industry in older areas" (Neill 1972: 38).

Core-periphery relationships are dynamic. Cores themselves rise and fall, and new ones, like Japan, do emerge, but the constant is the peripheries that supply staples are always in the shadow of the cores. Cores typically have access to a diverse range of peripheries, but the latter are often tied to one or very few cores. The trade between cores and peripheries is largely controlled by MNCs based in the cores. R&D occurs in the cores rather than the peripheries. Innis (1933, 1946) also applied the core-periphery framework within Canada. In this perspective, central Canada represents an internal metropole, while the maritime and western provinces, including BC, are doubly peripheral to the central Canadian core, the US, and the UK.

The second element in the triad is the institutional structure (Figure 1.4). Innisians argue that traditional economic theory, such as the comparative advantage model, does not have universal applicability. For this reason, the institutional framework presented within the traditional account of industrial centres need not necessarily hold in staple-producing areas. The lack of institutional congruence between Canada and industrial metropoles is manifold, but three themes are important.

First, in traditional theory, the prescribed role for the state as an institution is usually minimal (or simply ignored), but in staple-producing economies the state plays a vital role. Indeed, for any staple production to occur, the state must be willing to take an active role in upholding and partly providing the massive amounts of infrastructural capital necessary for staple production (Spry 1981). In fact, Watkins (1981: 25) argues: "Both the Act of Union and Confederation were eventually dictated by the need to create a larger state to provide security for foreign capital to build first the canals and then the railways to facilitate the movement of staples." The subsequent acknowledgment of provincial autonomy over resource development in Canada, following the Act of 1935, reinforced the interests of the state in staple production. In addition, Canadian governments have traditionally favoured an open door policy to foreign investment.

Second, the institutional structure of the firm, including issues of ownership, organizational control, internal structure, and competitiveness, is different in the Canadian periphery than in the industrial cores. Thus, in the periphery, foreign control of staples has been a dominating presence (Levitt 1970; Britton and Gilmour 1978). Moreover, Innis emphasized that the very nature of a staple economy – the need for large expenditures on fixed capital that required large firms involved in large-scale distant exports – precipitated this distinctive institutional structure.

Third, and related to the role of MNCs, the institution of the market is different at the margin than in the centre. In particular, the existence of MNCs, production indivisibilities, heavy state involvement, and unpredictability "seriously reduced the value of the price mechanism [in the periphery] as an effective means of adjusting supply to demand" (Spry 1981: 159). As a result, "market relations between the large central dominant areas and outlying areas at the periphery of larger economic systems will result in very different consequences for the dominant centre than for the outlying areas. Efficiency and stability in the central areas may be traded off for dependency and instability in the outlying areas" (Melody et al. 1981: 9).

More generally, Innisians explicitly interpret institutions to emphasize the political part of the idea of political economy. By including unequal relationships of power, which were instantiated within institutions, Innisians "explain theoretically why the external economy shaped, directed, and ultimately controlled the destiny of Canada as a hinterland, preventing it from becoming a fully-integrated, autonomous centre economy" (Drache 1982: 35).

The final element in the triad is technology. For Watkins (1963: 141), this element of staple theory is the most pivotal: "Methodologically, Innis' staple approach was more technological history writ large than a theory of economic growth in the conventional sense." Indeed, technology helps define both the kinds of staples available in the hinterland and the types of staples demanded by the core. From a broad historical perspective, Innis frequently emphasized the significant role of transportation technology. He revealed, for example, that the shift from waterborne transport to rail completely disrupted the inherited local social structures and relationships with massive long-term consequences; trappers and voyageurs became history, and prairie wheat farmers and railway workers became the future. Innis also frequently noted the intimate relationships among process technology, the staple way of life, and Canada's global role. Indeed, there remains a widely held view that Canada's goods-producing economy is based on a narrow technological foundation, specifically "mature" mass production methods to harvest and manufacture resources (Britton and Gilmour 1978; Hayter 1988).

In an Innisian perspective, staple development in Canada has unfolded in a more problematic way than is acknowledged in the comparative advantage model. As Porter (1993) observes, resource exploitation in Canada has been the source of substantial productivity improvements and high incomes. Canada is a rich periphery. The Innisian criticism is that the potentials for productivity gains and especially the use of resources for diversification have not been fully realized. As recent studies emphasize, the Canadian economy, among major industrial nations, remains unusually strongly tied to resources, especially from the perspective of trade (Britton 1996; Wilkinson 1997). Most Canadians may live in urban areas in occupations that are not resource-related, but resource-based company towns remain prevalent and still define Canada's global economic role (Randall and Ironside 1996). The remarkable decline in the value of the Canadian dollar in the 1990s is partly explained by global finance's perception of Canada as a low-value commodity supplier. In peripheral BC, this role is even more emphatic (Marchak 1983; Barnes and Hayter 1997).

In Watkins's terms, the Canadian economy is "locked-in" a staple trap: a state of dependence on resource exports from which it is extraordinarily difficult, if not impossible, to escape. Moreover, this trap cannot be solely explained in the terms of comparative advantage and disadvantage, distances from markets, small population size, and resource depletion. Rather, the staple trap is rooted in institutional structures and attitudes. In particular, the motivations of foreign-based MNCs and of government policies is to promote large-scale exports; diversification is of secondary or no importance. Foreign-owned MNCs do not have the mandate to add value or invest in R&D in Canada, thus reinforcing truncated branch-plant structures in the staple sector and throughout the Canadian economy (Britton and Gilmour 1978). Competition among rival MNCs and provincial governments to develop resources, along with significant indivisibilities in production, encourages cycles of excess capacity and deeply felt booms and busts. Unusually high levels of export dependence typically mean staple regions have little market power. The volatility of prices usually masks the downward trend in the real price of resources and the inevitability of the cost-price squeeze as the costs of resource exploitation inevitably rise to the extent that lowest-cost sources are exploited first (Freudenburg 1992). Ambiguous price signals often combine with ambiguities over depletion rates and resource availability to further entrench commitments to resource exploitation.

The idea of the staple trap implies Canada's economy is unusually resource-specialized and prone to crisis and booms and busts. Auty (1990, 1993) argues that resource-rich developing countries suffer from a "resource curse," while Freudenburg (1992) refers to resource towns across North America as "addictive economies." Trapped, cursed, or addicted as staple economies are, the staple framework remains an appropriate point of departure to investigate industrial restructuring in Canada, especially in peripheral regions such as BC. From this perspective, the volatility associated with the shift from Fordism to flexibility in Canadian resource production, is consistent with the boom-and-bust cycles inherent to staple theory. There are, however, new dimensions of contemporary restructuring not anticipated by staple theory, or indeed by the general literature on industrial transformation.

New Complexities in Contemporary Restructuring

In many parts of the world, the restructuring of industrial regions has faced new nonindustrial complexities scarcely apparent during Fordism (or before). Definitions of economic growth and development during Fordism were rooted in industrial values; even Baldwin's (1956) critique of "plantation economies" implied a commitment to higher per capita income through specialization, trade, and integration. In recent decades, however, these assumptions have been questioned by two powerful, global forces: environmentalism and Aboriginal rights. The emergence of environmentalism, which has developed rapidly around the globe since the early 1970s, greatly facilitated by the communications revolution made possible by microelectronics, has been noted. The implications of environmentalism are strongest in the context of resource regions such as BC. The same point can be made with respect to Aboriginal empowerment, a trend that directly confronts the empty land assumption of staple theory (Willems-Braun 1997a; see also Clapp 1998a and c).

Resource exploitation pushed industrialization to all corners of the globe, constantly forcing out indigenous or Aboriginal populations. During Fordism, it appeared that Aboriginal populations were to be pushed off the map altogether. In this context, Moore's lament offers a prescient, cultural definition of globalization (as "unification of the world"): "rapid incorporation of virtually every part of the world into the international and political 'community' marks the end, or the beginning of the end, for isolated and exotic tribal communities and also for complex and archaic civilizations. In this sense, and only in this sense, the unification of the world is complete" (1963: 89). Aboriginal communities have been at least touched if not transformed by globalization; Aboriginal consciousness has also reawakened in recent years, supported by growing claims for Aboriginal rights and empowerment, based variously on maintaining established ways of life or facilitating self-realization in various forms (Hecht and Cockburn 1989; Clapp 1998a). Confrontation between resource developers and Aboriginal peoples no longer has the same result as Aboriginal opposition is stopping major projects (Anderson and Huber 1988). Moreover, Aboriginal resistance to resource development is widespread; it is occurring in many developing countries, and in advanced countries, notably New World countries such as the US, Australia, New Zealand, and Canada.

The Empty Land Assumption

Both the conventional comparative advantage and the radical Innisian interpretations of Canadian resource development assume Baldwin's empty land as their theoretical starting point. While the vital role of Aboriginal peoples has long been recognized in Canada and BC, they essentially play no part in accounts of staple development since the onset of permanent settlement (Fisher 1977). Knight (1978) notes that Natives were employed for wages in the resource industries of BC in the late nineteenth and early twentieth centuries. In general, however, the empty land assumption reflects the reality that staple development since the fur trade was conducted as a capitalist enterprise in which Aboriginal peoples took little part. In much of Canada, the empty land assumption was given legal justification by treaties that placed Aboriginal peoples on reserves, typically remote places.

In BC, the treaty process was started but left incomplete, and in the terms of BC's entry into Confederation with the rest of Canada, Native interests and voices were excluded (Fisher 1977: 176). Indeed, with considerable suddenness after the decline of the fur trade in 1858, Native culture was undermined and "reduced from an integral to a peripheral role in British Columbia's economy" (Fisher 1977: 210). In BC, a system of reserves was established that, as elsewhere in Canada, oddly combined geographic isolation with an assumption of assimilation. The reserves were scarcely supportive, and the Indian population in BC, estimated at 25,661 in 1881, declined in number until the 1930s. The provincial government control over BC's forest resources and industrialization evolved as if BC were an empty land.

The completion of the treaty process, albeit 120 years or so late, has now been given priority by the provincial government to revisit and redress the empty land assumption. Historically, in BC as in much of Canada, forest land has not been privatized. In BC, all Crown land, and maybe private land as well, is still subject to land claims. Moreover, these claims have been given considerable impetus in recent years by the *Delgamuukw* decision of the Supreme Court of Canada, which has provided a stronger, albeit confusing, legal basis for Aboriginal peoples to pursue rights and title than had hitherto been the case (*Delgamuukw* v. *British Columbia* 1997). Indeed, the Supreme Court overturned BC's superior court judgment in this matter. Whether or not this judgment will redefine Aboriginal relations across the country is unclear.

For BC, a major implication of *Delgamuukw* is that Baldwin's scenario of potential capitalist development occurring in either an empty land or a region with a dense Aboriginal population has been turned upside down. In BC, the Aboriginal population of a sparsely populated area is looking for compensation and development in an already established capitalist system. Models for this scenario are not readily apparent. Aboriginal peoples, whose culture was discounted and irredeemably changed by the onset of industrialization in BC, will surely count in the future industrialization of BC's forests.

Conclusion

BC's forest economy has developed as a local model (Barnes 1993) or a regional world of production (Storper and Salois 1997), serving as an integral component of the Canadian political economy within the opportunities and constraints of global industrialization. But BC is also distinct in Canada. If the Canadian economy is a resource periphery to global industrialization, BC is peripheral within a Canadian context, too. Unlike other Canadian regions, BC's economy – and how its forest resource has been developed – has always been shaped by its links to the Pacific as well as to the rest of the continent and Europe. BC is clearly not a poor periphery; its forest resources are the richest and most diverse in Canada. The evolution of its economy and industrial forest policies has been powerfully shaped – not determined – by external forces and geographic situation. The next chapter examines this evolution.

2
Life on the Geographic Margin: The Evolution of British Columbia's Forest Economy from the 1880s to the 1970s

The geographic marginality of British Columbia, on the edge of an empire, a continent, and an ocean, is the defining context for understanding the evolution of its forest staple. In the last decades of the eighteenth century and first part of the nineteenth century, BC was peripherally engaged in supplying the British Admiralty, especially around the Pacific. The Admiralty's main needs for Canadian (colonial) timber were met by the so-called square-timber trade that centred on the central and eastern provinces and that waxed and waned according to the rhythm of Great Britain's wars and fluctuating access to the timber supplies of the Baltic and the US (Lower 1973). Consequently, BC was integrated into the world economy at a relatively late date. Although a few sawmills were constructed in the 1850s and 1860s, large-scale exploitation of BC's forests had to wait for the arrival of transcontinental railroads, beginning with the CPR in 1886 and the opening of the Panama Canal in 1914, which allowed access to continental and British as well as other European markets. As Harris (1997) emphasizes, the early economic development of BC, and its forest industries, was a problem in overcoming distance.

Yet, as the challenge of distance declined in the twentieth century, BC's role as peripheral supplier of forest-product commodities to the world's metropoles became more entrenched. Indeed, the massive expansion of BC's forest industries was predicated on export access, principally to the US, while in recent decades transatlantic trade has been surpassed by exports across the Pacific Ocean to Japan, a market on the other side of a great cultural as well as geographic divide. But attempts to overcome distance to link remote timber supplies with distant markets reveal a deeper relationship: from its beginnings, the evolution of BC's forest economy has been tied to the short- and long-term rhythms of global industrial transformation.

This chapter examines the evolution of BC's forest staple from the mid-nineteenth century to the early 1970s. It argues that, in accordance with global developments, the industrialization of BC's forests in this 100-year period was based on two major industrial transformations (see Table 1.1).

In particular, while there had been earlier, tentative developments, large-scale industrialization of BC's forest economy began in the 1880s, in association with a "global" boom centred on Europe and North America that Freeman (1987) cites as the start of the "electrical and heavy engineering" techno-economic paradigm. The effects of this phase of industrialization had more or less petered out by the 1930s, but in the 1940s, heralded by the Sloan Royal Commission (1945), the BC forest economy experienced a remarkable, sustained period of growth during the Fordist techno-economic paradigm.

In the BC forest economy, developments in the first transformation were fundamentally entrepreneurial in nature. In the second transformation, however, in a manner faithfully reflecting the central characteristics of Fordism, the large international corporation, employing unionized labour in big factories manufacturing vast quantities of relatively standardized outputs, became the organizational model of development. This chapter outlines the shift from entrepreneurialism to Fordism in BC's forest economy, noting the main institutional, geographical, and technological implications, and setting the stage for understanding the dynamics of contemporary change. Given that the chapter focuses on the 1880s to the 1970s, it uses 1873 and 1973 as convenient "markers." In 1873, just two years after BC entered Confederation with Canada, a global economic crisis began, and after 1973, the winds of change that so unsettled Fordism, beginning with an energy crisis and the termination of the Bretton Woods exchange-rate system, began to blow hard. The year also marked a change in the British Columbia provincial government as the left-wing New Democratic Party (NDP) was elected for the first time.

The First Industrial Transformation of BC's Forests

For 100 years or so, culminating in the gold rush era of 1858-79, outside powers expressed growing interest in BC and its forests (Galois and Hayter 1991). These early, tentative initiatives set the stage for the large-scale industrialization of the forest staple, led by the lumber industry, that began in earnest in the 1880s, a development much stimulated by the entry of the province into Confederation and the completion of the CPR.

The area that is now BC was integrated into the world economy in the second half of the eighteenth century as a result of political and economic competition among several European states, primarily Russia, Spain, and the UK, as well as the US and the colony of Lower Canada. European nations made the initial maritime approaches to the area, while the North American entrants, if later on the scene, had the advantage of land as well as sea routes. As a result of these activities, British Columbia, by the first decade of the nineteenth century, had become the intersection of complex spheres of influence. Three basic points about this activity may be noted. First, the establishment of British sovereignty conditioned the shape of subsequent economic and political developments over what was to become BC. Second, the tension between continental integration and maritime linkages, evident by 1810, persisted through much of BC's subsequent history. Third, implicit in the geopolitical considerations that attracted attention to the Northwest Coast was an interest in the resources of the region.

The maritime fur trade, built on sea otter pelts and access to Chinese markets, provided the initial impetus and was soon supplemented by the continental fur trade. By 1821, New Caledonia, as the region was then known, had fallen under the hegemony of the Hudson's Bay Company (HBC), the British fur-trading organization. The British had also begun to exploit the forest resources of the region. According to Taylor: "Captain James Cook in 1778 sparked the growth of the forest industry in British Columbia when his sailors replaced the rotting masts of their ships with Douglas Fir cut in Nootka Sound. Significantly, the first use made of BC timber by the explorers was as ship masts, which drew the attention of the British Admiralty and private investors, and became the primary use and main export of timber cut on the northwest coast well into the middle of the 19th century" (1975: 7).

In practice, this limited trade was primarily a maritime one, exemplified by the spars carried by Captain John Meares from Friendly Cove to China in 1788 (Taylor 1975: 7). But there were other developments. Thus, the HBC diversified its activities in BC beyond the confines of the fur trade to embrace other staples including lumber. HBC's first lumber shipments to Hawaii in 1829, and subsequently to California, South America, and even England, were actually from the lower Columbia in American territory. HBC sent its first lumber shipment from Vancouver Island to California in 1849, following the Oregon Treaty of 1846, which resulted in British territory being confined to north of the forty-ninth parallel and Vancouver Island being established as a colony (Galois and Hayter 1991: 193). The impacts of gold mining were also being felt.

The Gold Rush Era, 1858-79

In 1848-9, the rapid development of gold mining in California stimulated other industries and adjacent regions, including the forest industries of the Puget Sound and, to a limited extent, Vancouver Island. In 1853, despite a 20 percent tariff, eighteen vessels cleared Vancouver Island for San Francisco. Shipments of lumber, spars, and shingles were also sent to Hawaii, South America, and China. Then, in 1858, the mining frontier penetrated the Fraser Canyon and BC became a Crown colony, extended in 1867 to include Vancouver Island. The sudden influx of population that accompanied this advance of the mining frontier created instant local markets and aroused fresh interest in the timber resources of the new colony. While many small sawmills sprang up to supply the mining camps and service centres, a few bigger mills were established to serve export markets. Alberni, on the west coast of Vancouver Island, was the site of the first such "export" mill built in response to fears that timber supplies to Britain would be disrupted by the American Civil War (Hardwick 1963: 10). By the end of the colonial period, Burrard Inlet had developed as the centre of the export trade and two steam-driven sawmills operated on the future site of Vancouver.

Viewed from a broader perspective, these developments in the lumber industry of BC were modest. In neighbouring Washington Territory, with access to a more substantial domestic market, the industry was considerably more advanced. Even within BC, the economic significance of the forest industry was limited; in 1869, for example, the value of production was much less than gold, and in 1870, fur exports were greater than forest exports (Galois and Hayter 1991: 272). In the 1870s, after the initial flush of placer-mining activity had subsided, BC was a small, isolated settler colony with a non-Indian population of about 10,000.

In this context, BC entered Confederation in 1871, hoping to thus maintain ties with Britain by means of integration through the construction of a transcontinental railway. Unfortunately, the sharp global downturn beginning in 1873 heralded a long period of economic difficulty as the world's metropoles shifted to a new phase of development (see Table 1.1). By 1881, with construction of the Canadian Pacific Railway (CPR) finally under way, BC's population was just 50,000, of which slightly less than half were non-Indian. No reliable data are available on lumber production during this period, but the value of forest exports fluctuated between a high of $327,369 in 1878 and a low of $172,641 in 1881 when forest products comprised 7.2 percent of BC's exports. The large-scale industrialization of BC's forests was close at hand, however. An upsurge in the global economy and the completion of the CPR set the stage for a boom in the BC economy, especially by the lumber industry after the mid-1890s.

Lumber as the Lead Industry, 1880s-1930s

For BC, the construction of the CPR heralded a new pattern of integration into the world economy: maritime, continental, and industrial rather than mercantile. This integration was not smooth. From the 1880s to 1939, an overall massive increase in timber cut and lumber production reveals the effects of business cycles of varying intensity (Figure 2.1). From fluctuating beginnings, the boom between 1896 and 1913 was arrested by the Great Depression. Recovery and the Second World War then set the stage for the Fordist boom.

In 1895, a timber harvest of around 127 million board feet supplied seventy-seven sawmills operating in BC, of which forty-one were on the coast and thirty-six in the interior. By 1913, the total cut was thirteen times greater than in 1895; the number of sawmills had grown from seventy-seven to about 270 in the same period. In addition, the value of forest production reached $33.6 million in 1913, exceeding the value of mineral

Figure 2.1

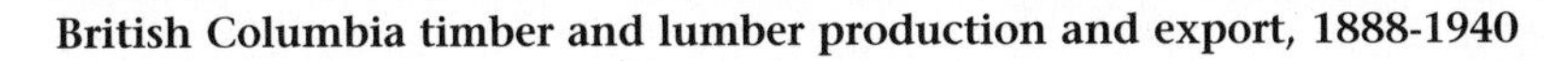

British Columbia timber and lumber production and export, 1888-1940

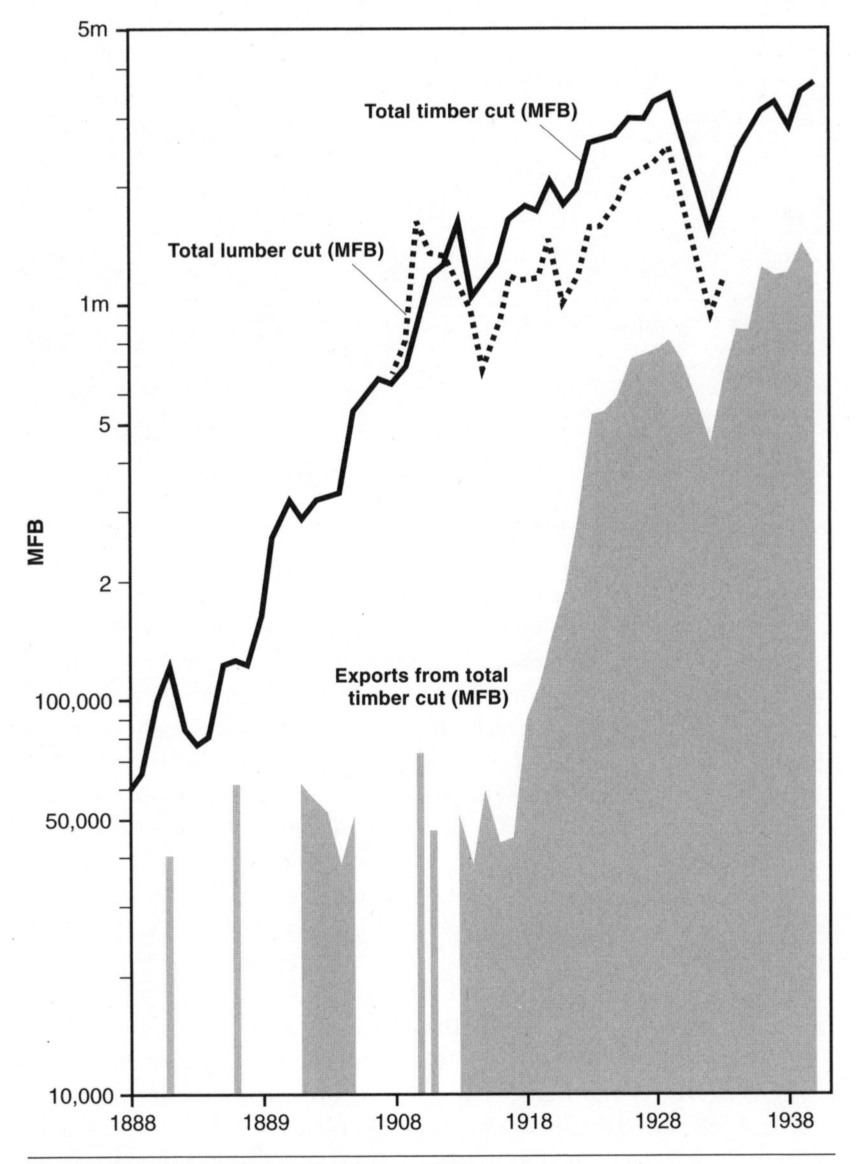

Source: Galois & Hayter 1991: 178-9 (columns A & C).

production for the first time: forest products had become the leading staple in the province and lumber the lead industry.

The pre-First World War boom faltered with the onset of war, but a sharp decline in production was soon offset by the demands of the war economy

Newsprint manufacture (and exports) at Powell River began in 1912.

and the total cut, if not lumber production, was at a record level by 1917. The trend of expanding production continued, apart from occasional recessionary years, until the peak of 1929 when the total cut and lumber production stood at levels approximately double those of 1913. The Great Depression was a severe blow to BC's forest industry; production declined precipitously from the peak of 1929 to the trough of 1932, and output was reduced to 1914 levels. Values declined even more dramatically. In 1932, lumber production yielded little more than one-quarter of the returns of three years earlier. Recovery from this collapse was slow but fairly consistent through the remainder of the decade. By 1939, lumber production and values were slightly in excess of the levels a decade earlier.

While the dramatic expansion of the forest staple after 1896 was led by lumber, other forest industries appeared, most notably pulp and paper; the first successful newsprint mill at Powell River opened in 1912. This mill, along with sulphite-pulp mills at Woodfibre and Port Alice, and a newsprint mill at Ocean Falls, survived the Depression and were operating in 1939. In addition, shingle-and-shake mills had been established as had the plywood industry. The timber cut in 1939 was about sixty times greater than in 1888 and twenty-seven times greater than in 1896. Such growth, in turn, given the great size of the trees, required considerable investment in transportation and logging technology.

Transportation and Technology

Initially, BC's huge trees were cut by axe and handsaw, and the distances and volumes that logs could be hauled were inevitably limited because they relied on manpower, oxen, and horses. The widespread application of steampower in the logging sector, however, was a crucial development. In the Vancouver area, around 1878, the spread of logging was facilitated by the use of a converted paddlewheel steamboat (Hardwick 1963: 50). On a bigger scale, railways changed the "whole scale of operation" (Hardwick 1963: 15). By 1913, railway route miles in the province had increased five-fold since the construction of the CPR. Company-owned logging railways also expanded; according to Hardwick (1963: 33), 170 operators laid 1,000 miles of track by 1925. The railways, in association with the Davis raft first used in 1911 to move logs across exposed coastal waters, allowed the industry to penetrate more distant areas, including the isolated coast, and to provide sawmills with sustained, large-volume supplies of logs. In the woods, the efficiency of logging was increased around the 1880s by the introduction of the crosscut saw, in the late 1890s by the steam donkey (a steam engine connected to a large drum and cable) that partially replaced horses and oxen for dragging and yarding logs, and around 1912 by high-lead logging systems (Cox 1974: 227-31). Even as horses continued to be used in coastal and interior operations, steampower in the woods reached its zenith in the 1930s.

The expansion of the industry also depended on new processing technology. The Alberni sawmill was the first to use steampower in 1860; bandsaws were adopted as a more efficient alternative to circular saws; and electric power was introduced in the early 1900s. Sawmills increased in size considerably and were more numerous in 1925 than in 1900. More, bigger sawmills in turn depended on the creation of markets.

The Search for Markets

If the evolution of BC's forest industries has depended on exports, as assumed by the staple thesis, the story is more complicated in practice. Export markets could not be taken for granted, waiting only for the establishment of transportation links. BC's lumber industry had to seek out and negotiate export markets. In the meantime, domestic markets played a decisive role in the industry's development, especially in its first boom of 1896-1913.

Before this boom, the lack of markets delayed development of BC's forest resources. BC was too far from Britain to serve the Admiralty, except around the Pacific, and American markets were served by American producers that were protected by high tariffs. BC's limited lumber exports were supplemented only by the limited local demands generated by fur trade and mining. Large-scale industrialization of BC's forests had to await the domestic demands created by railroad construction, Vancouver's housing boom, and the settlement of the Prairies. If many of these domestic demands were staple related, including the building supplies needed by western farmers to export wheat, the world's metropoles were not gasping for lumber from BC, at least not for a while.

In 1895, a little less than half the total cut was exported, mainly around the Pacific. Less than twenty years later, on the eve of the First World War, lumber exports had been reduced to insignificance: a mere 3.2 percent of the total cut, 4.4 percent of the lumber cut. In physical terms, the volume of lumber exported was lower in 1913 than in 1895. Moreover, a handful of companies was responsible for almost all lumber exports. The transformation of the BC forest industry at this time was driven by the expansion of domestic markets. Although data are not consistently available, in 1904 lumber shipments by the Pacific Division of CPR amounted to 119,761 tons, of which 8 percent stayed in BC and 85.7 percent went to Alberta and Saskatchewan (Galois and Hayter 1991: 180). A further 111,101 tons were shipped from the Kootenays by rail, probably mainly to the Prairies. Before 1913, perhaps the major change to this pattern stemmed from the localized demands of Vancouver's growth.

After the First World War, exports quickly assumed a much bigger role, and the year 1918 marked record levels. A variety of factors, domestic and international, contributed to this development. While prairie markets declined significantly, the completion of the Panama Canal in 1914, combined with subsidized shipping, offered BC access to the eastern seaboard of North America and to Europe, as well as to Pacific Rim markets. Inevitably, the Depression of the 1930s reduced export markets, but surprisingly perhaps, they survived better than domestic ones. Further, the recovery of the forest industry in BC after 1932, as measured by volume of production, was driven by exports. The export markets of the 1930s, however, were very different from those that contributed to the expansion of the 1920s.

Expanding after the First World War, American markets came to dominate the BC forest industry in the 1920s and did so until 1931. In 1932, however, the Smoot-Hartley Act imposed high tariffs on imported lumber and American markets collapsed. Pacific Rim markets increased but not by much. Markets had to be sought elsewhere, a goal made possible by the establishment of Imperial Preference, following the Ottawa Conference of 1932. Imperial Preference, combined with a rapid expansion of subsidized house construction in the UK, brought exports to record levels by the mid-1930s. By 1939, the UK was receiving almost 70 percent of all BC lumber exports.

The new export markets depended on new marketing efforts. Before 1913, BC's mills had been largely dependent on American brokers and carriers for exports, and they were in a fundamentally weak bargaining position. A number of measures aimed at market extension were undertaken by the federal and provincial governments. Probably the most significant measure was the appointment of H.R. MacMillan as a special trade commissioner charged with seeking out foreign markets, which he did. The experience MacMillan gained as trade commissioner also proved invaluable when he moved, in 1919, to establish his own export company. Within a decade, the

MacMillan Export Company assumed a dominant role in the export of BC lumber, acting on behalf of many sawmills in the province. By 1935, however, tension was evident between MacMillan and the producers concerning the division of benefits, and the producers formed their own marketing arm, Seaboard Lumber Sales Ltd., to break MacMillan's grip over export markets. In turn, to establish his own supply line, MacMillan began manufacturing by purchasing mills and timber limits. Since then, these sales organizations have dominated the offshore marketing of lumber.

Geography of Production

In the emerging geography of the forest staple within BC until the Second World War, two dominant and related themes are identifiable. First, the coastal region, and especially the Vancouver area, became the centre of production; Vancouver was a sawmill town (Table 2.1). Second, lumber, the leading industry of the period, related logging activities, the suppliers (that is, backward linkages) to these industries, and value-added activities were fundamentally entrepreneurial in nature. Apart from Ocean Falls, even the pulp-and-paper mills were individually owned and run.

Coastal Dominance

Before large-scale industrialization, sawmills had primarily followed local developments related to fur and mining, and mills were located in the interior as well as on the coast. In the 1880s and 1890s, lumber developments occurred in several places in the interior, notably the Kootenays, as well as

Table 2.1

British Columbia: Distribution of sawmills by region, 1914, 1928, and 1939

	1914			1928			1939		
Region	*N*	Capacity MBM	%	*N*	Capacity MBM	%	*N*	Capacity MBM	%
Vancouver	139	4,770	56.0	174	9,114	76.5	180	8,539	73.0
Prince Rupert	18	392	4.6	27	488	4.1	44	569	4.9
Total coast	**157**	**5,162**	**60.6**	**201**	**9,602**	**80.6**	**224**	**9,108**	**77.9**
Southern interior	107	2,312	27.1	83	1,626	13.6	90	1,020	8.7
Other interior	70	1,045	12.3	30	691	5.8	147	1,570	13.4
Total interior	**177**	**3,357**	**39.4**	**113**	**2,317**	**19.4**	**237**	**2,590**	**22.1**
Total	**334**	**8,519**		**314**	**11,919**		**461**	**11,698**	

Notes: The data for 1928 and 1939 refer to operating mills only. Capacity MBM: eight-hour daily capacity in thousands of board feet.
Source: Galois and Hayter 1991: 189.

the coastal regions. Interior locations were also in a good position to serve prairie markets. On the other hand, population growth favoured the coast. Thus, as the provincial population increased by more than 400 percent in twenty years to reach 392,480 in 1911, Vancouver developed rapidly to become a city of 100,000, outstripping Victoria in size. Moreover, as maritime-based export markets expanded, the coastal forest industry grew fastest. Thus, the coast's share of productive capacity increased from 56 percent in 1914 to 76.5 percent in 1928, and the share was still 73 percent in 1939. Meanwhile, the collapse of prairie markets and the reduction in merchantable timber through overcutting, combined to reduce the share of milling capacity in the Kootenays from 27.1 percent in 1914, to 13.6 percent in 1928 and 8.7 percent in 1939. In the interior, the main growth shifted to parts of the north where the Canadian National Railway had opened a railroad.

Within the coastal region, Vancouver had emerged by the 1880s as the most important lumber-manufacturing centre (Figure 2.2). Its first mill, an export mill built in 1863 by Captain Stamp (who had also built the Alberni mill in 1860) was located at Moodyville on the north shore of Burrard Inlet.

Figure 2.2

Log production and consumption

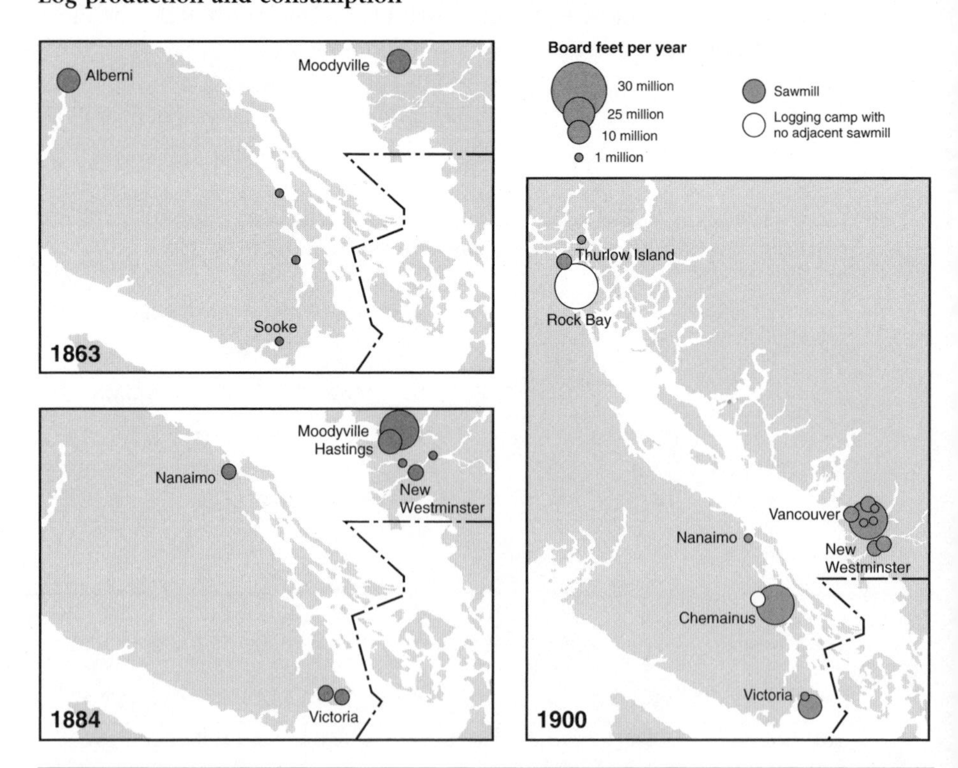

Source: Hardwick 1963: 16.

By 1891, the new city of Vancouver boasted a population of over 13,000 and contained nine sawmills, with another four in the immediate vicinity. The Moodyville mill was already obtaining logs from over 100 miles away via water transportation. In 1931, when the scale of activity had increased

Figure 2.3

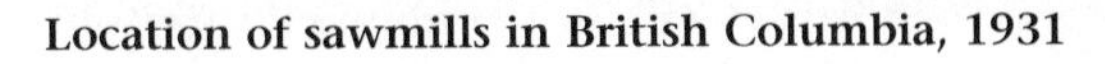

Location of sawmills in British Columbia, 1931

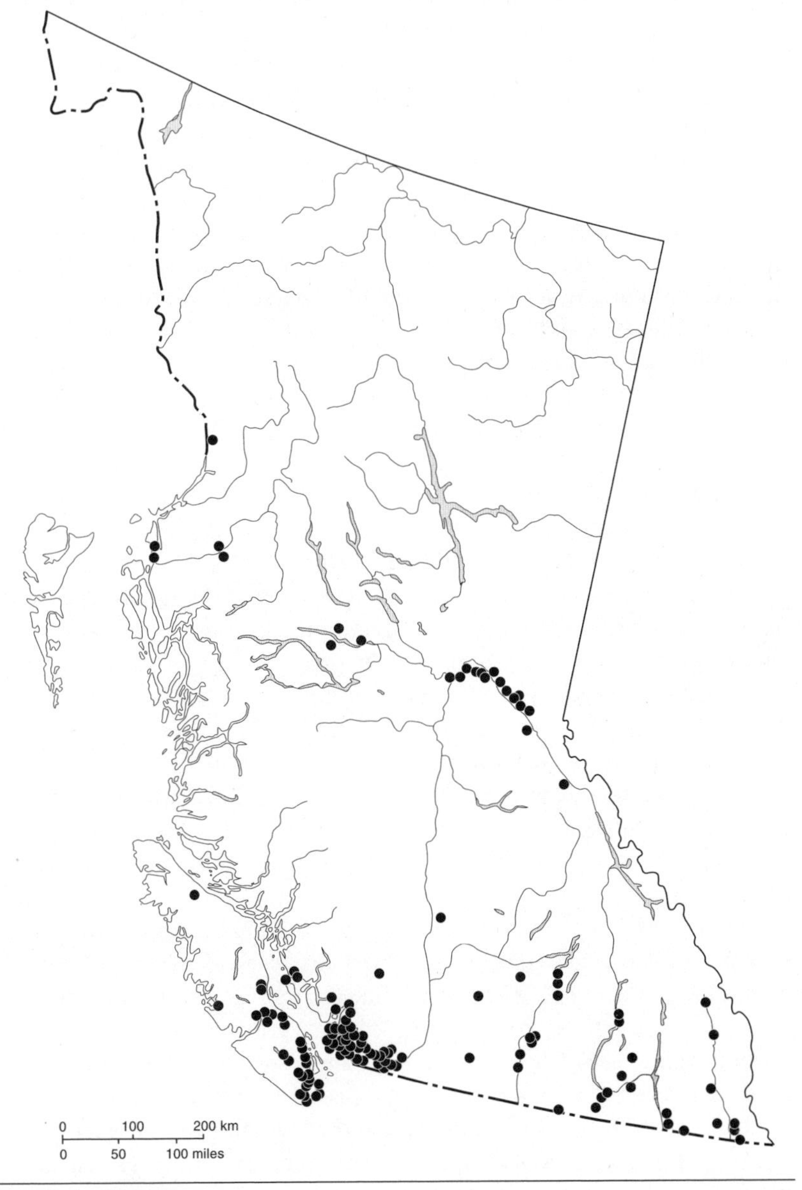

Source: Farley 1979: 65.

substantially, Vancouver was confirmed as BC's principal sawmilling centre (Figure 2.3). The second main concentration of sawmills was located across Georgia Strait in the Victoria-Chemainus region of southern Vancouver Island, with Port Alberni another established centre on the coast. In the interior, sawmill investments could be found among a few scattered pockets, such as along the Fraser near Prince George. Otherwise, the interior remained an empty land. The 1939 map is similar, except for changes in the Vancouver area that featured the closure of mills on Burrard Inlet and expansions along both arms of the Fraser, while False Creek remained Vancouver city's largest concentration of wood industries (Hardwick 1963: 20). The five pulp-and-paper mills were also located in the southern coastal region.

The Vancouver area obtained an early lead as the province's sawmill centre principally because of advantages in log-harvesting assembly and exporting. Slopes were gentler in this locale than at Alberni, for example, and the giant Douglas-firs more easily "skidded" to mills on the water's edge. The choice of Vancouver for the CPR's western terminus further boosted the region, giving it access to continental markets, and once established, the raw material supply hinterland was readily extended by rail and raft. Generally, lumber firms preferred to concentrate in the Vancouver metro, with its established infrastructure, labour, and suppliers (backward linkages), and to pay higher log transportation costs rather than disperse to more remote locations with lower transportation costs but with other disadvantages (Hardwick 1963).

The Entrepreneurial Character of the First Industrial Transformation

In Baldwin's terms, BC's forests were developed more along the lines of the (immigrant) farmer-entrepreneurial development in an empty land model rather than in the plantation model (Chapter 1). The entrepreneurial structure of the lumber, shingle, and logging industries, in which ownership and control were highly decentralized and held locally, further stimulated entrepreneurship through local linkage generation. Vancouver had become the administrative as well as the manufacturing and trade centre for the forest staple. Recognition of business opportunities and the consummation of business deals was further reinforced by concentration in the Vancouver area.

Moreover, distinctive problems arose in accessing and using the local resource, and they required distinctive local technical adaptations. Trees were enormous, not easy to log or move, while the problems of handling massive trees were not confined to the logging sector. As Carrothers (1938: 246) notes, sawmill equipment of "exceptional ruggedness and power" was necessary to handle the size and weight of logs in the coastal zone. Local suppliers were in the best position to meet these needs, and a Vancouver location combined access to imported materials, markets for equipment, and a labour supply. Indeed, by 1911, numerous backward linkages, principally comprising locally

Foreshadowing the industrialization of BC's forests. (This photo also reveals early links around the Pacific Rim through the supply of labour.)

owned small firms (notably, engineering and repair shops, logging and sawmill and shingle-mill equipment manufacturers, transportation and warehouse activities, and ship building and repair) had been established in the Vancouver area. In addition, some forward linkages (such as furniture, barrel, door, and coffin manufacturers) were formed by numerous small firms, also typically in the Vancouver area to be close to consumer markets as well as labour supplies and infrastructure.

It is reasonable to suppose that the forest industry and related activities created a flexibly specialized industrial district in the Vancouver area. Small firms were important: they had access to a common labour pool; numerous business transactions occurred; local consumer demands, most notably for equipment, were distinct; and if competition for business developed, cooperation and the sharing of information likely occurred as well. In the search for export markets, for example, H.R. MacMillan represented competitive entrepreneurialism as he sought to maximize the profits from markets he had identified and negotiated by acquiring his own productive base in BC. To compete with MacMillan, numerous other entrepreneurial firms formed Seaboard, a cooperative marketing agency.

If entrepreneurship injected vitality, risk taking, and diversification into the forest staple, there was also tremendous speculation that was not always well rooted in financial, organizational or technical ability, or social rationale. Timber speculation was rampant. As Hardwick (1963) notes, after 1900 many of the entrepreneurs were newcomers who had previously helped mine midwestern (and other American) forests, and they now applied the same philosophy to BC. Before BC's entry into Confederation, land and forests could be purchased outright. After 1870, an evolving system of long-term leases, short-term licences, and long-term licences was introduced in return for rental income and royalties (Cail 1974; Pearse 1976; Schwindt and Heaps 1996: 24-6). The government retained ownership of most forest lands in the province and increased fees, but forest management issues were practically ignored. In some cases, lease and licence holders were required to harvest and invest in manufacturing facilities; in other cases, forests could be held as a long-term investment. A policy of long-term, twenty-one-year licences introduced in 1905 stimulated particularly frenzied speculation, and the first provincial Royal Commission on timber and forestry, chaired by F.J. Fulton, was established in 1909 to assess timber-leasing and -licensing policies. This commission found that enough forests had already been alienated to sustain industrial forestry in the foreseeable future. Its recommendation that no further allocations be made, apart from timber sale licences, needed to rationalize logging operations, was accepted.

By the 1940s, BC's forest policy set by the Fulton Commission, faced two major concerns. First, industry needed more timber to expand, and the increasing size of processing facilities required larger supplies of raw material. Second, forest liquidation, was a problem in the areas already under lease and licence, specifically for Douglas-fir logs. Under an entrepreneurial regime, on public and private lands, forest management was clearly deficient. These issues stimulated a second Royal Commission of inquiry into forest policy in 1945, headed by Chief Justice Gordon Sloan, which laid the foundations for another transformation of BC's forest industries, now summarily labelled the shift to Fordism (Sloan 1945).

The Sloan Commission and the Transformation to Fordism

The Sloan Commission's mandate was to recommend policies to establish sustained yield, rather than liquidation, as the basis for forest policy in BC, and to provide industry with secure supplies of timber for large-scale operations. To quote Sloan, the policy challenge was how to replace "a system of unrestrained and unregulated forest exploitation regarding the forest as a mine to be exhausted of its wealth ... to a system based on sustained yield, wherein the forest was to be considered as a perpetually renewable asset" (quoted in Mackay 1982: 153). In his follow-up Royal Commission, Sloan (1956: 40) stated the goal more specifically as providing a "perpetual yield of wood of commercially usable quality from regional areas in yearly or periodic quantities of equal or increasing volume." If by the 1980s public policy considered sustainability of the forests to be a priority, the Sloan hearings are an important precedent.

In the 1940s, Sloan equated forest policy with industrial policy. Sustainable or renewable forests were needed for sustainable or stable industrial growth and community development. Sloan believed that this goal could best be achieved by the creation of extensive timber leases that would be granted over long terms in exchange for large-scale industrial development, preferably by large corporations with the appropriate industrial and financial capabilities. The logic is simple enough. Long-term leases covering huge forest areas allow permanent large-volume supplies of wood while permitting harvested areas sufficient time to renew. Large firms with major investments would be committed to forest renewal and would have the capability to follow through.

Bearing in mind that expansion of the capital-intensive pulp-and-paper industry was foreseen, the assumptions underlying Sloan's recommendations were not unreasonable. Forest management was in bad shape, and the forest industries had recently experienced the upheavals of the Great Depression. The global violence of the Second World War was also fresh in memory. Small-scale loggers and lumber operators had not fared well in the face of market and supply uncertainties (nor had they demonstrated much interest in forestry), and signs of corporate integration were already evident in the 1930s (Hardwick 1963: 21). The impact of the policies adopted from Sloan's recommendations on the demise of the entrepreneur was debated and criticized. Ironically, a chief critic, H.R. MacMillan, presided over the firm leading the charge to corporate integration.

Sloan's recommendations became the basis of forest policy in 1947 which provided BC's recipe for Fordism. As Sloan prescribed, BC's forest economy was to be dominated by Big Business that, in close alliance with the provincial government, emphasized the forest as timber supply for mass production. Moreover, labour unions quickly accepted this recipe to create an implicit (and classic Fordist) deal among (big) business, (big) government, and (big) labour that Wilson (1987/88: 9) calls the "wood exploitation axis." The basis of the deal rested in forest policy.

Forest Policy

Two major new forms of tenure were created: private and public working circles, which respectively became known as tree farm licences (TFLs) and public sustained yield units (PSYUs). Both forms were "regulated," ultimately according to the allowable annual cut (AAC) set by the forest ministry. The ostensible principle underlying an AAC was that harvesting rates would not jeopardize the flow of timber in the long run. Harvesting in any given year was allowed to exceed the AAC but not over a longer, usually five-year, period. In practice, AACs were revised in light of new information and technological change that permitted full utilization of timber. During Fordism, and until relatively recently, AACs were revised only in an upward direction.

In the TFLs, firms were required to have logging plans approved and were responsible for forest management (notably, replanting, preventing and controlling fire, and ensuring that AACs were not exceeded) to meet sustained-yield principles. For companies with private lands and older leases and licences (old temporary tenures or OTTs), the government sought to encourage sustained yield by offering to combine such holdings with unencumbered tracts of Crown land as part of a TFL. MacMillan Bloedel's TFL 20 on Vancouver Island, for example, was created in 1954 by such a combination. TFLs had varying time horizons, although twenty-five years became standard. In 1961, the government introduced pulpwood harvesting areas (PHAs), huge areas within which pulp-and-paper mills had rights to pulpwood and chips from sawmills (Hayter 1978c). On the PSYUs, the government had responsibility for forest management, and wood was allocated by competitive bidding. PSYUs were modified in 1967 by timber sale harvesting licences, which were usually ten-year agreements in which firms took over forest management responsibilities.

For timber cut on Crown land, companies paid stumpage fees to the provincial government. Stumpage is essentially a tax on logs harvested, and the formula for its calculation reflected a market-based system that remained in effect until 1987. As practised in BC, stumpage calculations have been regarded as a rather mysterious process, although the principle is simple enough (Schwindt 1987). Thus, stumpage was calculated as a residual of the difference between revenues less costs in which the latter involved an uncertainty factor associated with the difficulties of logging in a particular site or with marketing problems. In practice, stumpage decisions therefore involved judgment and close negotiations between companies and the provincial government. A fundamental implication of the market-driven formula was that stumpage fluctuated considerably, and it could be both high and extremely low, even lower than the government's own costs of forest management. Whenever booms were greater than busts, however, the political damage of such occurrences was mitigated.

The award of harvesting rights to companies was tied to investment proposals, the larger the scale, the better. Large companies were also able to finance the construction of logging roads, which had replaced rail as a means of access to trees, as well as to invest in forest management. Moreover, as improvements in wood utilization were made, for example, by the introduction of chippers, chip 'n sawmills, small-log sawmills, and sawdust-pulp refiners, the provincial government passed increasingly strict wood-harvesting and manufacturing utilization standards, in terms of how much of the tree could be left in the forest. Stumpage formulas were also modified to assume trees were effectively used, specifically so that sawmills would provide "waste" wood as chips for pulp mills, as bark for fuel, or as waste for other uses.

In more general terms, the provincial government sought to facilitate the development of the forest industries by providing a stable and positive business climate. Thus, considerable investments were made in infrastructure, especially extensions of rail, road, and power networks; new town legislation further facilitated "instant resource towns" in remote areas of the province (Bradbury 1978); and the government welcomed foreign direct investment (FDI) without restriction. These policies further reinforced preferences for large firms in BC's forest economy. Thus, massive infrastructure investments were readily justified by large investments in productive facilities whereas the FDI process was largely dominated by large MNCs. Moreover, a major goal of forest policy was to disperse forest industries throughout the province to achieve "regional self-sufficiency" in the production and consumption of logs (Hardwick 1963: 73; Bradbury 1978).

Public Forests

Crucially, the government chose to maintain BC's forests overwhelmingly in public ownership. Such a choice may be considered surprising. After all, private ownership of forest lands is a widespread feature in other industrialized societies, and the provincial government in BC was controlled by a single, right-wing party committed to free enterprise throughout the Fordist period. As Drushka (1993: 5) points out, a land ordinance in 1865 allowed the Crown to grant timber-harvesting rights on public land while maintaining public ownership of the land itself. In the 1880s, land grants were made to promoters as an inducement to build railways, the most important being 1.9 million acres on the east side of Vancouver Island for construction of the Esquimalt and Nanaimo Railway. In 1896, however, legislation was passed prohibiting privatization of timber land, although timber speculation continued until 1912. In other words, the principle of public ownership of forests was introduced during the first industrial transformation of BC's forests.

This principle has proven resistant to change. Drushka (1993: 1-11) reveals that in 1942 the chief forester privately urged the government to adopt a sustained-yield policy based on the creation of sustained-yield management units and the retention of Crown ownership. This proposal was known to Sloan, and it became policy, apparently without much public debate. Drushka (1993: 7-11) is probably right when he says that the key decision makers in both the public and the private sectors believed that the public would not accept the idea of privatization (even though they were never asked). The rationale for privatization, and the problems of public ownership, was discussed, however, especially regarding the greater incentives of the former for forest management. Drushka notes that the government may have been concerned about losing revenues or about public opinion, which may have been aroused if privatization also implied foreign ownership.

Kraft Pulp as the Lead Industry, 1947-73

During the 1950s and 1960s, forest-product commodities expanded impressively in BC, faster than national averages (Figure 2.4). Thus, the long-established lumber industry grew at an annual average of 4.4 percent in the twenty-one years from 1950 to 1970, more than doubled in production, and retained its almost 70 percent share of the industry in Canada. Newsprint,

Figure 2.4

Output of selected forest-product commodities, 1950-70, Canada and British Columbia

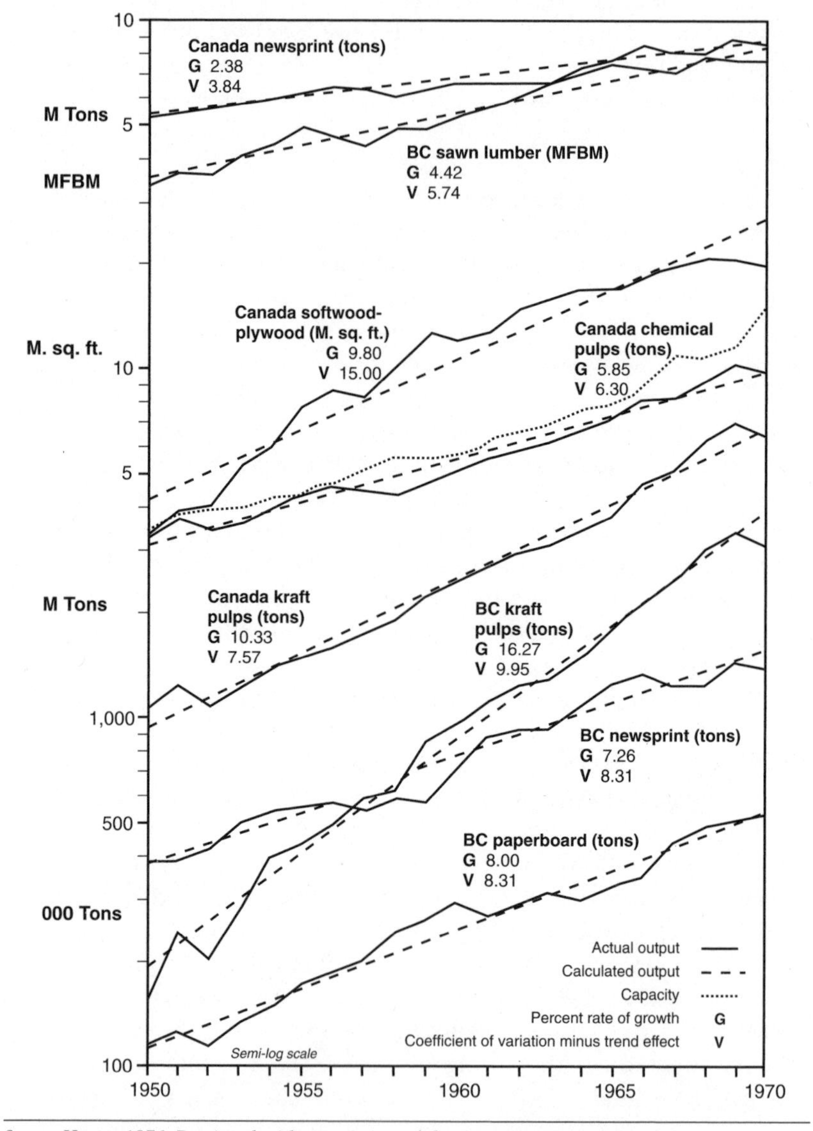

Source: Hayter 1976. Reprinted with permission of the American Geographical Society.

softwood-plywood, and most notably kraft pulp grew even faster. Indeed, newsprint and kraft pulp in BC increased their respective share of national production from 7.3 percent and 15.1 percent in 1950 to 16.2 percent and 46.2 percent by 1970 (Hayter 1976). There was also growth of paper-converting operations, notably box, folding carton, and bag plants, in the

1950s and 1960s. Such forward linkages were typically controlled by the major forest-product firms, which were situated mainly in the Vancouver area and served local markets. Meanwhile, modest growth of secondary wood-product manufacturing tended to be entrepreneurial, small scale, and local market oriented.

The BC forest economy during Fordism, however, was principally a commodity industry and kraft pulp was its lead industry. Indeed, kraft pulp production grew at the extraordinary annual rate of more than 16 percent between 1950 and 1970 as fifteen new, large-scale, export-oriented mills were built between 1947 and 1970, and two more startups occurred by 1973 (Barr and Fairbairn 1974; Hayter 1978c). Kraft pulp, stimulated by technical advances in the bleaching processes in the 1930s, began to replace sulphite pulp for converted paper products, and then the chemical pulp input for newsprint manufacture in the 1950s. The cost advantage of kraft over sulphite pulp resulted because a wider fibre base may be used, and lignin, comprising 25 percent to 30 percent of wood volume, is recovered and used as fuel, reducing costs and stream pollution (Guthrie and Armstrong 1961: 117). The first two kraft pulp mills in BC (at Port Alberni in 1948 and Harmac near Nanaimo in 1950) were both designed to use wood formerly wasted in wood-processing mills, thus initiating the technical integration of pulp and paper with wood processing in the province. Even so, the kraft pulp process only recovered a little over half of the wood fibre used, and its chemical (chlorine dioxide) requirements are considerable, a threat to the environment and health.

Kraft pulp was the signature industry of the Fordist period in BC. In a time often labelled the long boom, this industry grew the fastest among BC's forest commodities, reflecting powerful competitive advantages based on low-cost and high-quality fibre supplies, an increasing proportion of which were supplied as chips from sawmill operations. In organization, the kraft pulp industry was a homogeneous commodity manufactured in large, capital-intensive mills operated by unionized labour and controlled by large, integrated corporations. In fact, these characteristics were strongly evident in softwood-plywood, paperboard, and newsprint, and increasingly in sawmilling. If the industrial use of BC's forests was enlarged and diversified, industrial organization was clearly focused on the Fordist model.

BC's Forest Economy as a Fordist Model

The Sloan-inspired forest policy of 1947 anticipated a forest industry dominated by big firms and factories. This anticipation in turn was justified by the supposed stability and efficiency of large-scale operations. Indeed, productivity increases were widely associated with the idea of economies of scale (see Table 1.1). Admittedly, the extent to which large size implies *increasing* economies of scale (decreasing average costs of production with increasing size) cannot be taken for granted. Indeed, Schwindt (1977)

Table 2.2

Output by major commodity groupings for three large forest-product corporations in British Columbia, 1950 and 1970

Company	Lumber (MFBM)	Plywood (M.sq.ft.)	Newsprint (000 tons)	Market pulp (000 tons)	Papers, paperboard (000 tons)	Fine papers (000 tons)
MacMillan Bloedel						
1950	375	127	–	21	–	–
1970	1,180	450	1,094	483	349	37
Crown Zellerbach						
1950	–	–	86	16	22	22
1970	373	195	300	220	100	100
BCFP						
1950	249		–	–	–	–
1970	484		187	218	–	–

Note: MFBM refers to million board feet; M.sq.ft. refers to million square feet, 3/8" base.
Source: Hayter 1973:112

argues that the largest forest-product mills in the early 1970s were probably bigger than what could be justified solely by factory-level economies of scale. It may well be that the biggest firms were also bigger than anticipated on the basis of firm-level economies of scale.

Yet, scale economies are not easy to measure and are to some extent a matter of judgment, for example, in terms of how a large size helps compensate for uncertainty in serving distant markets and negotiating with powerful consumers. Economies of scale in production at the factory level and economies of scale at the firm level (in marketing, finance, research, and labour relations) are also not easy to integrate and are unlikely to dovetail neatly with one another. Moreover, in the real world, size confers advantages other than average cost (notably, bargaining power). In practice, the search for scale economies in the forest industries inevitably meant building the largest mills possible. Implicitly, the virtues associated with the model of flexible specialization, as well as criticisms rooted in diseconomies of scale, abuses of power, and a loss of competitiveness, were discounted.

Scale, Size, and Integration

To a significant degree, the Fordist boom in the BC forest industries was dominated by large corporations pursuing strategies of horizontal and vertical integration (Table 2.2). The largest had their parent head offices in BC (MacMillan Bloedel and Canadian Forest Industries), in the US (Crown Zellerbach Canada), and elsewhere in Canada (British Columbia Forest Products). These firms became major producers of pulp and paper as well as wood products, and all had their own long-term timber licences.

Table 2.3

Corporate concentration in BC's timber harvest, 1973

	Leading firms	
Provincial harvest	Top 4	Top 10
TFLs	71%	97%
PSYU	27%	54%
Total	33%	54%

Source: Schwindt 1979.

As expected, concentration levels are greater in pulp, paper, and plywood than in lumber, reflecting the greater capital intensity and entry barriers of the former. Thus, in 1975 the four largest firms accounted for 22 percent of lumber production, but 75 percent of plywood, 52 percent of market pulp, and 94 percent of newsprint production (Schwindt 1979). Clearly, numerous other industries exhibit greater degrees of concentration, and given the strong export orientation of the BC forest industries, the concentration of production within BC does not equate to market power. Indeed, producers on the geographic margin scarcely enjoy such influence, and the BC forest industries have consistently emphasized their roles as price takers rather than setters. The great fluctuations in price that occur in forest commodities support this view. On the other hand, the largest firms are typically important in both wood products and pulp and paper (and in the corrugated-box industry). In addition, corporate concentration extended in 1973 to control over the wood harvest, especially for TFLs (Table 2.3). Outside of the TFLs, however, corporate concentration was much less.

Intimately linked to the benefits associated with size and scale were those stemming from integration. The "ideal" firm was not only big, but it was integrated across all forest-product commodities. Thus, horizontal integration enhanced market power and facilitated firm-level economies of scale in the use of marketing networks, production know-how, and R&D, whereas vertical integration enabled firms to gain control (and thereby security and quality control) over lines of supply and market links. Integrated firms, in comparison to specialized firms buying and selling to one another, could arguably convert logs to their highest and best use. Moreover, integrated corporations would be more stable to the extent that the fluctuating fortunes of different forest commodities offset one another. As noted, Sloan also hoped that integrated firms would have a stronger commitment to sustained-yield forestry.

During Fordism, growth and size themselves were objectives, a desired characteristic of industry culture. New mills were typically extolled as "state-of-the-art," a label that in BC rarely meant a global-first in product or process terms, but often meant the biggest size that could be engineered with the latest, proven technology. Kraft pulp mills led the march to increasing size. In 1948, Bloedel, Stewart, and Welch's new kraft mill at Port Alberni was designed to produce 200 tons per day; Eurocan's Kitimat mill was designed at 1,000 tons per day in 1970; the new mills in between those years had become progressively bigger. Paper machines similarly increased in size. At Powell River, the new No. 7 newsprint machine established in 1948 had designed speeds of 1,650 feet per minute; its new machine No. 10 was built in 1968 to produce 3,000 feet per minute (Hayter and Holmes 1993: 24). Its first paper machine, which opened in 1912, had a designed speed of 650 feet per minute. For wood processing, although plywood mills were all relatively large, sawmilling after 1960 was concentrated rapidly in fewer, larger mills, many producing over 100 million board feet per year (Dobie 1968). Hundreds of mills were closed in this period, virtually all manufacturing less than 5 million board feet per year.

Among the dominant commodity industries, production philosophy rested on the economics of mass production. Stimulated by substantial capital investments in specialized, indivisible machinery as well as by increasing labour costs, factories in these industries emphasized cost minimization by maximizing "throughput," in which speed and volume were the commonsense keys to increased productivity. Counterparts to highly specialized pulp mills and paper machines were readily found in wood processing, notably the so-called spaghetti factories of the interior, the sawmills that cut two-by-fours for the construction industry.

To some extent, the specialized and rigid structures characterizing the BC forest industries during Fordism were offset by the diversified range of commodities produced and by flexibility through technical integration, especially if achieved on specific sites. The epitome of such integration at this time is MB's Port Alberni forest-product complex (Hardwick 1964). In the early 1970s, annual product capacities at this site amounted to 410,000 tons of newsprint, 99,000 tons of market pulp, 112,000 tons of paperboard, 478 million board feet of lumber, 187,000 square feet of plywood, and 272,000 squares of shingles. Logs were mainly obtained from nearby TFL 44, which were sorted (at that time mainly in the Alberni Inlet) and directed to various operations; for example, the best grades of Douglas-fir supplied the plywood mill, and the lowest-quality logs were sent to the groundwood pulp mill. The pulp-and-paper operations obtained chips from the wood-processing mills; bark and waste wood was used as an energy source for the entire complex; kraft pulp and groundwood (or mechanical) pulp was manufactured to supply the newsprint mill. Located along the Alberni Inlet, the complex enjoyed sea access to global as well as continental markets.

In effect, the entrepreneurialism associated with the early industrialization of BC's forests had been replaced by the large, integrated corporation as the ideal and dominant model of organization. In contrast to Fordist structures in the US, however, many large corporations were partially or wholly foreign owned, a distinctive characteristic important to understanding the structure of BC's forest economy (Chapter 5).

Keeping Suppliers Cost-Competitive

Although integrated corporations dominated the industrial organization of the forest industries during Fordism, external suppliers (such as logging contractors, engineering consultants, and equipment manufacturers) continued to play important roles. In some instances, supplier relationships were stable. H.A. Simons, the engineering firm, for example, long enjoyed a close relationship with MacMillan Bloedel. Indeed, it is likely that many interfirm relationships enjoyed considerable longevity. At the same time, formal, long-term cooperative relations were unusual, and firm-supplier relationships reflected the independence of arm's-length transactions to encourage suppliers to remain competitive.

Indeed, in the critical area of processing technology, forest-product firms in BC, as throughout Canada, insisted that their export competitiveness depended on being able to purchase machinery at the lowest possible price, wherever the source. For this reason, forest-product firms lobbied the Canadian government to have tariffs on imported machinery reduced, if not eliminated. Local suppliers enjoyed the advantages of proximity and an understanding of local conditions, along with considerable advantages in logging and wood-processing equipment. But these advantages were not cemented by long-term cooperative understandings by which forest-product firms would guarantee to at least partially underwrite technology development and to purchase innovations. In Sweden, Finland, and elsewhere, on the other hand, such liaisons have been fostered, and equipment suppliers were more willing to take risks in developing new products and to acquire the kinds of expertise that encouraged exports and other forms of internationalization (Chapter 11).

During Fordism, BC forest-product firms adopted conservative strategies to technology choice, preferring the latest, *proven* equipment (Hayter 1973: 217). In 1972, a manager of the new kraft pulp mill in Mackenzie summarized the views of the entire industry by noting that a principal "concept of the mill's design is that the selection of equipment has been limited to only those ideas that have proved themselves elsewhere. Enough problems can arise out of the mill's relative isolation and northern location ... without adding those of bringing untried equipment on stream" (Starks 1972: 30).

Taylorized Work and Collective Bargaining

The Fordist boom witnessed a considerable expansion of employment. Logging and the export industries (notably, sawmills and pulp-and-paper mills as well as plywood) added employment throughout BC, while secondary manufacturing, such as paper-box and door manufacturing needed access to local markets, principally in the Vancouver area. Direct jobs increased from over 30,000 in 1945 to over 90,000 in 1970. Despite this rapid growth, employment in pulp and paper remained well below levels in the other major sectors, a direct reflection of variations in capital intensity.

By the 1940s, the industry was comprehensively unionized. In the early 1900s, various attempts to unionize, including thoughts about "one big union," had met with only temporary success (Seager 1988: 123). The Great Depression further undermined labour power while reinforcing worker interest in unionization. In the late 1930s, at one mill after another, unions were again created, and after Canada had passed the necessary legislation, collective bargaining rapidly became the norm in the postwar forest industry. The main unions (the International Woodworkers of America [IWA], the International Union of Pulp, Sulphite and Papermill Workers, and United Papermakers) were affiliates of American parents, until 1974 when the two pulp-and-paper industry unions broke away to form the Canadian Papermakers (CPU), now part of the Communications, Energy and Papermakers Union (CEPU). Unions encompassed virtually all workers in the main commodity industries and paper-box manufacture, as well as some smaller-scale activities.

In practice, collective bargaining in BC closely followed the American Fordist model of industrywide or pattern bargaining in which large groups of workers were represented by a single union, itself organized into locals usually affiliated with a particular mill. Especially at the more diversified sites, two or three unions might be present. Both management and labour agreed that labour markets should be structured around the basic principles of seniority, job demarcation (including a sharp dividing line between workers and management), and job grievance procedures. For unions, these principles were vital to reduce competition among workers for jobs, to establish stability and dignity in the workplace, and to reduce discretionary powers of management based on favouritism and whim. For management, these principles confirmed Taylorism as the basis for work organization in which managers retained decision-making responsibilities, and stable and specialized workforces provided the basis for productivity improvement.

Until the mid-1970s, labour supply in the BC forest industries was a more important problem than labour demand, a situation that inevitably helped the bargaining strength of the union (Hayter 1979). Collective bargaining evolved as an adversarial process, a characteristic publicly confirmed by occasional strikes and lockouts of varying length. In recessions, management retained the right to lay off workers, but layoffs were temporary, at least in the large mills, and layoffs and rehiring were themselves structured, specifically by the seniority principle. Workers achieved improved wage and nonwage benefits to become among the highest-paid workers in North America. During Fordism, forest communities became high income and relatively stable.

The Geography of Fordism

During Fordism, industrial forestry spread to all regions of BC, as the government had hoped. By 1970, the size of the forest industry in the interior rivalled that of the coast. The number of sawmills in the central and southern interior mushroomed, reaching an all-time peak in 1961 (Figure 2.5).

Figure 2.5

Location of sawmills in British Columbia, 1961

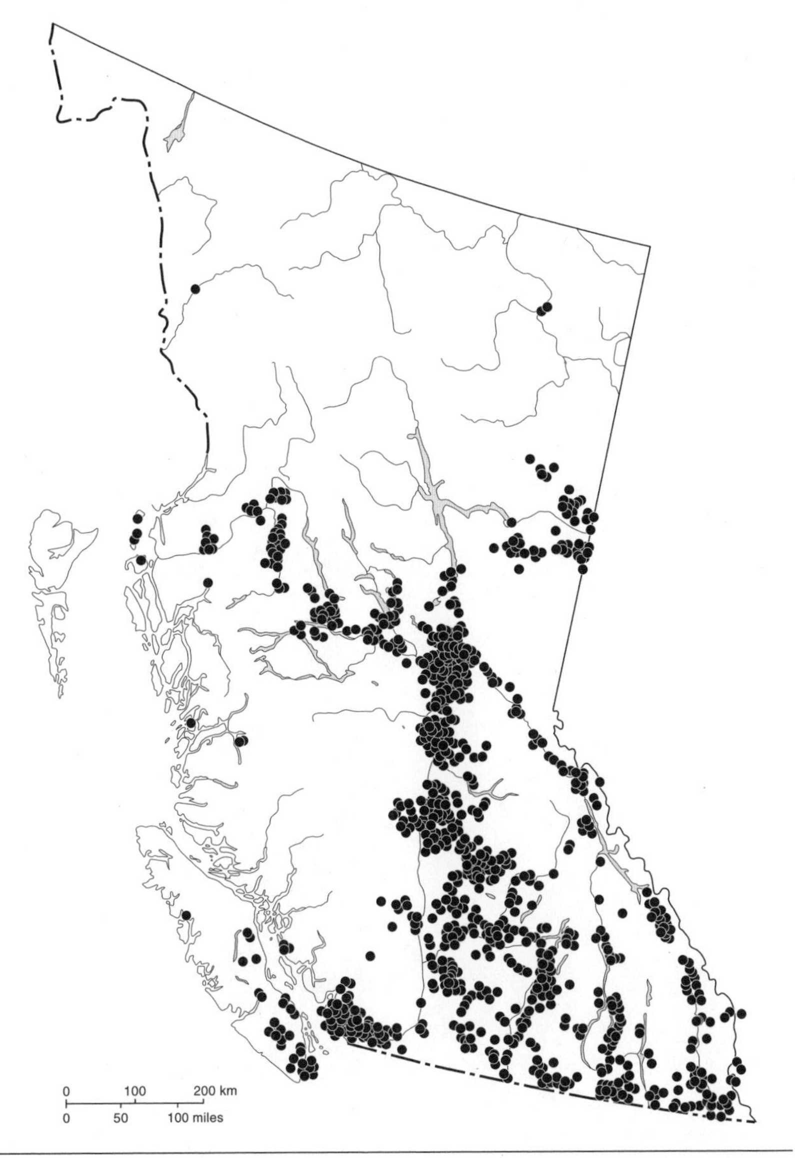

Source: Farley 1979: 65.

This number then declined rapidly, especially for small-scale operations, but the overall pattern in 1971 remained much the same (Figure 2.6). The rapidly developing softwood-plywood industry was concentrated in the Vancouver area, but it was also established in the interior. Most importantly, kraft pulp mills anchored forestry development throughout BC, creating a pattern of "dispersed concentration" (Hayter 1978a).

Figure 2.6

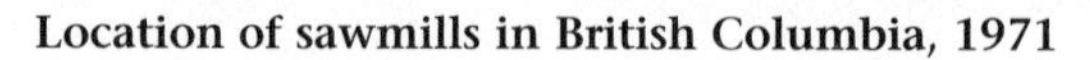

Location of sawmills in British Columbia, 1971

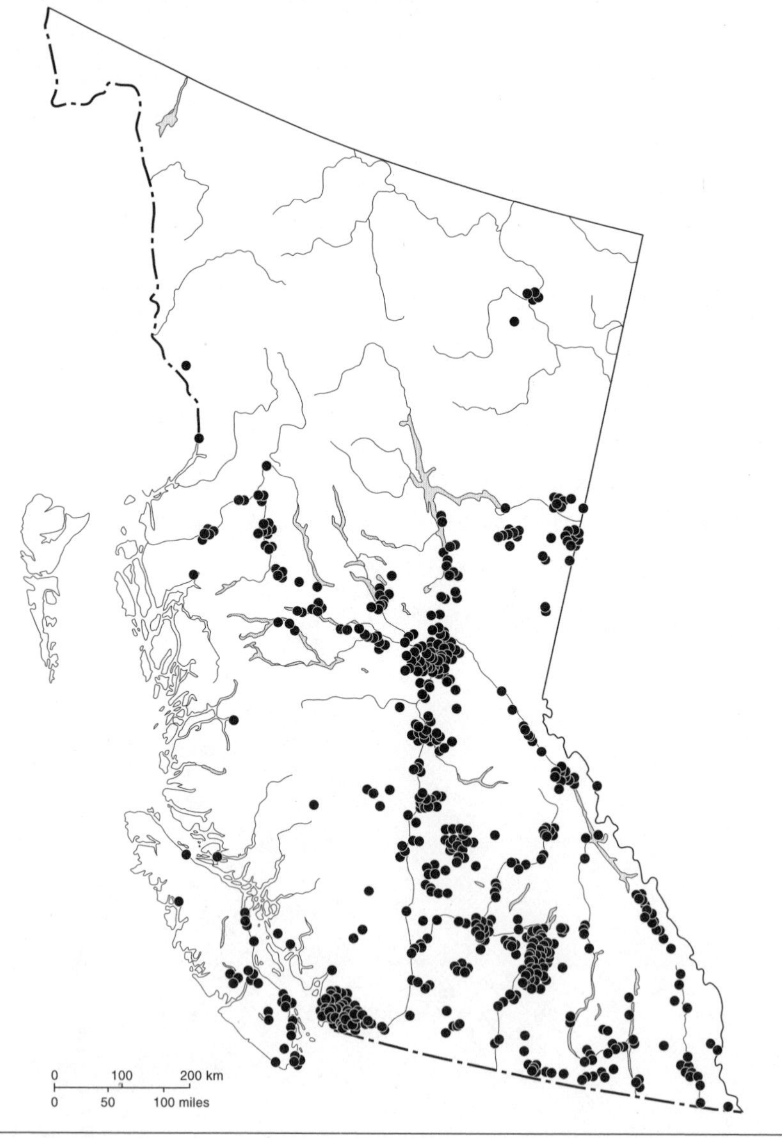

Source: Farley 1979: 67.

Thus, kraft pulp mills were first built in the coastal region, beginning with the MacMillan Export Company's (Harmac) mill near Nanaimo in 1947, and after the Castlegar mill opened in 1961, throughout the interior as well. In all, between 1947 and 1973, new site kraft pulp mills were built on the coast at Port Alberni (1948), Prince Rupert (1949, 1967), Harmac adjacent to Nanaimo (1950), Crofton (1958), Elk Falls adjacent to Campbell River (1954), Gold River (1967), and Kitimat (1970), and in the interior at Castlegar (1961), Prince George (1966, 1966, 1968), Kamloops (1965), Skookumchuck (1969), Quesnel (1972), and Mackenzie (1970, 1973). In most cases, these mills were based on newly awarded timber supplies in the form of TFLs, PHAs, or PSYUs. At several of these sites during this time, newsprint machines were built (notably, Crofton, Campbell River, and Port Alberni), and existing paper mills, especially at Powell River and New Westminster, were greatly expanded. Paperboard machines were also added to the Port Alberni, Campbell River, Kitimat, and one of the Prince George mills.

Moreover, BC's forest economy under Fordism shifted from an entrepreneurial model to a plantation model of development. Admittedly, BC's version of the plantation model was not the extreme type outlined in Baldwin's theory. Thus, some MNCs were locally based, the population highly educated, and the culture of entrepreneurialism of the previous transformation still existed. Even so, most forest-product head offices in Vancouver were themselves subsidiaries and subject to control from bigger centres such as New York and Tokyo. In addition, the kraft pulp boom did not develop backward linkages to the same extent as the previous boom in lumber. Pulp-and-paper machinery was largely imported, and BC never attracted branch plants in this industry.

By 1970, the geography of the forest industry within BC revealed two broad dimensions: coast-interior contrasts and core-periphery contrasts, the latter institutionalized by the organizational structures of large corporations.

Coast-Interior Contrasts

Fundamental differences between the coastal and the interior forest regions of BC relate to a substantially different species mix and productivities (Figure 2.7). Thus, the coastal regions are more productive for tree growth, and hemlock and cedar are extremely important, but they are scarcely present in interior regions. The largest quantities of Douglas-fir harvested are concentrated in the southwest coastal region, although it is also harvested in the interior (and not on the northwest coast). In the interior, spruce is the dominant species, along with important quantities of lodgepole pine. In all regions, except for the northeast, various other species make important contributions to species mix.

Coast-interior contrasts have been reinforced by the timing of forest exploitation which initially emphasized the coastal forests. Coastal forests are accessible via water transportation. Although interior resources became widely

Figure 2.7

The AAC and harvest by region and species, 1969

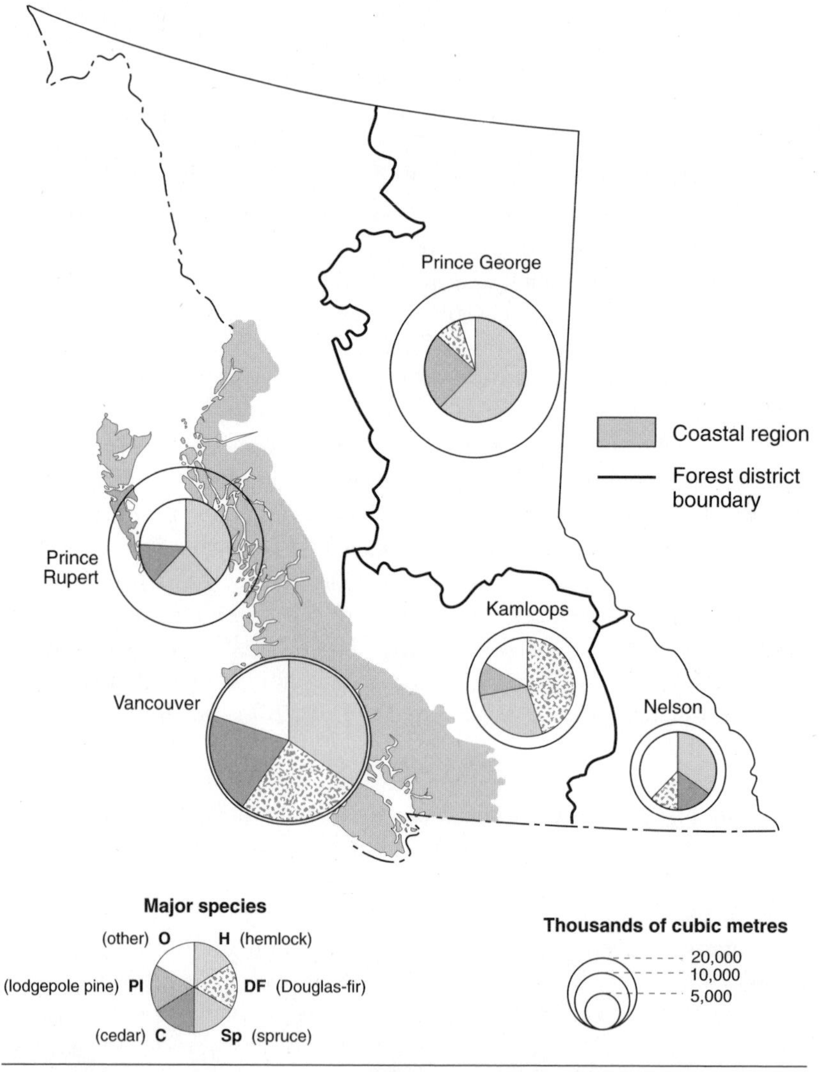

Note: Diameter of pie-graph symbols is proportional to volume of cut for each forest district. Inner circle represents 1969 cut. Outer circle represents possible allowable cut. Size of pie-graph segments is proportional to species cut, by forest district.
Source: Farley 1972: 99.

accessible in the 1950s and 1960s via road and rail transportation, the coast still had advantages in reaching a wider range of global markets via sea transportation. The interior has been much more tied to North American markets. Moreover, developing later, with less valuable timber resources and more limited market connections, the interior became more locked in to basic commodities than did the coast. By 1969, however, the allowable annual cut was already close to its potential in the southwest coastal region (Figure 2.7).

Institutionalization of Core-Periphery Contrasts

Within BC, Fordism created strong core-periphery contrasts between the metropole of Vancouver and the communities of the hinterland within both the coastal and the interior forest regions. Vancouver had always been at the centre of the forest industries in terms of manufacturing, distribution, and forward, backward, and final demand linkage effects. Before 1945, however, decision-making control of forest-product activities was relatively dispersed, even if many entrepreneurs did live in the Vancouver area. This situation changed under Fordism. In particular, the rise of corporate concentration in the 1950s and 1960s was equated with the geographic concentration of head offices of forest-product corporations whose control covered the province (Hayter 1978a). The leading forest-product corporations typically developed large, strongly hierarchical organizational structures, anchored by head offices in Vancouver. In MacMillan Bloedel's case, the head-office staff reached over 1,000 by the 1970s. For a time, forest-product firms built and occupied the biggest high-rise office towers in Vancouver's downtown.

If Vancouver by the 1970s was an emerging metropolis, the forest industry was the principal economic basis for this status. The decision-making influence of corporate head offices in some instances extended globally as well as throughout the hinterland, and some backward linkages, especially logging and wood-processing equipment and engineering expertise, had established exports. In the periphery, forest communities were highly specialized and increasingly dependent on decision makers in Vancouver. Many tiny mills and settlements failed as growth focused on a regionally dispersed system of larger factories and towns. But these towns offered both high wages and employment opportunity. Indeed, Lucas's (1971) classic sociological model of the evolution of Canadian mill towns, outlined at the end of the Fordist period, emphasizes the stability not the ephemerality of resource towns. In this model, development terminates in mature but stable communities (Hayter and Barnes 1997). As it turned out, this stability was not to last.

Conclusion

The large-scale industrialization of BC's forests faced the significant problem of finding markets to justify production. This initial challenge and transformation of BC's forests was met by an entrepreneurial form of development. As this first surge of industrialization expired in the Great Depression, a new way of exploiting BC's forests was promoted. In particular, after the Second World War, BC's forest economy became a willing partner in the system of production now labelled Fordism. Thus, the forest economy became dominated by large, integrated corporations and large-volume production for distant markets. In this development, provincial government aspirations were fully complementary with global markets.

BC's situation on the geographic margin inevitably shaped its particular expression of Fordism. In particular, the BC forest economy developed in tandem with an unusually strong commitment to exports. In addition, both national and provincial attitudes encouraged high levels of foreign involvement. In effect, both exports and foreign direct investment, along with high-volume production, reinforced BC's role as a geographically marginal producer. Certainly, no concerted attempt was made to develop either an indigenous technological or an entrepreneurial capability that would have provided a basis for some kind of diversification. Indeed, in the area of technology, the forest industries chose to be deliberately conservative; despite the massive growth of the pulp-and-paper industry, it is hard to think of any example of a global-first for BC producers in this period.

The approach to forest management in BC during Fordism was entirely consistent with a situation on the geographic margin, at least as predicted by conventional land use theory. Thus, this theory predicts and prescribes that producers on the margin can only be profitable by using land "extensively" – that is, with minimum inputs of labour, fertilizer, management, or capital per unit of land. In BC, sustained yield was literally defined as extensive forest management because, following Sloan, management units were huge to allow harvesting to be replenished by limited inputs of forestry but most of all by natural regeneration. At least, that was the hope. This hope was also convenient because it implied minimal forest management costs at a time when production philosophy emphasized cost minimization. In these ways, BC's situation on the geographic margin shaped both the industrial and the resource dynamics of the forest sector, and in turn it provided the basis for contemporary patterns of industrial and resource policy.

3
Booms, Busts, and Forest Reregulation in an Age of Flexibility

The technological, geographical, and institutional conditions underlying the Fordist model of BC's forest economy in the 1950s and 1960s began to change in the 1970s (Hayter and Barnes 1990). These winds of change placed BC's forest economy at a double crossroads created by both industrial and resource dynamics. In the context of industrial dynamics, technological change spearheaded by the revolution in microelectronics has exerted pervasive impacts on the structure of the forest industries in BC and elsewhere, offering all kinds of possibilities – in BC and rival regions – for efficiency gains, quality improvements, and product differentiation. In the context of resource dynamics, the end of the era of extensive growth in which forest-product growth is based on old-growth timber, if not entirely over, is close at hand. Easily accessible old growth is gone. Moreover, this loss of BC's natural advantage has occurred when rival regions that had already established second-growth forest management were lower cost and becoming more important (Marchak 1991).

The implications of resource and industrial dynamics in the BC forest industries are intimately associated with a search for flexibility. Industrial flexibility in the operations of factories, firms, and production systems is demanded to increase productivity, to respond to consumer demands, and to cope with a rapidly changing fibre-supply situation. At the same time, changing forest management regimes, especially the need to incorporate nonindustrial ("environmental") values, is requiring flexibility in forest policy. Flexibility is an ambiguous imperative, however, and the double crossroads are not well signposted. Industrial volatility has proven an unfortunate context for the rethinking of forest policy.

This chapter provides an overview of the restructuring of BC's forest economy from a Fordist to a more flexible production system in the last quarter of the twentieth century. Two themes are pursued. First, the BC forest industries since the early 1970s have been marred by profound volatility, specifically by a series of booms and busts. Thus, while record-setting booms occurred in the late 1970s, late 1980s, and 1993-5, sharp recessions in 1971 and 1975 were followed by the extended, deep crisis of the early

1980s, a significant downturn in the early 1990s, and the dramatic "bust" in the second half of the 1990s – this last event made all the more painful because elsewhere in Canada the forest industries have enjoyed relative prosperity. Second, industrial volatility has been accompanied by a growing, if uncertain, appreciation of the need for a fundamental change in forest policy, beginning with the establishment of the Pearse Royal Commission in 1975 by BC's first-ever New Democratic Party (NDP) government. While this commission led to important modifications in forest policy in the early 1980s, the onset of American protectionism, based on the charge that BC's forests were undervalued, and the environmentally led "war in the woods" protests indicated powerful criticisms remained. In response, especially after the 1991 election of a second NDP government, a barrage of policies was rapidly introduced that collectively constitutes a major reregulation of BC's forests. In this chapter, this reregulation is labelled the "high-stumpage regime." This reregulation has proven controversial; if volatility is an unfortunate context for rethinking forest policy, the reregulation may have added to the volatility.

The organization of this chapter follows these themes. Thus, the first part of the chapter focuses on volatility in the forest economy; the second part outlines the reregulation of BC's forests, culminating in the high-stumpage regime of the 1990s; and the last part notes the paradoxical relationships between volatility and reregulation. Indeed, the reregulation of BC's forests, rather than setting the stage for, has been continually compromised by, industrial dynamics. Certainly, the reregulation has been prolonged, and while gaining much momentum in the 1990s, some recent initiatives have already been modified. If the Sloan Royal Commission outlined the policy context for subsequent industrial development for over two decades, the relationships between forest policy and industrial dynamics since the mid-1970s have been less clear. Indeed, there is a widely held view that forest reregulation is itself a source of instability (*Vancouver Sun* 1999).

Volatility in British Columbia's Forest Industries

In the latter part of the 1990s, a stream of announcements about layoffs, plant closures, and corporate losses has been paralleled by pleas to restructure BC's forest industries. Pearse, for example, has called for another Royal Commission, noting that the dynamics affecting the forest industries are structural, not temporary (*Vancouver Sun* 1998). In fact, the present crisis is part of a longer period of volatility that began in the 1970s, and the need for structural change was starkly revealed by the recessionary crisis of the early 1980s. Indeed, a crucial question is why the forest industries have failed to effectively restructure before the present crisis. At least part of the answer lies in the nature of volatility itself, rooted in industry culture and the "staple trap."

It is worthwhile recalling that the 1970s was a period of globally fluctuating fortunes featuring two energy crises, the first in 1973, and recessionary crises, notably in the mid-1970s, then considered the worst for several decades. In BC, the latter in particular affected the forest industries, the poor fortunes of which helped derail the new NDP provincial government's ideas for forest reform, resulting in part in its 1976 election loss (Wilson 1987/88, 1997). Since then, the volatility of the forest industries has become even more marked; thus, record-setting booms in the late 1970s and 1980s have been separated by severe, prolonged recessionary crises in the early 1980s and during both the beginning and the end of the 1990s.

The turnarounds from booms to busts have been startling. In 1979, the forest-product industries generated profits of $500 million; in 1981, the loss was similarly $500 million. By 1982, debt-equity levels were unusually high, stimulated by record levels of capital expenditures initiated in 1979; production levels were cut; and immediate and massive layoffs occurred. A federal government study estimated that 21,341 jobs were lost in the provincial forest-product industries between 1979 and 1982, with the bulk of the layoffs taking place in late 1981 and 1982 (Grass 1987). Correspondingly, official unemployment rates in these industries rose dramatically from 6.4 percent in 1979 to 19.2 percent in 1982 (Grass 1987: 32). The recession was also unusually long lasting. Between 1981 and 1984, the industry lost $1.1 billion, and as late as 1985, several major forest-product corporations were still experiencing income losses. Industry unemployment remained at 13.7 percent.

The problems of the early 1980s were then followed by a sharp boom, and in 1989, the forest industry achieved all-time peak levels of production and profit. In 1987-9, profits amounted to over $3 billion, and a decade after the first public announcements of the falldown effect were made, harvest levels reached an all-time record high. Yet, industry fortunes by 1993 were again sliding, and in 1994 and 1995, steep annual losses of around $1.3 billion were reported, followed by further losses totalling $412 million in 1996 and 1997, and a staggering $1.1 billion in 1998 (Price Waterhouse 1998: i). In mid-1999, job losses of over 15,000 were reported in the previous two years, while planned investment in plant and equipment, so essential for long-term viability, remain at low levels. As in the mid-1970s, and especially the early 1980s, the same pleas for restructuring were being made in the late 1990s.

Production and Harvesting Trends

Apart from plywood, the major BC forest industries have experienced significant expansions since the mid-1970s (Figure 3.1). In comparison to 1950-70, however, growth rates between 1974 and 1996 have declined, and plywood production is less than in its peak years in a short-lived late 1970s boom (compare Figure 2.4 with Figure 3.1). Moreover, as average annual ("trend-line") growth rates have levelled off, production in the main commodities has become more volatile, as measured by the coefficient of

Figure 3.1

Production trends in BC's major forest commodities, 1974-96

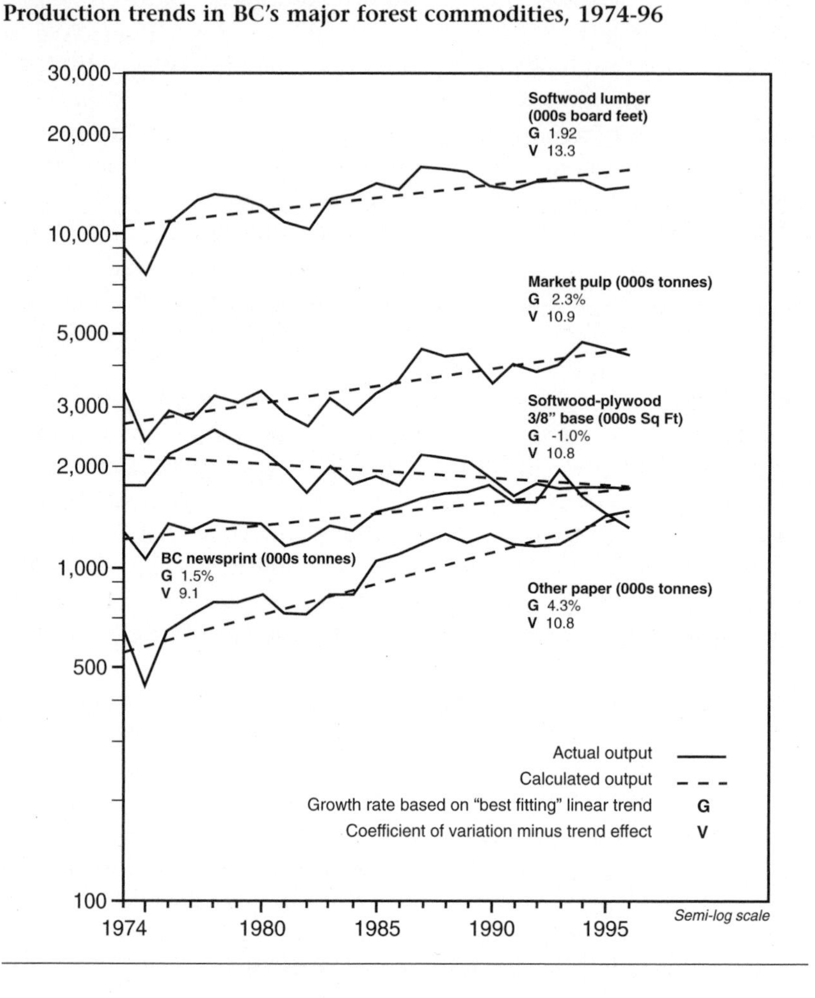

variation (based on percentage variations around the trend line). With lumber, for example, the growth rate of 4.4 percent between 1950 and 1970 declined to 1.9 percent between 1974 and 1996, while the coefficient of variation increased from 5.7 to 13.3. Indeed, whether the criterion is lumber, pulp, or plywood production, log harvesting, capital investment, profitability, or jobs, a lower growth rate (or even decline) combined with much greater instability is the statistical signature of the age of flexibility in BC's forest economy.

Within the lumber and paper industries, there is a shift towards more differentiated production that has helped maintain overall growth but not reduce overall volatility. Thus, "other paper" products, in part created by the conversion of newsprint machines to higher-value papers and in part

Vertical (Fordist) integration at Port Alberni. Cedar logs will be first cut in the sawmill and waste wood chips and sawdust will provide raw materials to adjacent pulp-and-paper operations.

the result of new plants, have grown much faster than newsprint, but stability remains a problem. Although not reflected in the aggregate statistics, the lumber industry has also experienced differentiation of its product range, especially on the coast and in association with Japanese markets (Chapter 7). The lumber industry is further differentiated by the growth of secondary (wood-using) activities in BC, often labelled "value-added" production, the output of which is not included in these data.

In employment terms, all three broad statistical categories (logging, wood industries, and paper and allied) have experienced permanent downsizing in 1979, 1989, and 1996 (Table 3.1). In the last few years, job losses have continued (Price Waterhouse 1998). The biggest losses are in the main commodity industries (logging, sawmills, plywood mills, and pulp and paper, as well as shingles and shakes); in wood processing, several mills ("establishments" in Table 3.1) have been closed. While many more sawmill closures took place in the 1960s, those mills were primarily small operations. Closures since the early 1980s have involved large factories and have not only included sawmills but plywood mills, a rapid-growth industry of Fordism. Many other big mills, including pulp and paper, downsized between 1979 and 1996 (Hayter 1997a).

Table 3.1

British Columbia's forest industries: Employment levels and number of establishments, 1979, 1989, and 1996

	Employment			Establishments		
	1979	1989	1996	1979	1989	1996
Logging	24,940	22,437	22,775	1,554	3,801	3,468
Wood products	51,369	42,416	38,973	717	687	714
Sawmills/planing mills	38,937	31,287	28,647	442	353	351
Veneer/plywood mills	7,930	4,571	3,786	27	26	22
Sash, door, millwork	3,483	4,985	4,611	192	220	250
Wooden box, pullets	202	n/a	n/a	15	n/a	n/a
Coffin and casket	37	n/a	n/a	4	n/a	n/a
Other wood	780	1,344	1,741	37	74	78
Paper and allied	20,998	18,643	17,423	74	69	73
Pulp and paper	18,557	16,723	15,893[1]	–	15	18
Boxes and bags	1,938	n/a	1,041	23	–	18
Other converted paper	n/a	n/a	186	24	–	12

1 Statistics are for 1995.
Source: Statistics Canada Cat. 31-203 1979, 1989, and 1996; Cat. 25-202-XPB, 1996.

A few new sawmills and plywood-and-pulp mills have been opened, while new mills and expansions of existing mills in secondary wood-based industries, especially particleboard, engineered woods, and door, window, and related component manufacture (which are classified as "sash, door, and other millwork" and as "other wood industries"), have partially offset the employment declines in the commodity sector. Although these data do not reveal the extent of fluctuations in employment, in both union and nonunion mills employment is typically rapidly adjusted to operating levels. In addition, job declines in large-scale wood processing and pulp and paper have emphasized union employees, whereas job increases among smaller firms have tended to involve nonunion employees. Although job flexibility is a traditional characteristic of the nonunion sector, unions in BC's forest industries in the 1990s have been faced with increasing pressure to accept more flexible work practices (Chapter 8). To some extent, the shift towards secondary manufacturing has required smaller firms and flexibly specialized production (Rees 1993; Chapter 6). Especially when considered in relation to the downsizing of large operations, the trends of the 1950s and 1960s towards fewer, larger mills have at least been modified.

Harvesting levels on regulated Crown lands since the early 1970s largely reflect trends in production (Figure 3.2). Thus, harvesting levels were generally higher in the 1980s than in the 1970s, while levels in the 1990s have been modestly reduced from those in the 1980s. Fluctuations have been considerable, the peak harvest (on Crown lands) occurring in 1988 at almost 80 million cubic metres, compared to a 1983 low of around 59 million

Figure 3.2

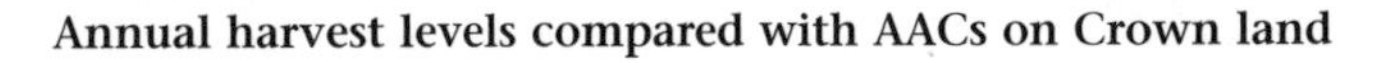

Annual harvest levels compared with AACs on Crown land

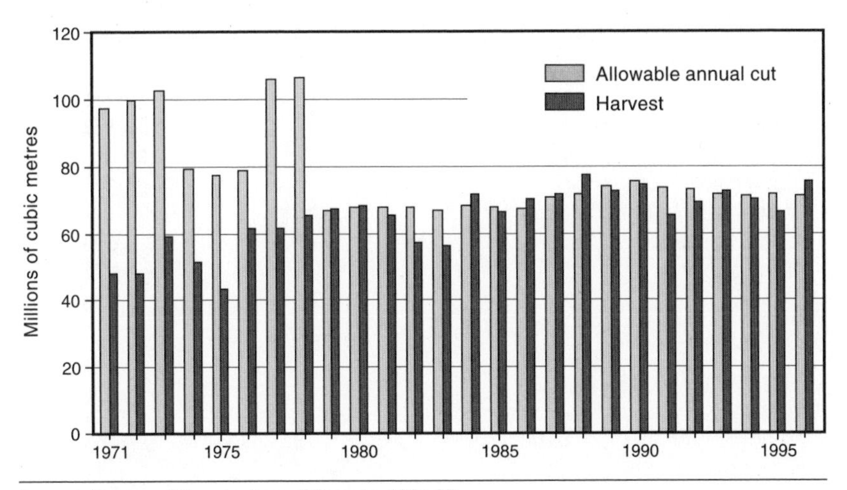

Source: Ministry of Forests, Annual Reports, 1971-97.

cubic metres. Estimates of the AAC became more stable in the 1980s, corresponding closely to the actual harvest. In fact, in the late 1990s, despite fears of wood-fibre shortage, not all the AAC has been consumed for market and cost reasons, specifically that part of the AAC allocated to the Small Business Forest Enterprise Program (SBFEP).

If the AAC appears to have become more stable, however, there are widely differing estimates, each based on government data, of what the long run-sustained yield (LRSY) or the more conservative long term harvesting level (LTHL) should be. Thus, one report (BC Wild 1998) identified a LTHL of 53 million cubic metres, a widely quoted figure for the LRSY is 59 million cubic metres, which the government has also referred to as the "long run sustainable cut," while the government recently estimated 71 million cubic metres as the "post-timber supply review" AAC (Ministry of Forests 1996b). Reed's (n.d.) suggestion for future annual harvests of 90 million cubic metres, if intensive forestry is practised, further widens possible scenarios.

In private unregulated lands, not subject to AAC calculations, harvest volatility is also evident. Thus, Travers (1993: 189) notes that private forests have provided a "relatively stable" cut of between 12 million cubic metres and 15 million cubic metres, often around 15 percent to 20 percent of the total harvest. In the 1995-6 fiscal year, however, private forests supplied 8.8 million cubic metres (or 11.8 percent) of the total provincial cut of 75.4 million cubic metres (Ministry of Forests 1996: 81). As the costs of logging Crown land increased rapidly after 1995, some firms with access to their own land (such as MB) have given more emphasis to harvesting private

Table 3.2

Forest tenure types, 1993

Tenure type	Number	AAC	%	Harvest	%
Regulated harvest					
Tree farm licence	35	17,044	24	15,935	20
Forest licence	183	39,655	56	39,171	50
Timber sale (SBFEP)	1,370	9,986	14	10,216	13
Timber sale (non-SBFEP)	13	328	<1	298	<1
Timber sale (pulpwood)	21	1,624	2		
Woodlot licence	500	503	1	603	1
Licence to cut	658			971	1
Other		1,250	2	1,126	1
Total (regulated)	**2,780**	**70,390**			
Unregulated harvest					
Timber licence				1,862	2
Private land (Crown grant)				7,583	10
Federal lands				247	<1
Total (regulated and unregulated)				**78,012**	

Note: AAC and harvest expressed in thousand cubic metres.
Source: Schwindt and Heaps 1996: 28.

timber. Not much is known about the status of forestry on private lands, although private forests have historically provided a bigger share of the harvest than their share of forest area. This trend could imply better forest management or the onset of a substantial falldown. On all lands, public and private, the peak harvest was over 90 million cubic metres in 1987, subsequently dropping to just under 80 million cubic metres in 1990 and 68.6 million cubic metres in 1997.

Large corporations continue to control significant timber rights. Despite fears of some observers (M'Gonigle and Parfitt 1994), however, corporate concentration has not increased, at least until 1996. While TFL rights are dominated by the largest firms (Schwindt and Heaps 1996: 30), the largest ten firms accounted for 53.1 percent of allocated timber rights in 1995, the same as in 1975 (see Table 2.3; Wilson 1998). In fact, in 1993, SBFEP licensees held 14 percent of the AAC and they are now the third most important type of tenure, after forest licences and TFLs (Table 3.2). In 1995, firms smaller than the top fifty controlled 30 percent of harvesting rights, a situation inconsistent with the claim of unusually high levels of corporate concentration. The trend in harvesting seems to be modestly more in line with the growing role of SMEs.

Export Trade: Marching to the Tune of Core Countries

BC's forest industries have long been based on high levels of exports. Since the early 1950s, when the export sales ratio was around 60 percent, the

Table 3.3

Total and export values of British Columbia forest products 1952, 1966, 1987, and 1996

	1952	1966	1987	1996
Net value of shipments ($M)	483.9	1,037.0	11,602.0	16,466.0
Export share percentage	60.2	80.6	80.3	90.3

Source: British Columbia, Government of, 1967: 47; COFI 1997: 69 (and Table 2.E).

industry has consistently exported around 80 percent of its shipments (Table 3.3). In 1996, the export sales ratio reached an astonishing 90.3 percent of sales. In the 1950s and 1960s, expansion of BC's forest industries was largely predicated on accessing American markets. The US accounted for the lion's share of forest-product exports, and Europe, especially the UK, constituted an important secondary market, especially for lumber and pulp. Japan was a relatively small market at this time (Table 3.4). In the 1980s and 1990s, while the US remained the dominant market, if less so than before, Japan and other Pacific Rim markets greatly increased in importance; in 1996, they were three times more important than Europe.

Export trade is therefore the lifeblood of the BC forest economy, as it was during Fordism. Indeed, the reliance on export trade, specifically the markets of major industrial powers, is extreme and signifies a highly open economy susceptible to price fluctuations in markets over which BC producers have little control and to competition from alternative supply areas or "intervening opportunities." Typically, commodity markets are more competitive than higher-value markets. Moreover, export dependence on core countries has exposed BC producers not only to the fluctuations of the "laws" of supply and demand but also to changes in trade policies in countries with considerable power.

Table 3.4

British Columbia exports of forest products to major markets, 1952, 1966, 1987, and 1996

	1952 (%)	1966 (%)	1987 (%)	1996 (%)
United States	57.1	68.9	50.6	56.0
European Community	1.5	6.6	12.6	9.5
United Kingdom	29.8	10.8	5.2	
Japan	2.0	6.9	19.0	23.7
Other Pacific Rim			9.6	9.2
Australia	2.9	2.4	(2.2)	
Other	6.7	4.4	3.1	1.5
Total ($M)	**291.5**	**836.0**	**9,311.9**	**14,863.0**

Source: British Columbia, Government of, 1967: 134 and COFI 1997: 66.

During Fordism, BC's commodity trade was buttressed by trade policy. The major industrial powers, led by the US, eliminated tariffs on these commodities (but not for more processed products) to facilitate trade. Thus, the US (and Europe) allowed the import of basic commodities duty free, except for plywood in the case of the US, while secondary manufactures faced restrictive tariffs. Apart from raw logs, Japan severely restricted the import of all forest products until the 1980s when it, too, sought to increase the import of lumber. In other words, under Fordism, the BC forest economy developed under a partial free trade regime that enshrined its role as commodity supplier. Since the early 1980s, however, despite the North American Free Trade Agreement (NAFTA) and its 1989 forerunner signed by Canada and the US, trade policies in the US towards BC have changed drastically (Chapter 7). Specifically, the US since 1986 has restricted BC's lumber exports to protect its own industry. Meanwhile, European consumers have threatened boycotts because of environmental objections to BC forestry practices. Moreover, especially after the decline in Asian markets, BC producers clearly need exports more than the US or Europe need BC's imports, a dependency directly reflecting BC's role as periphery to the world's cores.

Changing Cost Competitiveness: The Long Shadow of the Cost-Price Squeeze

Within the context of increasing volatility, the competitive position of BC's forest industries has changed fundamentally. In particular, BC is no longer among the low-cost producers in the world. In this context, Marchak (1991) has rightly pointed to developing countries that have developed low-cost forest plantations, especially for pulp and paper. Technology has both widened the fibre base for pulping processes (Westoby 1989) and reduced the quality advantages associated with traditional coniferous producers such as BC. The quality of plantation-based wood products is improving as well (Clapp 1995). In addition to increased competition from developing countries, of even greater significance is the emergence of the American South over the past several decades as a large-scale, low-cost producer of forest products, especially pulp and paper but also lumber. The massive growth of the pulp-and-paper industry, and to some extent lumber, in Alberta constitutes further competition for BC mills.

Cost comparisons of forest industries between regions are not easy to make, and they are usually expressed as estimated averages based on the operations of existing mills or hypothetical estimates of state-of-the-art mills that have yet to be built. Cost structures comprise numerous components (fibre, chemicals, energy, labour, capital, taxes, and transportation), each of which can show distinct variations. Foreign-exchange rates also dramatically affect comparative costs. Nevertheless, there is strong evidence that BC is now among the highest-cost producing regions in the world. In the case of kraft pulp, Price Waterhouse noted that BC's costs in 1990 were lower

Table 3.5

Market kraft pulp: International cost comparisons 1997 (Cdn $ per tonne)

	US South	US West	Eastern Canada	BC Interior	BC Coast	Sweden/ Finland
Fibre	227	179	256	192	244	316
Labour	100	83	98	105	131	58
Chemicals	70	90	64	73	73	60
Energy	35	51	33	36	53	8
Other mill	76	81	82	115	150	65
Corporate, sales	16	35	28	30	20	5
Delivery	93	106	64	88	79	55
Total	**617**	**625**	**625**	**639**	**750**	**567**

Note: Costs exclude depreciation, interest. Data refer to softwood kraft market pulp. Local currency converted into Canadian dollars as of April 1998.
Source: Price Waterhouse 1997: 10.

only than those of Finland and that the American South had a big competitive advantage. By 1997, kraft pulp's cost situation in BC, particularly on the coast, remained dire, especially in view of the low value of the Canadian dollar (Table 3.5). Virtually, the same generalizations apply to newsprint (NLK and Associates 1992; Hayter 1997a: 130). Moreover, in pulp (especially) and newsprint, developing countries such as Brazil and Chile, fundamentally based on low wood-fibre costs of plantation-grown wood, have developed significant export-oriented industries since the 1970s.

Similarly, for lumber, the American South has become the benchmark region, at least in North America. An American study in the mid-1980s, for example, noted that cost estimates for state-of-the-art mills of different sizes were much lower in the South than in the Pacific Northwest states (Table 3.6). BC's lumber costs are probably as high or higher than in the Pacific Northwest, especially given that wood costs dominate the overall cost structure of sawmilling. Thus, a recent global comparison of the total harvesting costs of softwood saw logs, not including stumpage, revealed the BC coast to be the highest-cost region in the world (Ernst and Young 1998: 36). On the BC coast around 1997, the "delivered cost of sawlogs" was about US$59 per cubic metre compared to the approximately US$43 per cubic metre of the second-highest-cost producer, Germany. All major rival regions enjoyed an even bigger cost advantage. In this study, interior BC costs were the same as American costs in the Pacific Northwest (around US$32 per cubic metre) but much higher than in the American South, which again is the global low-cost point at US$18 per cubic metre. Higher costs in BC are further complicated by problems of access to American markets that the producers in Washington or Oregon do not face. In addition, Binkley (1997a: 22) reports that labour costs in four North American regions were highest on the BC coast. Finally, as will be discussed later, stumpage and regulatory costs of harvesting have increased rapidly in BC in the 1990s (Cafferata 1997).

Table 3.6

Annual operating costs: Lumber mills in the US, c. 1984

	Southeast region		Pacific Northwest region	
Cost	Large	Medium size	Large	Medium size
Wood	67.7%	59.2%	85.0%	78.0%
Labour	3.4%	7.0%	2.1%	4.6%
Maintenance	9.1%	10.6%	5.6%	6.9%
Overhead	2.2%	7.8%	1.4%	5.1%
Energy	17.6%	15.4%	5.9%	5.4%
Total ($000)	11,135	3,195	18,014	4,916
Cost/MBF	$111	$128	$180	$196

Note: The medium-sized mill has an annual capacity of 25 million board feet; the large mill has an annual capacity of 100 million board feet.
Source: Tillman 1985: 196.

For individual industries, significant losses were recorded in the early 1980s in wood processing and in pulp and paper. The most significant losses in the 1990s are in the market pulp and newsprint industries, especially the former (Table 3.7). Lumber lost substantial income in 1990 and 1991, but since then, the industry has generated profits, and in the mid-1990s, substantial profits. Market pulp, however, has been a consistent money loser since 1990, apart from 1995. Newsprint and other operations have not fared much better. In financial terms, the government's decision in 1998 to subsidize the market kraft pulp mill at Prince Rupert, to the tune of $280 million, does not make much sense, especially given that this mill is marginal even within a BC context. The 1998 announcement by Bowaters to close its pulp mill at Gold River, built as recently as 1967, is not surprising in view of these figures. Howe Sound Pulp and Paper has reported losses for the past ten years, and even the new mills at Taylor and Chetwynd have faced difficulties. In kraft pulp, a commodity industry, BC has excess capacity.

Clearly, the major BC forest industries no longer enjoy the cost leadership that was provided by nature's bounty of high-quality, large logs. This bounty has been substantially exploited at a time when forest plantations in rival regions have increased their wood supply. Moreover, BC's cost-competitiveness, especially in relation to the US, would have been much worse by 1998 if not for the unprecedented low value of the Canadian dollar, because income is largely received in American dollars and payments are largely made in Canadian dollars. As a recent Price Waterhouse study (1998) indicates, a one cent change in the exchange rate for Canadian and American dollars translates into a gross sales change of around $200 million for the BC forest sector. A substantial increase in the exchange rate could be catastrophic to BC's forest industries.

Table 3.7

British Columbia's forest economy: Net earnings by industry, 1988-97 ($M)

Year	Lumber	Plywood	Market pulp	Newsprint	Other operations	Total
1988	175	18	654	163	330	1,340
1989	77	35	694	36	309	1,151
1990	-281	-4	227	-33	82	-9
1991	-334	-2	-250	-152	-131	-869
1992	182	24	-178	-220	-70	-262
1993	1,235	70	-479	-188	-118	520
1994	1,136	65	-103	-64	326	1,360
1995	479	46	393	112	250	1,280
1996	392	-2	-587	90	-183	-290
1997	259	3	-284	-65	-105	-192

Note: In 1995, 1996, and 1997 the effects of "unusual" items have not been shown.
Source: Price Waterhouse 1988: Table 9.

Unfortunately for most of BC's forest-product firms, caveats to the exchange rate aside, increased costs have not been offset by sustained price increases. In general, reductions in the quality and accessibility of the resource, competition from other regions, and overcapacity impose significant long-term downward pressures on forest-commodity prices. Moreover, the technology for commodity production is well known and readily transferred to rival regions in developing countries where fibre costs are much lower. Consequently, a cost-price squeeze has been apparent for some time (Woodbridge, Reed 1984). Thus, Woodbridge, Reed (1984: 57-8) show that commodity lumber prices, specifically for spruce-pine-fir, fluctuated widely from the late 1960s to 1983 but had generally declined to the point where 1983 prices were 20 percent below 1969 levels in real terms. For the same period, this study also noted that while wood costs have constantly increased, the real price of market pulp remained constant and the real price of paper declined (Woodbridge, Reed 1984: 87). In the second half of the 1990s, as costs increased, prices for forest commodities have been soft, and the cost-price squeeze for lumber, newsprint, and pulp commodities is a long-term reality.

Compared to new supply regions, BC has significant advantages in terms of the existence of established infrastructure, plant capacity, and pools of skilled labour and engineering expertise. The key to survival in the long run is for BC producers to use this accumulated expertise to shift to higher-revenue-generating products that can compensate for the high costs. However, producers in the US and Europe, where pulp-and-paper costs are the highest in the world, have been more successful than BC firms in innovating grades that command price premiums (Woodbridge, Reed 1984; Croon 1988; Ernst and Young 1998: 39). In BC, forest-product firms remain too

Figure 3.3

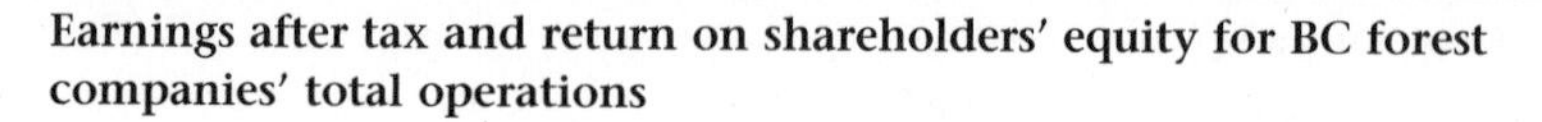

Earnings after tax and return on shareholders' equity for BC forest companies' total operations

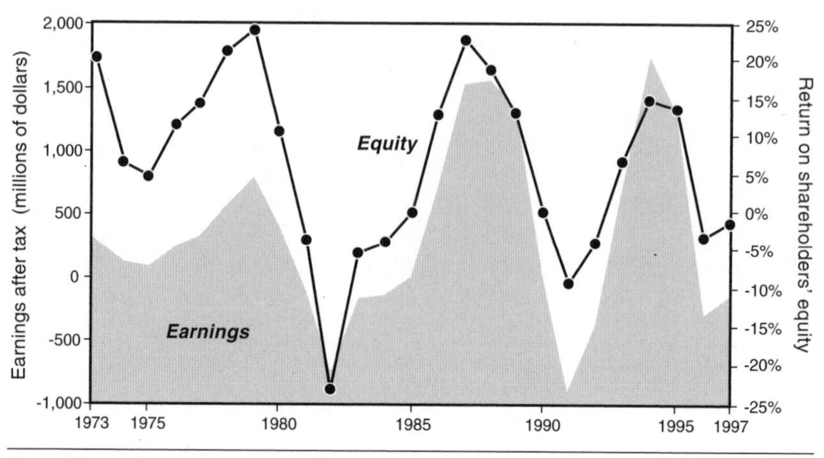

Source: Schwindt and Heaps 1996: 83 and Price Waterhouse 1988.

heavily concentrated on commodity grades. In the late 1990s, low prices and high costs have combined once again to create losses and another crisis.

Fluctuating Rates of Return and Capital Investments

Volatile patterns of production have been accompanied by fluctuating rates of return on capital invested (Figure 3.3). Thus, earnings after tax, as measured in actual dollars, and the percentage return on shareholder equity, both reflect the sharp booms that occurred in the late 1970s, late 1980s, and 1993-5, and the equally sharp busts that immediately followed the booms. As Schwindt and Heaps (1996: 83) note with reference to the 1980s and 1990s, "when times are good, they are very good; when times are bad, they are very bad." In fact, in terms of number of good years versus number of bad years, the overall performance can be said to be very bad. Thus, according to after-tax earnings, BC forest industries did not generate any profit for five full years (1981-5), while in the eight years from 1990 to 1997 inclusive, net earnings have been positive in just three years. In these years, the earnings have almost doubled the losses (+$3,160 million compared to -$1,609 million), but overall the rate of return on investment in BC's forest industries has been modest to say the least; the return on Canada Savings Bonds since 1975 has generally been better (Forgacs 1997: 169).

The volatility in earnings in BC's forest industries directly reflects a combination of dependence on export markets and a still too strong orientation towards commodities. The commodity segments are extremely vulnerable to price changes over which they have little control, and much of the forest sector is vulnerable to exchange-rate fluctuations.

Strongly fluctuating rates of return, which are often negative, have important implications for capital investment. Profits directly fund investment and indirectly help firms attract funds from other sources. Losses reduce funds for investment and potentially add to debt burdens generated by borrowing from previous rounds of capital investment. In this latter regard, high levels of capital investment encouraged by the boom in the late 1970s contributed to the forest industries' financial woes in the early 1980s with high debt-equity ratios that reached above 1.2. Subsequently, the forest industry reduced its debt-equity ratio to a low of 0.5 in 1987. Following the boom of the late 1980s, it then rose again to 0.8 in 1995, declining slightly since then as the industry has reduced levels of capital investment.

Typically, capital expenditures in BC's forest economy have increased during booms or in the later stages of booms and declined during recession (Figure 3.4). Because investments already implemented cannot be automatically adjusted to the realities of demand and profitability, some time-lag effect inevitably occurs between investment and business cycles. Moreover, investment in BC's forest industries is a feature of oligopolistic rivalry, so that firms often collectively decide to go ahead with investments to compete for markets, a pattern of behaviour that in turn reinforces tendencies towards overcapacity and recession (Schwindt 1979). In these respects, the substantial increases in capital investment that occurred in the late 1980s and early 1990s are consistent with the optimistic expectations generated

Figure 3.4

Capital expenditures (on machinery) in BC's paper and allied and wood-processing industries

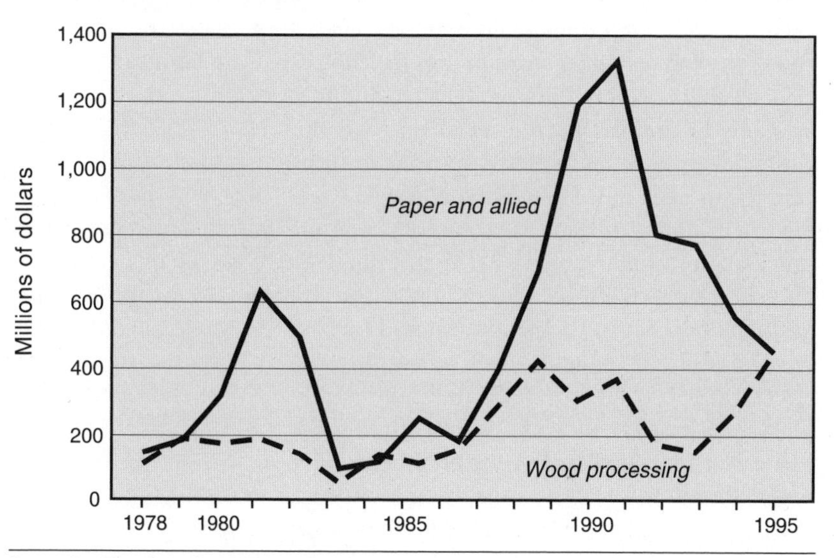

Source: Statistics Canada 61-205 and 61-206.

by the boom and the desire to keep pace with the plans of rivals. Significant expenditures at this time were also designed to reduce environmental impacts and to meet environmental regulations. Indeed, according to Price Waterhouse estimates of capital expenditures (lower than those estimated by Statistics Canada and shown in Figure 3.4), in 1990 and 1991 expenditures for environmental purposes were in excess of $.5 billion (Price Waterhouse 1992b: 20) and respectively accounted for 24.8 percent and 34.5 percent of total expenditures. Since 1993, environmental expenditures have been lower, ranging from a low of $69 million in 1994, or 6.5 percent of total expenditures, to $167 million or 14.4 percent of total expenditures (Price Waterhouse 1997: 20).

Capital-investment expenditures provide a crude indicator of the demand for technology, and they are essential for the long-term health and vitality of the forest industries (Hayter 1988). Investment provides the basis for expansion, modernization, and diversification; it is a requirement for enhanced efficiency and higher-value production. As noted, investment is also vital to meeting environmental goals. SMEs often scavenge old equipment, thereby adding to the region's capital-output ratio. Larger firms, and sophisticated operations, need to invest in new machinery and equipment to realize productivity gains.

If capital investment is vital to industrial viability in the long run, the relationships between investment and productivity are not straightforward. During Fordism, BC's forest industries performed well by increasing the efficiency of large-scale production; according to this cost-minimizing imperative, increasing capital intensity was based on increasing throughput or volume of fibre processed per unit of time, labour, or machine. Thus, Travers (1993: 206) notes that the ratio of harvest cut to labour inputs steadily declined since 1960 from about two jobs per 1,000 cubic metres in 1960 to about one job per 1,000 cubic metres in 1989. There are limits to such efficiency gains in commodity production, and in recent years, the productivity gains of the past have levelled off for pulp and declined for logging, while increasing only modestly for lumber (Ernst and Young 1998: 24-32).

In the present age of flexibility, however, the "productivity problem" facing the BC forest industry is primarily one of maximizing value per unit of time, labour, or machine, rather than minimizing costs. In this argument, BC's costs are so high that cost-minimization strategies alone cannot ensure profitability. That is, the priority should be based on value maximization, after which cost reduction can be emphasized. In recent years, Binkley (1997a) has become an articulate spokesperson for just such efforts. For lumber, his analysis shows improvement in productivity from 1976 to 1989, as measured by cubic metres per hour of production, the increase occurring entirely after 1981. Value-added trends, however, have fluctuated and in 1989 were not different from in 1976, as measured in 1986 dollars per hour of production. Binkley's (1997a: 23) fear is that while noteworthy increases in productive efficiency are evident, "it is not clear that this increase in technical efficiency was adequate to offset higher production costs, competitor responses, and market place effects." Bluntly put, the industry needs to generate higher values. As Travers (1993) reveals, compared to many other jurisdictions within Canada and elsewhere, the BC forest industry generates less value-added activity and fewer jobs per unit of harvested wood. To an important extent, the inheritance of Fordism constrains the ability of BC's forest economy to fully engage the imperatives of a flexible age. Both attitudes and existing machinery and infrastructure reflect the history of the cost-minimizing priority. In some cases (for example, kraft pulp), possibilities for higher value exist, but their application in BC is extremely limited (Tillman 1985). It may be that their potential in BC is small. Meanwhile, if the kraft pulp industry can neither drive down its costs much further nor rely on higher-value output, more mills are likely to close. Indeed, closure of large-scale mills is already a part of the geography of production in the present age of flexibility.

Geography of Production

In 1970, the coast and interior industries were about the same size in terms of log harvest, but the species mix was substantially different. By the 1990s, the distinctions in species mix remain: hemlock, Douglas-fir, and cedar

Figure 3.5

Log production in British Columbia by species and region, 1996

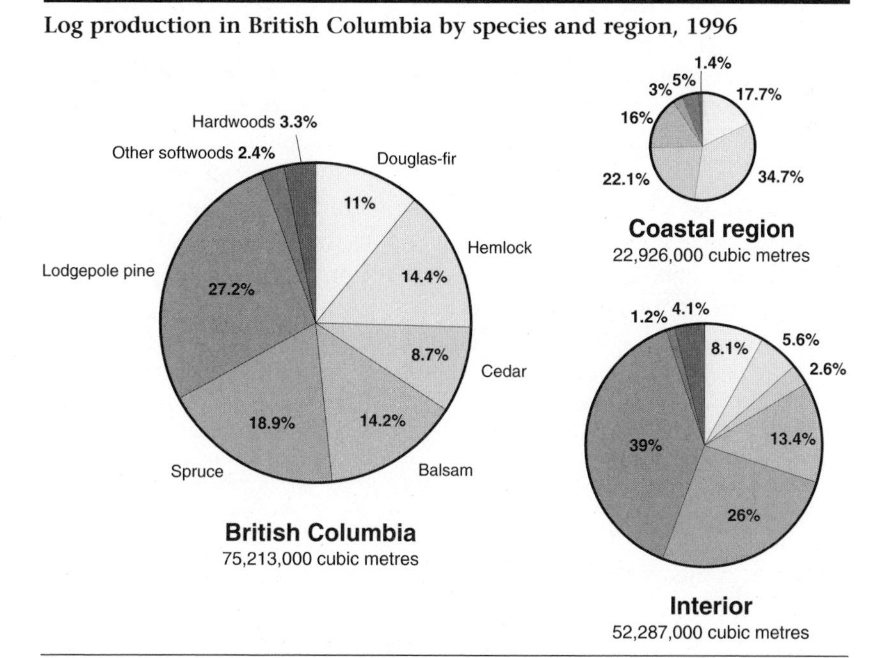

Source: Based on data in Council of Forest Industries, 1997.

The old Somass mill at Port Alberni. Industrial restructuring now occurs in situ.

dominate the coastal cut, and lodgepole pine, spruce, and balsam dominate the interior cut (Figure 3.5). However, the interior harvest in 1996 had grown to almost two-and-one-half times the size of the coastal harvest. Coastal harvests may well have declined further if private forest lands had not been concentrated in this region. Yet, if the wood harvest is greater in 1996 than in 1970, especially on the coast, wood supply is a problem; large mills may now employ fewer people, but they still consume large timber inputs, timber quality is declining, and the creation of parks and related legislation has had the impact of reducing cedar availability on the coast.

In the Fordist period, the spatial dynamics of the forest industries were dominated by the building of greenfield mills (mills on new sites) and by diffusion throughout the interior. In the present age of flexibility, spatial dynamics are dominated by in situ change. For the established commodity industries, expansion, modernization, diversification, and contraction have primarily occurred on existing sites. For newly emerging industries (such as engineered woods and oriented strandboard or OSB), many new mills have been built, including small-scale operations controlled by small and medium-sized enterprises (SMEs). For the most part, however, these new mills have been built in long-established forest centres, especially in the Vancouver area. Thus, in 1993 the greatest concentration of wood remanufacturing plants was in this area, with important outliers in towns such as Prince George (Figure 3.6; Chapter 6).

Similarly, forestry is beginning to tap second-growth forest; that is, to reharvest established timber supply areas. Indeed, over the next decades,

Figure 3.6

Distribution of remanufacturing firms in British Columbia, 1989

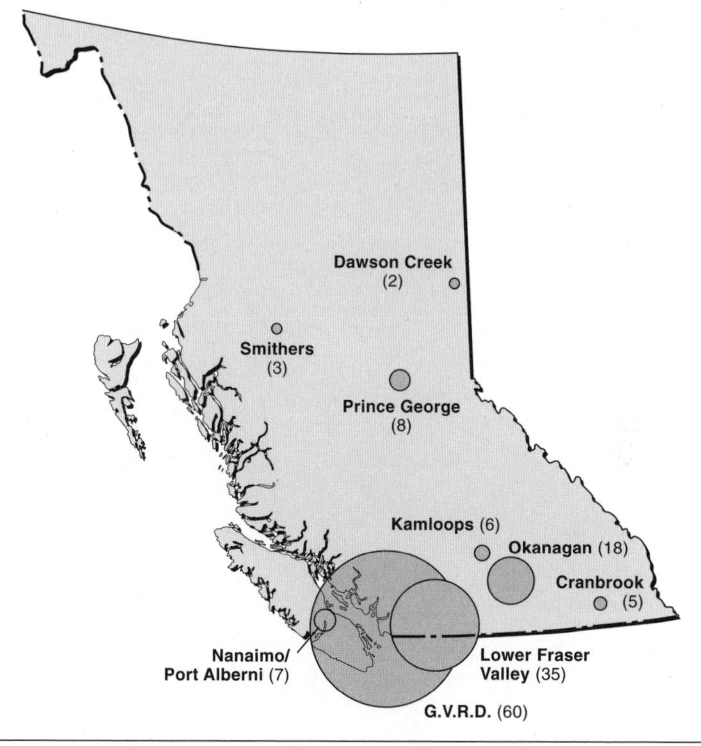

Note: Numbers in brackets indicate total number of firms.
Source: Rees 1993: 67.

the proportion of the harvest accounted for by second-growth forestry will increase dramatically. The in situ nature of contemporary industrial and resource dynamics is a fundamental dimension of the flexible crossroads.

A caveat to the theme of in situ restructuring of long-established production sites and forest-harvesting areas is provided by northeastern BC, where in 1985 Louisiana Pacific of Portland, Oregon, built an OSB mill in Dawson Creek and a bleached chemi-thermo-mechanical pulp mill (BCTMP) in Chetwynd (west of Taylor). These facilities pioneered the use of aspen in BC, hitherto considered a weed species and, with a few other developments (such as Slocan's new CTMP mill at Taylor, built in 1986), extended the geographic margins of large-scale export-oriented pulp production in BC (Figure 3.7). The other new pulp mill built in 1981, also a BCTMP mill, was located at Quesnel, a forest community that had boomed during Fordism.

Given that in situ change has provided the predominant context for industrial (and resource) dynamics in the present age of flexibility, the forest-production landscape of BC has become increasingly differentiated. The established (standardized) commodity industries are in particularly serious

Figure 3.7

Annual capacity of pulp-and-paper mills in BC, 1994

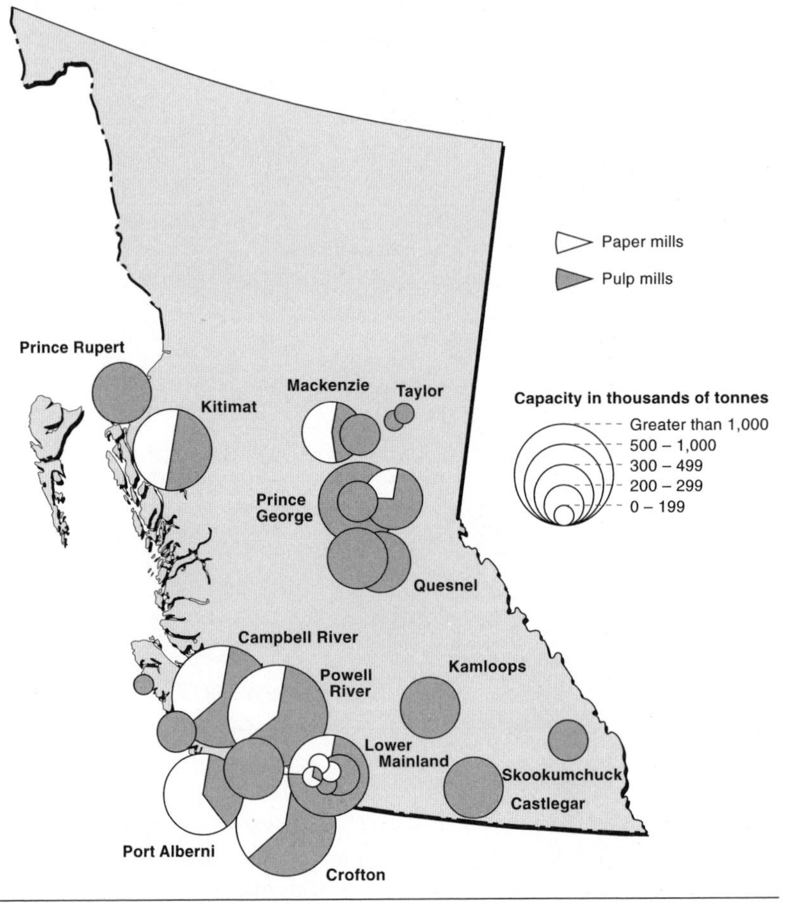

Note: Annual capacity is based on 365 operating days per year.
Source: British Columbia, Government of, 1995.

trouble. Commodity-based plywood mills and sawmills, along with kraft pulp mills and newsprint machines have been closed, especially in the coastal region. In other cases, old mills have been modernized to produce, in still large volumes, a wider and distinctive range of products using strategies of flexible mass production. MacMillan Bloedel (MB) is a case in point: it has removed some commodity lines entirely while restructuring virtually all its remaining sawmills and pulp-and-paper mills to manufacture a wider range of higher-value outputs with each mill playing a distinct role (Chapter 4). New large-scale mills typically focus on new products. The new pulp mills at Quesnel, Taylor, and Chetwynd, for example, all produce high-yielding pulps (around 95 percent wood-fibre recovery) in contrast to kraft pulps (around 50 percent wood-fibre recovery); all are environmentally friendly –

the Chetwynd mill claims to be the world's first to be based on "closed loop" (no effluent) principles; all three mills realize economies of scale at lower levels of output than kraft mills; and the Chetwynd mill uses a hitherto "waste" species. In wood processing, large firms have invested in distinct, new OSB- and engineered-wood mills, further differentiating the forest landscape within the context of large-scale production.

Meanwhile, the growth of SMEs in recent years has raised the possibility of the development of flexibly specialized production networks within BC's forest economy, adding to its diversity (Rees 1993; Reiffenstein 1999; Chapter 6). As the flexible specialization thesis predicts, SME-based growth favours the Vancouver area, but it has not excluded smaller urban centres or remote, tiny places. Moreover, as the corporate head-office hierarchies have been downsized, the Fordist underpinnings of core-periphery relations are undermined (Chapter 9).

In BC's forest economy, however, the age of flexibility is still associated with boom and bust. The new in situ geography is a volatile geography. Most mill closures and employment downsizing has occurred in the coastal region, but much value-added activity has been concentrated in this region, too. Most mill closures and job loss have also occurred at older facilities, but not exclusively so. The Gold River pulp mill was built in 1967, and it is now gone. Even new developments do not guarantee success. Thus, in 1999, the new pulp mills at Taylor and Chetwynd both requested financial assistance to maintain or restart operations, and both are looking to sell (in the case of Slocan's mill at Taylor, successfully). Small firms throughout the province are constantly threatened by wood-fibre shortages and problems of market access (Chapter 6). These are still perilous times, and in the crisis atmosphere of the late 1990s, provincial forest policy has itself become the centre of debate over whether it is resolving or worsening the situation.

Reregulation: Towards the High-Stumpage Regime of the 1990s

During the Fordist period (1950-73), the provincial government was formed by one free enterprise party (Social Credit) and one strong-minded premier (W.A.C. Bennett). Forest policy and attitudes towards the forest industries remained generally consistent with the philosophy of the 1945 Sloan Royal Commission. The Social Credit government was briefly replaced in 1973 by the more left-wing New Democratic Party (NDP), which had long been critical of forest policy and, led by its forest minister, Bob Williams, immediately sought radical changes (Wilson 1997). In 1975, Williams established a Royal Commission and a single commissioner, Peter Pearse, to direct the rethinking of forest policy. While the NDP were ousted in 1976 and Social Credit returned with a new leader, W.A.C. Bennett's son Bill, many of the subsequent recommendations of the 1976 Pearse Commission were accepted. Yet, in contrast to the Fulton and Sloan Royal Commissions of 1909 and 1945, respectively, the Pearse report, rather than bringing closure, marked the beginning of debate on the forest economy that culminated in the barrage of policies in the 1990s that forms the basis of the high-stumpage regime. There is still no end in sight for the debate.

The Pearse Commission: Towards the Forest Transition

The Pearse Commission was born out of the new provincial NDP government's concern for the unevenness of forest management in TFLs and PSYUs and for the corporate concentration of harvesting rights. Following the election of a new provincial government, the Forest Act of 1978 included Pearse's main recommendations. Thus, the act harmonized the TFLs and replaced the quota system used to allocate wood from the PSYUs by a new form of harvesting licence, the forest licence (FL), which provided licensees with stronger entitlements to wood and greater forest management responsibilities. In addition, the new policy provided wood fibre to SMEs, simultaneously revealing the government power to change timber licences that critics had suggested were de facto in perpetuity arrangements that favoured large corporations. SMEs that qualified for the new Small Business Forest Enterprise Program (SBFEP) were allowed to obtain a timber sale licence by bidding on timber made available from the nonquota wood in the old PSYUs. A decade later in 1988, 5 percent of the cut in TFLs was also redirected to the SBFEP.

Moreover, the Pearse Commission stimulated better forest management, most obviously in the effective stocking of logged-over land. Thus, reforestation efforts in BC since the late 1970s have escalated significantly, are much bigger than those of any other province, and reforestation by the late 1980s was more or less on the same scale as harvesting (Table 3.8). In support of reforestation, nursery activities have also grown impressively. In the crude terms of areas harvested and planted, the Forest Act of 1978 marks the beginning of an effective grappling with the forest transition in BC. Indeed, a recent study by the Ministry of Forests concluded that second-growth forests in BC are growing faster than expected, a finding with important implications for the timing of future harvests. For lodgepole pine in the interior, for example, at age fifty trees are now expected to be 20 metres in height rather than 16 metres, and to be ready for harvesting in ninety-three years rather than 124 years. Not all experiments in reforestation have worked as well as expected, including MB's pine plantation near Port Alberni. Nevertheless, the scale of planting is impressive and raises hope that at some time sustainable harvests could actually increase; in Reed's view (c. 1996), a future harvest of 90 million cubic metres is not out of the question.

In its emphasis on forest sustainability, support for SMEs, and reference to wider environmental values, the Pearse Commission was consistent with the needs of the emerging age of flexibility. In 1975, however, the full implications of industrial dynamics and resource dynamics, not to mention their interrelations, were not fully understood. Two years after the Forest

Table 3.8

British Columbia forest area harvested, area planted, and number of trees planted, 1970-96 (selected years)

	Area harvested (million hectares)		Area planted (million hectares)		Trees planted (million)	
1970	150		30		35	
1975	157	(680)	63	(128)	73	(128)
1980	188	(873)	64	(147)	75	(147)
1989	199	(1,018)	172	(447)	210	(447)
1995	190	(1,011)	206	(436)	262	(436)
1996	200				261	

Note: The figures in parentheses are Canadian totals.
Sources: Canadian Council of Forest Ministers 1997; Canadian Forest Industries Council 1986; and Council of Forest Industries 1997.

Act of 1978, BC's forest economy was shaken to its roots by the deepest recession since the 1930s. In turn, amid financial loss and job loss, this recession stimulated American protectionism against BC lumber producers and modifications in BC's forest policies that infuriated environmentalists. The volatility of the 1970s became real trouble in the 1980s (Barnes and Hayter 1997).

The 1980s Recession: Industrial and Resource Dynamics Join Together

The basic cause of the early 1980s recession in BC's forest economy was industrial. Overcapacity existed and global demands declined alarmingly while technological change, especially that driven by microelectronics, was changing production, work organization, and market possibilities. Similar problems faced American lumber producers elsewhere in the Pacific Northwest, but they had an easier scapegoat in the form of exports from BC. Accordingly, on the basis that stumpage in Canada was so low as to constitute a subsidy, American producers began protectionist action against Canadian lumber that haunts BC to this day (Chapter 7).

By 1980, on the eve of recession, falldown effects had also emerged as an issue within the Ministry of Forests (Percy 1986). The Ministry of Forests (1984) emphasized that harvests will decline at some future date because of falldown, while noting that past estimates of the AAC had been "bullish" by assuming optimistic scenarios of continuing improvements in wood-utilization technology. By 1990, published estimates of the long run-sustained yield (LRSY) for regulated lands were in the order of 59 million cubic metres, about 20 percent less than the estimated AAC. An estimate for the long-term harvest level (LTHL) was less (BC Wild 1998). Surprisingly, the 1996 estimates again appear "bullish" and partly because timber supplies previously excluded from calculations are now included and partly because of "improved information" and "changes in technology," the new

Table 3.9

Estimates of allowable annual cut and long-run sustainable cut, 1990 and 1996

	1990		1996	
Region	AAC (1,000 m³)	LRSC	Pre-TSR (1,000 m³)	Post-TSR
Cariboo	7,903	5,567	8,204	8,168
Kamloops	7,985	5,922	7,699	8,185
Nelson	6,633	5,114	6,048	5,485
Prince George	19,189	16,552	18,368	19,400
Interior total	**41,710**	**33,155**	**40,319**	**41,238**
Prince Rupert	9,345	6,266	9,189	9,317
Vancouver	23,663	19,771	21,674	20,300
Coast total	**33,008**	**26,037**	**30,863**	**29,617**
Total	**74,718**	**59,192**	**71,182**	**70,855**

Notes: AAC is allowable annual cut and LRSC is long-run sustainable cut. TSR refers to Timber Supply Reviews commissioned in 1992. These estimates are for regulated lands only.
Source: Ministry of Forests 1990 and 1996b (web page: http://www.for.gov.bc.ca/tsb/other/review/review.htm).

AAC, at around 71 million cubic metres, is only modestly less – much bigger than the previously determined LRSY (Table 3.9). Such estimates are sensitive to many factors that are hard to assess, such as growth estimates, silvicultural treatments, land use changes, and technology. Moreover, the falldown effects predicted by the 1990 figures were projected to unfold over time horizons from twenty to 200 years depending on the region. Clearly, BC still has a huge forest resource and in one opinion is not suffering from shortages of wood (Runyon 1991: 64). Yet, the falldown is a reality that specifically signals the decline of old-growth forests and fundamental changes in forest management.

For industry, initial announcements of falldown meant a reduced AAC at some future date; for environmentalists, they provided clarion calls to save remaining vestiges of old growth, and they suggested the irrevocable loss of a wide range of ecological values. At the same time, the industrial problems reflected and caused by recession invigorated environmental criticism that portrayed the forest industry as part of a sunset sector with no economic future. Under the pressure of crisis, the Ministry of Forests had secretly introduced a policy of "sympathetic administration" in the early 1980s, to allow companies leeway in meeting forest regulations. Once this policy was discovered, however, environmental anger deepened, further encouraging "a war in the woods" that in the 1980s led to logging blockades and even tree spiking in such pristine areas as Meares Island, the Stein, the Carmanah, Clayoquot Sound, and the Walbrand (Blomley 1996). Environmentalists' arguments typically conflate their interests with those of Aboriginal peoples

(M'Gonigle and Parfitt 1994), although Aboriginal spokespeople have distanced themselves from environmentalists in making their own claim on the forests of BC (Nathan 1993).

It is a little ironic that as forest exhaustion has declined as a threat, controversy over forest policy has escalated. Globally, this controversy reflects growing environmental consciousness and the more varied needs of the postindustrial society (see Table 1.1). In BC, if the Forest Act of 1978 did introduce change, the long-established tradition of secrecy in decisions about forest use in BC maintained by what Wilson (1987/8: 9) labels the "wood exploitation axis" of government, business, and labour, remained intact. Similarly, the provincial philosophy of leasing large tracts of timber over long time horizons to corporations in return for relatively small royalties and/or stumpage payments, but large-scale investment in industrial facilities, did not appear to be violated. By the late 1980s, public scepticism about existing forest policy was on the rise, triggered in part by announcements of the falldown effect. Even if falldown effects can be rationalized as a natural evolution in forest management, the implication that sustained yields should decline was a new policy implication in BC that had not been publicly anticipated. There was also growing recognition that forestry practices were not well scrutinized in the past, with too much casual reliance on "natural regeneration," the cheapest form of forestry. Unfortunately, the failure rate of natural regeneration can be high (Smith and Lessard 1970).

The heat generated by environmental criticism and conflict, as well as the reemergence of Aboriginal land claims in the 1980s, revealed that the Forest Act of 1978 had not gone far enough in meeting a wider range of values demanded by the public related to alternative uses such as fishing and tourism as well as the concerns associated with aesthetics, biodiversity, and land claims. Moreover, the recession of the 1980s raised the spectre that neither jobs nor communities could be sustained, thus contributing to an image of the forest industries as a declining force. To further complicate matters, American protectionism after 1983 towards BC's lumber industry was based on the accusation that provincial stumpage – the tax charged on companies for cutting trees on Crown land – constituted a subsidy, thus supporting the view that BC was not placing proper value on its forest resource.

Environmentalism, Aboriginalism, protectionism, falldown effects, sharp booms and busts, and technological change conspired to make BC's forest economy a deeply troubled landscape in the 1980s, a situation that demanded new policy initiatives. The very considerable response of the NDP, partly elected to resolve these troubles, can be labelled the high-stumpage regime.

The High-Stumpage Regime and Industrial Restructuring

In a climate of growing uncertainty, conflict, protest, and civil disobedience, the new NDP provincial government established the Peel Commission (1991)

Table 3.10

The high-stumpage regime: Reregulation for the 1990s

Policy	Comment
Revised stumpage formula, 1988	Shift from market-based system to waterbed system (introduced 1988, retained by new provincial government)
Stumpage level ratchets, 1992-	Ratcheting up of stumpage on yearly basis, after 1992
Pulp Mill Effluent Standards, 1992	New targets for all pollutants; AOX levels to be reduced to zero by 2001
Forest Practices Code, 1995	Reform forestry to meet environmental values: size of clear-cuts reduced, continuous clear-cutting eliminated; wildlife, biotic, and aesthetic values incorporated in forest plans
Timber Supply Reviews, 1992-	Reassessment of allowable annual cut (AAC) for industry, to assess falldown effect, and effects of provincial policies
CORE, 1992	CORE (Commission on Resources and Environment) established to develop regional land use plans for entire province
Parks, 1990-	2.5 million hectares added to the provincial protected area system between 1990 and 1995, with goal to have 12 percent of provincial land base in parks to conserve environmental values; 108 new parks in 1995
Clayoquot Sound Compromise, 1993	Committees representing numerous interest groups proposed (first) land use plan, in 1993; a scientific advisory panel established
Community forests	Proposals requested in 1997, with three to be awarded
Small firms, 1988-	In 1988, 5 percent of wood fibre of TFLs diverted to small firms, more wood fibre diverted in 1990s; industry association activity funded
Forest Renewal, 1994	New agency, funded from "super stumpage," with BC, broad mandate to invest in silviculture, help forest communities, workers, firms, and other interest groups, and to fund research projects and organizations. NewFo established (1997) to redirect laidoff union workers to silviculture
Jobs Accord, 1997	Agreement-in-principle with industry, to promote jobs, especially by small firms and in "value added"; subsidies for new jobs provided
Treaty Process, 1993-	Canada-BC Memorandum of Understanding creates a five-stage treaty process to resolve Aboriginal land claims that cover entire province
Tenure Reform?	Devolution? Market-driven timber pricing?

to formally recommend a new policy of "enhanced stewardship" designed to secure environmental as well as economic values from provincial forests. The provincial government committed to substantive changes in forest policy as well as to settling land claims. This recommendation has been pursued and indeed, the breadth and scope of the forest policy initiatives introduced in the 1990s are extraordinary (Table 3.10). The overall objective was to fully realize the economic and environmental values of the forest resource, while also finally settling Aboriginal land claims.

At the core of the high-stumpage regime is stumpage itself; the new regime comprised a change in both formula and levels (Figure 3.8). Both the new formula and higher stumpage were first introduced by a Socred provincial government in 1988, primarily stimulated by American countervail threats, while offering the perk of additional revenue. The NDP government retained the new formula and higher stumpage, and in 1992, the government introduced a built-in escalator so that stumpage would progressively increase each year. The stumpage formula was changed from a market-based system to a waterbed system. In the former, stumpage was a residual of the difference between costs and revenues, with costs incorporating a risk factor. According to the waterbed formula, overall stumpage does not fall below a level set by government, although stumpage can vary according to location and species, reflecting different cost, risk, and price factors.

Figure 3.8

Provincial government rationale for higher stumpage

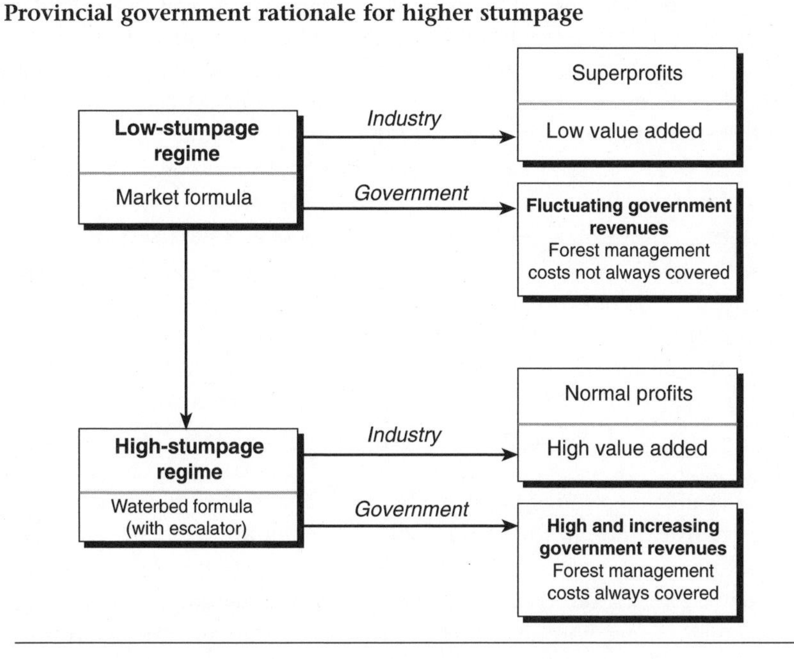

Table 3.11

Average log costs ($/m³) on Crown lands, 1988, 1992, and 1997

	Province			Coast	Interior
	1988	1992	1997	1997	1997
Total costs	$42	$46	$88	$107	$79
Stumpage	$8	$8	$27	$23	$29

Source: Price Waterhouse 1997: 15.

Numerous other initiatives, several of a substantive nature, supplement changes in stumpage formula and levels within the overall framework of the high-stumpage regime. Environmentally, the main policies are the Forest Practices Code (1995), designed to ensure environmentally appropriate forest practices; the Commission on Resources and Environment (CORE), established in 1992 to develop regionally based comprehensive land use plans for the entire province; the creation of new parks that have taken land out of forestry; the Clayoquot Sound Compromise, set up to maximize multiple values in a highly sensitive region of Vancouver Island; and a provincial timber-supply review, initiated in 1993 to revise the AAC allocations, a revision which more or less justified, in aggregate terms, the March 1994 AAC level of 71.3 million cubic metres when industry had expected a downward revision, possibly to 59 million cubic metres (Price Waterhouse 1995: 6; see Table 3.9). The government also set up Forest Renewal BC, funded out of "super stumpage," to invest in forestry, help retrain laidoff forest workers, help forest communities diversify, and fund research in support of value-added initiatives. Through the Jobs Accord, the government also affirmed its interest, in broad terms, in promoting value-added activities and small firms, including providing incentives and wood fibre for new investments. As well, the treaty process was introduced to resolve all land claims in the province, beginning with the Nisga'a. Moreover, forest plans and contracts are now public documents, and in general, it has been made clear that BC's forests are no longer the sole priority of industrial forestry, especially large corporate forestry. What this battery of policies lacked, however, was any explicit comprehensive plan for tenure reform (Wilson 1997).

With this important caveat to tenure reform aside, the battery of policies that constituted the high-stumpage regime represents the widely held Innisian view that BC's forest resources are undervalued and that public policy should seek higher rents (in the form of stumpage) and meet the full range of public goals for a publicly owned resource. In this view, stumpage is the key institution in the pattern of forest exploitation (Figure 3.8). From this perspective, low stumpage allows industry to earn "superprofits" while maintaining a low-value commodity production, a strategy in which the public loses many of the environmental benefits of the forest and the government

Table 3.12

Crown revenues from the timber harvest ($000)

Year	Ministry of Forest data	Price Waterhouse data	Schwindt and Heaps data
1978/9	254		
1979/80	571		
1980/1	361		442
1981/2	107		172
1982/3	93	142	146
1983/4	137	191	196
1984/5	187	187	187
1985/6	209	209	209
1986/7	246	222	246
1987/8	536	671	536
1988/9	624	627	622
1989/90	652	648	650
1990/1	573	605	571
1991/2	608	574	607
1992/3	711	616	710
1993/4	1,025	849	1,023
1994/5	1,884	1,430	
1995/6	1,666	1,798	
1996/7	1,852	1,733	
1997		1,733	
1998		1,415	

Notes: The ministry data are taken from the financial and statistical tables section without modification. The single most important revenue contribution is stumpage from major licences and the second most important is stumpage from the SBFEP. For example, in 1996-7, the former constituted $1,451 million and the latter $348 million, for a total of $1,799 million. The Price Waterhouse data are for calendar years and only refer to stumpage.
Sources: Ministry of Forests annual reports 1979 to 1997; Price Waterhouse 1988 and 1998; Schwindt and Heaps 1993.

loses a source of rent. In contrast, the high-stumpage regime increases revenues to society by eliminating superprofits, provokes industry into shifting towards higher-value activities, and realizes environmental and social goals.

The extent of the stumpage increase, of overall logging costs, has been considerable (Table 3.11). From 1988 to 1997, average stumpage costs per cubic metre increased more than threefold while total costs doubled. Moreover, the cost escalation has primarily occurred since 1992 when average provincial stumpage was around $8. By 1997 stumpage was $27 per cubic metre, on average more in the interior than on the coast, although overall log costs are much greater on the coast. Even as harvests have declined, the absolute increase in Crown revenues from the timber harvest, the central source of which is stumpage, has been considerable (Table 3.12).

In terms of industrial dynamics, the high-stumpage regime is a deliberate attempt to kick-start or at least boost restructuring processes by emphasizing the need for industry to transform itself from standardized, low-value commodity production to high-value, differentiated production. Stated simply, the high-stumpage regime assumes that with a smaller supply of higher-priced timber, industry is forced to create higher values. In turn, higher values rationalize the waterbed formula and the argument, long advocated by foresters, that stumpage should at least cover the forest management costs incurred by the government – a principle that was not met in practice three times in the early 1980s when low prices reduced stumpage to a bare minimum (Schwindt and Heaps 1996: 46).

Indeed, higher-value activity is the theoretical basis for a positive-sum game in which both economic and environmental goals are fully met. Thus, investment in value-added activities compensates for job loss in commodity industries, especially to the extent that value-added activities are associated with labour-intensive small firms. Because value-added activities are associated with creating more jobs per cubic metre, more of the forest is available for environmental purposes (or for Aboriginal land claims), without loss of jobs in the industrial forestry workforce. Similarly, the new Forest Practices Code, CORE, the creation of parks, reductions in the AAC, and the Clayoquot Compromise were all introduced to enhance the environmental stewardship of the forest, including logging itself (Gunton 1997; Chapter 10). While these environmental initiatives impose significant extra costs on industry, their effect is to reinforce higher stumpage levels in stimulating high-value-added activities. A higher-value resource also (potentially) implies more intensive forms of forestry on lands available to industry. Indeed, a shift from basic forest management to more intensive forestry (with greater investment in research, seedling production, planting, precommercial and commercial thinning, pruning, fertilizing, and so on) offers a beguiling, but legitimate, hope that in the future the AAC might actually increase (Haley 1985; Percy 1986: 27; Binkley 1997b). Finally, as already noted, the high-stumpage regime counters the threat of American protectionism based on the argument that low stumpage is a subsidy. In fact, successive provincial governments "used" American countervail threats as a justification for higher stumpage.

According to the high-stumpage regime, "value added" is a magic wand that will resolve the problem of industrial and resource dynamics to the benefit of all concerned, reward the government with higher rents, and allow a wide range of social goals to be met, including addressing Aboriginal land claims (M'Gonigle and Parfitt 1994).

The High-Stumpage Regime as a Leap of Faith

Changes in BC's forest policy were necessary and environmental accountability was an absolute priority. The high-stumpage regime has certainly provided a massive shove in this direction. Yet, by 1999, the reality of the unfolding of the high-stumpage regime has proven problematic to say the least; highly touted expectations of the early 1990s have ended in the crisis of the late 1990s, one that is as severe, if not more so, than the recession of the early 1980s. Thousands of workers have been laid off; corporate losses have mounted; the remaining industry R&D centres, so vital to innovation and value added, have been closed; log exports have increased; and acrimony is growing among and within interest groups. Booming and busting is still the modus operandi of BC's forest economy. Moreover, the implication of this policy that decision-making processes regarding forest use would be more democratic and representative has in practice meant increasingly fierce acrimony. The high-stumpage regime, designed to resolve BC's troubled economy, is itself controversial, thereby adding to the troubles.

Two criticisms based on the underlying assumptions and formulation of the high-stumpage regime may be noted. The first criticism is that the high-stumpage regime is based on questionable assumptions that amounted to a leap of faith. Thus, the high-stumpage regime assumed superprofits existed in the industry. Yet, despite attempts to do so, no superprofits have been empirically revealed (see Schwindt and Heaps 1996); rather, evidence suggests that while rates of return have been volatile, average returns are not high by virtually any standard. Moreover, the argument that the deliberate escalation of stumpage and cost-increasing forest regulations in the 1990s would, by itself, stimulate higher-value production (and more jobs) involves further leaps of faith. In particular, the higher-stumpage/stronger regulation policy assumed that value-added opportunities on a scale sufficient to pay for the increased costs exists; that existing or new firms in BC would be able to exploit these unspecified opportunities, more or less immediately and comprehensively (while adding to the industry's employment base); and that in the absence of private-sector responses, the government would be able to stimulate them. A related assumption is that the provincial government knew precisely by how much to increase costs to generate the desired value-added responses. It might also be noted that while positive benefits to more intensive forestry can be expected, their calculation is not straightforward, and increasing benefits cannot be simply assumed to cover ever-increasing costs (Percy 1986: 25-9).

In practice, the main effect of higher-stumpage and related initiatives, especially the Forest Practices Code, has been to raise costs well beyond levels in the rest of Canada and similar to levels in the US where other costs are much lower. In the case of the US Southeast, the cost of harvesting sawlogs, excluding stumpage, is more than three times less than on the US coast (Ernst and Young 1998: 36). These other regions are also competing for the same higher-value markets, and any advantage BC once had in terms of wood quality has been substantially reduced. In addition, industry in BC has not only paid higher stumpage but also higher environmental costs and higher costs of forestry management that were once the government's responsibility, while losing part of its tenures. In the meantime, the increased costs are even difficult for the small, so-called higher-value producers to bear. Moreover, in competing in US markets, BC firms face important problems of market access in comparison to US rivals.

The government has conceded to this criticism, and in 1998 it reduced stumpage somewhat, although levels remain comparatively high, and modified aspects of the Code. However, there is no simple answer to the stimulation of a higher-value-added industry. Furthermore, the changes to stumpage and the Code underline the experimental nature of the high-stumpage regime. In fact, virtually all other components of the regime have been modified or delayed so that the battery of policies is taking on an ad hoc, highly politicized nature. Such experimentation may have made sense in the late 1940s, when the provincial government chose the path of corporate-dominated forestry, and even in the early 1970s, when there was still first-growth wood fibre available. Social experimentation of this kind, however, is far more difficult in an already settled land where changes have to be in situ.

A second set of criticisms relates to the issue of provincial autonomy and can only be intimated at this point. In the high-stumpage regime, the provincial government saw itself as conducting an entirely new vision of the future for BC's forest economy. It has unwittingly undermined its own ability to orchestrate this vision, however. While the government did not outline a comprehensive plan (or "map") of tenure reform or land use in general, it did open up debate over BC's forests to interest groups hitherto excluded, notably environmentalists and Aboriginal peoples. The government did not provide or effectively negotiate any lids to the influence of these interest groups, while government control over forest policy continues to be further complicated by increasing interference by the US. In effect, the high-stumpage regime opened up a Pandora's box of voices that are expressing powerful views far beyond what the government anticipated. Thus, in 1999, there were pleas from industry to privatize forest lands (Stephens 1999), proposals vehemently opposed by environmentalists, and an environmentally sponsored plea to basically shut down the coastal forest industry (Marchak et al. 1999), a proposal vehemently opposed by industry. Both proposals may run aground on Aboriginal land claims, stimulated by the provincial government's reference to "inherent rights."

Reregulation as an Expression of Flexibility

The rethinking of forest policy begun by Pearse (1976), and gaining momentum in the 1990s, can be broadly seen as seeking flexibility in three fundamental ways. Thus, forest-policy rethinking seeks to develop greater flexibility in the use of BC's forests: first, by accommodating a wider range of values, notably by giving more emphasis to the so-called nonwood benefits of the forest than in the past; second, by reallocating forest rights to more varied constituencies than in the past; and third, specifically with respect to industrial use, by promoting a shift from commodity-based activities to more value-oriented activities that can profitably maintain and add jobs by product strategies that are highly responsive to a more differentiated range of consumer needs.

In other words, the rethinking of forest policy is extraordinarily complicated. Nonwood benefits of the forest refer to a wide mix of values related to recreation, erosion control, climatic benefits, wildlife, biodiversity, hunting and fishing, health, research, and other activities. That is, in contrast to previous times, the idea of sustainability is a more complex, multidimensional concept, and forest management is no longer exclusively about maintaining timber volumes for industry (see Figure 1.2; Baskerville 1990). More varied constituencies imply a shift in forest rights to small firms, municipalities, and Aboriginal peoples, the latter including the treaty process that itself entails principles of equity and in some sense the resolution of historical wrongs. Meanwhile, established industry is expected to reformulate its basic attitude and structures developed over the past 100 years within the context of a lower AAC.

Whether the provincial government's forest-policy initiatives of the 1990s effectively reflect the imperatives of flexibility in a socially beneficial way is problematic. There is much criticism of policy and evidence of inconsistency, and arguments can be made that provincial policy in some ways has been counterproductive (Chapter 11). The causes of the industry crisis in the late 1990s are complex, and they are not simply the result of government policy. Ultimately, however, provincial forest policy exercises a defining influence on the path of industrial evolution, even if many consequences of such policy are unintended.

Reflections on Recession and Restructuring

If industry, government, and labour anticipated the severe recessions of the early 1980s and 1990s, little was done to avoid them. Possibly, these recessions, at least in their degree of severity, were not anticipated as well. Moreover, the extent of the crisis in the 1990s suggests that the decision makers for BC's forest economy, public and private sectors, did not learn enough from the crisis of the early 1980s, or at least have not been able to develop a sufficiently coherent response to ensure effective restructuring. There are new, complex considerations, relating to American protectionism, environmentalism, provincial forest policy, Aboriginal land claims, and the Asian economic crisis. Even so, the pleas for restructuring in the late 1990s are strikingly familiar to the pleas of the mid-1980s. Restructuring appears to have been slow and haphazard.

Brunelle (1990) noted similar restructuring lethargy in Washington and Oregon, and other parts of the American Pacific Northwest, with respect to responses by forest-product firms to recessionary conditions in the early 1980s. Brunelle reveals that the firms slowest to react were the largest "national" firms. In his view, the relative decline in the importance of the largest firms in forest production of the Pacific Northwest between 1979 and 1985 did not reflect an inherent lack of loyalty to the region, compared to "regional" and small companies. Rather, he suggests that, despite their control over resources, professional management, vertical integration, R&D groups, and planning departments, the major corporations were simply "too large and cumbersome to respond quickly and did not see the changes coming in the economy and industry and reacted slowly" (Brunelle 1990: 119). Brunelle quotes an industry spokesperson: "Let's be very candid about it. We'd been coasting along, making a lot of money and doing what we thought was a reasonable job of innovative management. When we were really faced with the reality, we found that we had almost been asleep at the switch. We found ourselves trying to start a recovery marathon with a pocket full of rocks. We had over-aged plants. We had the highest forest-product wage scales in the world in the Pacific Northwest with no tie-back to productivity. We had some 'yesterday' management attitudes" (Brunelle 1990: 119).

This view is undoubtedly relevant to BC, where large corporations, including some of the same ones operating in the Pacific Northwest, gave little evidence of impending crisis. The fact that BC forest-product firms embarked on record-setting capital-expenditure programs in 1979 and 1980 testifies to this lack of anticipation or appreciation of the forces of change. Indeed, by the 1980s, for some observers, large corporations in BC were dinosaurs. Schwindt (1979: 19), for example, agrees with Rumelt's (1974) characterization of forest-product corporations as mature firms locked in to old technology and ideas within a "low-performance strategy" based on high levels of vertical integration and slow to react to market signals.

Yet, the underlying issue in this context is not simply the extent to which the impact of recession, and more importantly the need for long-term restructuring, was unanticipated. As Freeman (1987) emphasizes, recessions and crises facilitate change that is overdue (see Table 1.1). From this point of view, forest-product firms were so locked in to ingrained managerial attitudes, labour relations, and technologies, and to ways of organizing business, that the impact of recession could not have been avoided, even if it had been anticipated. Alternatively put, in the absence of crisis, vested interests met with insufficient incentive to change established practices and values, even if some in the industry did anticipate the need for change. Clark (1986) makes this point in the context of the auto industry in the American Midwest. Unfortunately, the implications of crisis for restructuring are not straightforward.

The Paradox of Busts (and Booms) for Corporate Restructuring

As a context for restructuring, recession posed a double-edged sword for forest-product firms in BC. On the one hand, the recessionary crisis encouraged radical thinking about corporate strategies and structures throughout the industry. On the other hand, the same crisis imposed strong restraints to the implementation of such thinking. These restraints are primarily twofold: fiscal and human.

Typically, the immediate response to recession is to reduce costs by laying off employees and management. In the early 1980s, these layoffs frequently became permanent (Grass and Hayter 1989; Hayter and Barnes 1992; Chapter 8). However, if such layoffs reduce costs, they simultaneously reduce the human resources available to plan and implement initiatives. Moreover, among the unionized workforce, it is extremely difficult for firms to be selective over layoffs because of the seniority principle. Admittedly, firms can choose more freely the professional positions they wish to eliminate so that, in theory at least, perceived problems of antiquated attitude and deadwood can be more readily addressed at the management level. Yet, fewer managers, as well as workers, reduce the organizational slack available to cope with a crisis; any shift to new managers involves learning-curve costs; and to the extent that firms wish to implement flexible labour relations, shifts in managerial attitudes have to be matched by shifts in worker attitudes. The human-resource constraint to coping with recession may well have been underestimated; the shift towards labour market flexibility and to "high-performance organizations" in general has proven time consuming and contentious (Chapter 8).

A second major constraint to responding effectively to a sustained recessionary crisis is fiscal. Thus, the recession of the early 1980s provided a powerful signal to modernize and develop new product lines while simultaneously compromising the financial ability of the industry to do so. Indeed, forest-product firms lost a lot of money and had high debt-equity loads as a result of capital-investment programs begun in the late 1970s and the capital-intensive nature of the primary forest industries, especially pulp and paper. Many large corporations, including MacMillan Bloedel, faced desperate situations: they were losing money but needed cash both to cover losses and make investments. Faced with this predicament, many corporations, including MB, chose to sell some assets to acquire cash quickly. Such divestments, however, were not necessarily consistent with long-term needs, serving to offset losses and debts, as much as to redirect the corporation.

Although the decision makers in BC's forest economy began to recognize the need for fundamental change during the early 1980s recession, the subsequent boom appears to have soon retrenched traditional attitudes. Indeed, if the paradox of recession is to encourage firms to rethink strategy but to limit the resources to do so, the paradox of booms is to provide resources for change but to encourage status quo thinking. Remarkably, almost two decades after the recession of the early 1980s, the same concerns about corporate strategy in BC's forest industries are still being voiced.

Thus, using similar language, sometimes exactly similar language, of earlier studies (Schwindt 1979; Woodbridge, Reed 1984; Hayter 1987), the Science Council of BC (Ernst and Young 1998) has recently argued that forest-product corporations are "locked-in" to a narrow range of mature technologies and commodities. This study, like its predecessors, wants to see more commitment to value-added strategies, whether incorporated in flexible mass production (Chapter 4) or flexible specialization (Chapter 6).

A Model of the Restructuring from Fordism to Flexibility

In BC, in the decades immediately following the Second World War, the forest industries evolved as a distinct, coherent production system which is now (but was not then) labelled Fordist. The features of Fordism were in turn shaped by BC's location on the geographic margin. Thus, patterns of production, industrial organization, employment, forest management, and technological innovation were collectively organized to export commodities to global industrial markets as efficiently as possible. Low stumpage reflected the low commodity values placed on the resource. BC achieved rapid growth and consumers received cheap wood.

This production system is presently in transition, in Freeman's (1988) terms as part of the ICT paradigm, or more generally, in response to the

Figure 3.9

The BC forest economy: Directions of change from Fordism to flexibility

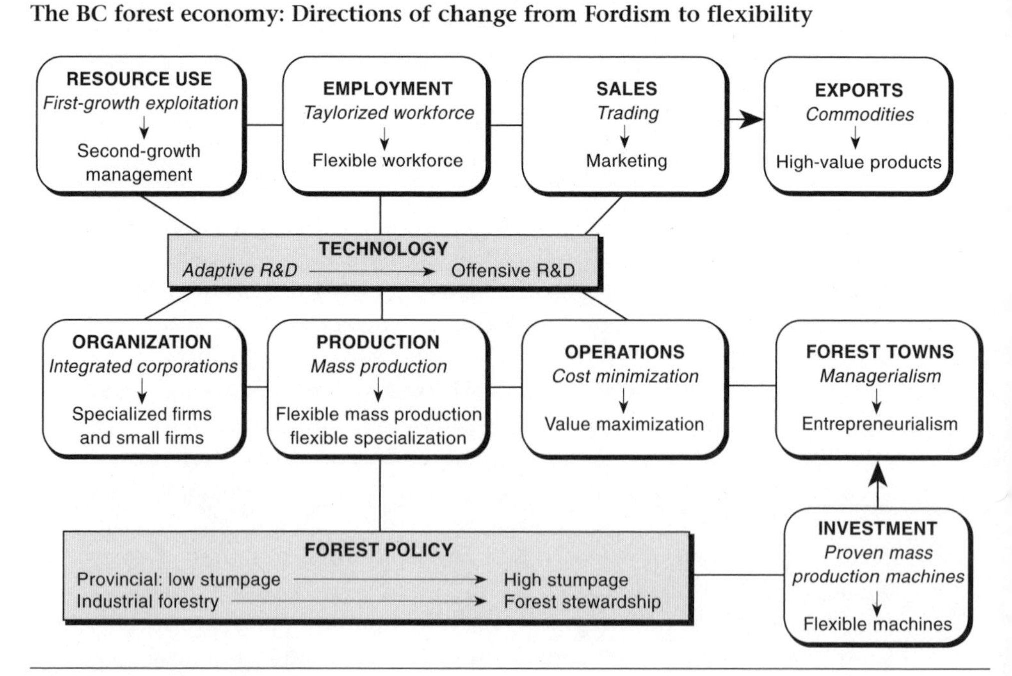

Source: Based on Hayter 1996: 102.

imperatives of flexibility. In this regard, microelectronics, and other high-tech activities, have spearheaded the shift with significant impact in all economic sectors. In BC's forest industries, new technologies have been the basis of tremendous efficiency gains, but also new forms of quality control, environmental protection, and new ways of designing and manufacturing products and services. In short, the new technologies are not simply about automation (and labour saving) but also about the creation of knowledge that can be used in the creation of new endeavours. In turn, the new technologies have allowed producers to serve a more differentiated range of demands, a trend evident in the BC forest economy. In BC, pressures for change have also arisen through natural resource dynamics.

A model of the crossroads explicitly recognizes the interrelated nature of the activities comprising the BC forest-production system (Figure 3.9). It prescribes directions of change in these activities that are deemed necessary for a flexible production system consistent with high wages and high value. The actual nature and strength of these interrelated shifts, or restructuring processes, are problematical, and they are systematically explored in the second part of this book. From a prescriptive or policy perspective, the model argues that to remain viable, forest industries must shift from cost minimization to value maximization, from specialization to diversification; commodity trading must give way to product marketing; the emphasis on proven equipment should be modified by more innovative machinery; dependent technology strategies need to be replaced by offensive technology strategies; corporate concentration must change to a more balanced size distribution of firms; forest management ought to become less extensive and more intensive; the environmental implications of technology and product choices need to become more central; and forest communities need to shift from managerialism to entrepreneurialism (that is, become more proactive in their own development).

If a globally oriented forest industry is considered desirable, forest policy needs to reinforce these shifts, fully recognizing their interrelated nature, as it replaces its narrow (Fordist) concern for the industrial values of the forest with the (flexible) concerns of enhanced stewardship, which gives full recognition to non-industrial values as well. Enhanced stewardship will not pay for itself in the absence of innovation. Policy also needs to recognize that if restructuring is not new to BC's forests, its contemporary form is.

Transformations or restructurings of BC's forest economy, broadly conceived as redefinitions of its competitive foundation and global role, are not new. The global crises of the 1880s and 1930s had severe local impacts while subsequent global booms also featured similar expansions in BC and the emergence of new lead industries (see Table 1.1). Clearly, each of these transformations in BC's forest industries has been different, distinguished in the first instance by the emergence of technological and institutional innovations that create new activities and modify existing structures and ideas. Each transformation is also distinguished by significant changes in the global economy and in local resource dynamics. From these two perspectives, two general points need to be emphasized about the present transformation, involving a unique intersection of industrial and resource dynamics.

First, in contrast to previous transformations, the BC forest industries now face competition that is much more global (Marchak 1991, 1997). Indeed, even on the North American continent, BC's coniferous forests no longer are the low-cost location. Technological change has broadened the scope of faster-growing hardwoods, and it is the American South that defines low-cost forest-product locations in North America, and possibly the world, although plantations in Brazil, Chile, Spain, Portugal, South Africa, New Zealand, and elsewhere, also supply very low-cost wood fibre, especially for pulp and paper. In addition, in sharp contrast to previous transformations, environmental concerns are global rather than local. In the past, environmental issues were typically subsumed within locally based professional forestry, which in BC developed close alliances with industry so that environmental legislation was always seen as a friendly amendment to private-sector behaviour. This is no longer the case; environmentalism is placing global pressures on BC's forest industries, with or without the support of locally based professional forestry.

Second, in previous transformations and crises, subsequent forest industry booms in BC were based on accessing new abundant and cheap fibre supplies. In the past, there were always empty lands. While old-growth areas still remain, even in the coastal region, there are no new empty lands in which to invest. Moreover, with a population of over 3 million, BC's territory is occupied, and although this population is concentrated in the southwest corner, there are settlements, transportation networks, and property rights, recognized and otherwise, throughout the province.

Essentially, the present transformation of BC's forest industries is taking place in situ. This restructuring – labelled in this book as the change from Fordism to flexibility – is inevitably more difficult because it is in existing factories and communities where the friction between vested interests is strongest. In addition, in situ restructuring among forest-based activities has to be considered within the context of other demands on forestry and in light of the needs and values of a population who, for the most part, are not directly linked to the forest economy. It is not clear that the provincial government has understood the implications of rethinking forest policy in situ.

Conclusion: The Contemporary Significance of BC's Forest Industries

There is no doubt that the forest-product industries have declined in importance. The prevailing wisdom of the Fordist period that fifty cents out of every dollar made in BC stems from the forest sector is now discredited. As

Schwindt and Heaps (1996) document, the relative importance of the forest sector for direct and indirect employment in BC or for its share of gross provincial (domestic) product has declined.

Whether or not the forest industries should be thought of as a sunset sector is an entirely different matter. Depending on estimates, the forest industries still directly account for around 90,000 to 100,000 jobs – about 80,000 to 90,000 in the wood-processing, paper and allied, and logging industries (see Table 3.1), and another 8,000 in government and silviculture. In 1996, the forest industries accounted for $16.5 billion (or 50 percent) of total manufacturing shipments. These jobs further support additional jobs through multiplier effects throughout the economy, including the retail, business service, construction, transportation, and government sectors. The Council of Forest Industries (COFI) generally uses a multiplier of two, indicating that for every direct job, the forest industry generates an additional two indirect jobs. Such a multiplier is probably a little high; even so, the forest industry remains a force in the province for employment and wealth generation. The fortunes of many communities remain closely tied to the forest industries, and even in Vancouver, the forest industry is surprisingly important.

Moreover, forest-product exports continue to define BC's global role (Hayter and Barnes 1990). Thus, in 1995 and 1996, forest-product exports amounted to $16.8 and $14.8 billion, respectively, and overall forest products account for over 75 percent of provincial visible exports.

Based on a renewable resource, the forest economy does not have to wither away. Some environmentalists may wish to see the deindustrialization of the forest economy, but global demands for forest products will only shift production elsewhere, with far worse levels of forest management than BC. There is legitimate hope in BC for a viable forest industry that is environmentally acceptable. Some planners may see BC's future in high-tech activities which at the present time are growing rapidly, albeit from a small base. Yet, high-tech and forest industries should not be automatically allocated to two separate parts of the economy. The forest economy offers opportunities for sophisticated research and development resulting in the manufacture and production of complex equipment and software. These opportunities extend to solutions to environmental problems. In short, the forest industries can continue to play a vital role in the provincial economy. A key issue is its ability to innovate and develop as a learning system.

Part 2
The Anatomy of Change

4
MacMillan Bloedel: Corporate Restructuring and the Search for Flexible Mass Production

MacMillan Bloedel (MB) has long been recognized as the leader or champion of British Columbia's forest economy. During the industrialization of BC's forests, H.R. MacMillan, as much as anybody, embodied the entrepreneurial spirit of the times. In often-quoted remarks, MacMillan, before the second Sloan Commission, bemoaned the passing of the entrepreneur in BC's forest industries: "It will be a sorry day for ... British Columbia when forest industry here consists chiefly of a few very big companies, holding most of the good timber – or pretty near all of it – and good growing sites to the disadvantage and early extermination of the most hard working, virile, versatile, and ingenious element of our population, the independent market logger and the small mill man" (Schwindt 1979: 34-5).

Yet, MacMillan was entrepreneurial enough to recognize the winds of change towards corporate integration and to be on its leading edge in creating the biggest forest-product corporation in the province and Canada (Hayter 1976). More than any other forest-product corporation, MB has been at the centre of public controversy in BC. Thus, mergers of its founding companies in the 1950s fuelled public outcries about corporate concentration; its international expansion of the 1960s raised questions about its loyalty to BC; in the 1990s, MB was targeted by international environmentalism; and its most recent restructuring plans, which have encompassed major divestments, a commitment to stop clear-cut logging practices, espousal of privatization of log markets (and land itself), a willingness to give up land for tenure reform, and, finally in mid-1999, its proposed sale to the American MNC Weyerhaeuser have provoked eye-catching headlines.

MB's most recent (1997-8) plan for restructuring, closely associated with its new chief executive officer, Tom Stephens, is explicitly designed to enhance shareholder value by creating a leaner, more focused corporation. This plan extends the corporate restructuring that began in the depths of the recession of the early 1980s, while recognizing that the first phase of restructuring did not go far enough. In 1980, MB was quintessentially Fordist. As a horizontally and vertically integrated MNC, MB mass produced low-value commodities in big factories employing Taylorized labour relations within a strongly hierarchical multidivisional (M-form) corporate structure. By this time, there were already warning signs about the need for change in corporate structures and strategies. MB incurred its first recorded losses in the mid-1970s, and for some observers, MB had become a dinosaur rather than a champion, a metaphor that applied widely throughout the coastal region in particular (Schwindt 1979; see also Brunelle 1990). It was the recession of the early 1980s, however, that confirmed the need for restructuring.

For MB, "flexible mass production," in which high volumes are associated with a more differentiated, higher-value (innovative) product mix, has been a central theme of its restructuring since 1980. For BC, the implications of this strategy, ultimately based on innovating products that command a price premium and cannot immediately be copied by less capable, lower-cost rivals, are substantial. Thus, in 1980, MB diversified across the basic forest industries (lumber, plywood, shingles and shakes, pulp, newsprint, paperboard, and containers) and within each industry production specialized in mature commodities and was designed to maximize throughput (volume) at the lowest possible cost. By 1991, MB had moved out of some commodities (such as plywood, shingles and shakes, and paperboard), almost out of market pulp, and significantly reduced its production of standard newsprint, as well as standardized lumber products. With fewer employees, MB increasingly emphasized value maximization in the manufacture of a wide range of specialty papers and building products, including engineered woods it had developed.

Yet, in the mid-1990s, MB again faced significant problems of profitability, and in 1997, led by CEO Tom Stephens, embarked on another phase of restructuring affecting all aspects of operations. Thus, within two years of Stephens's arrival, MB chose to focus on building products, and in BC, the pulp-and-paper business was spunoff, mill and logging employment downsized, the head-office staff halved, a shipping transportation subsidiary was sold, and in a puzzling move, its R&D laboratory closed. Other radical changes have occurred in MB's operations outside of BC. This shareholder value driven restructuring plan implies a strong commitment by MB to "lean production," in which human and capital resources are fully used in support of flexible mass production. Moreover, this "new" plan recognizes that in a period of product dynamism and differentiation, MB needs to focus sharply on particular segments of the forest sector rather than try to compete across the forest-product spectrum – a strategy that was possible during Fordism when the focus was relatively homogenous commodities.

This chapter examines the restructuring of MB's operations within BC as the corporate exemplar of a shift from Fordist to flexible mass-production strategies, until its takeover by Weyerhaeuser in October 1999. The chapter is in four main parts. First, the evolution of MB is outlined, especially with

Table 4.1

MacMillan Bloedel: Selected aggregate trends, 1979-97

Year	Sales ($M)	Employment	Earnings before taxes ($M)	Net earnings ($M)	Capital investment ($M)	Debt equity ratio (%)
1979	2,180	24,730		155	280	27
1980	2,436	24,505		113	328	35
1981	2,210	22,049		3	308	39
1982	1,843	18,581		-58	207	46
1983	2,044	15,472		24	104	42
1984	2,128	14,994		19	138	46
1985	2,336	15,139		43	97	42
1986	2,512	15,102		178	100	28
1987	2,863	15,226	508	271	249	19
1988	3,037	15,384	336	327	350	18
1989	2,923	15,094	211	245	553	27
1990	2,818	15,036	-69	50	363	35
1991	2,477	13,905	-226	-93	275	46
1992	2,918	13,203	-64	-49	214	52
1993	3,739	12,258	134	54	224	50
1994	4,417	12,549	235	180	288	49
1995	4,327	12,886	359	279	758	50
1996	4,267	13,497	43	51	497	47
1997	4,521	10,592	-138	-368	257	54

Note: In 1995, MB slightly revised some of its accounting formulas, and the data from 1991 reflect these changes. All dollar figures are nominal.
Source: MacMillan Bloedel selected Annual Reports.

respect to its expansion as a BC-based MNC in the Fordist period. The second and third parts examine MB's search for flexible mass production between 1980 and 1996, respectively for changing product (and market) dynamics and the adjustments implemented at specific manufacturing sites. The last part of the chapter reflects on MB's role in BC's forest economy, a theme that has taken on a significant new twist as a result of the acquisition of MB by Weyerhaeuser.

As a case study, MB is a distinctive corporation, by virtue of its size, location, ownership structure, innovativeness, and history. MB is the largest forest-product corporation in BC, with sales of over $5 billion in 1996 (the equivalent of 390th on the Fortune 500 list), and MB ranked as the eleventh largest forest-product firm in North America (the biggest forest-product corporation was International Paper of New York with sales of over US$20 billion). MB has long been the central player in BC's forest economy in the coastal region, where its private forests and forest tenures are concentrated (see Figure 10.1). Its roller-coaster fortunes since 1979 are a key dimension of the contemporary transformation of this region's forest economy. Selected aggregate indicators of performance in terms of sales, employment, earnings, and capital investment provide context for these fluctuations (Table 4.1).

MacMillan Bloedel's Strategies During Fordism

MB originated in three long-established, family-run enterprises in BC: the MacMillan Export Company, incorporated in 1919 as a lumber-trading concern; Bloedel, Stewart and Welch, founded in 1911 as a logging company; and the Powell River Company, incorporated in 1910 to manufacture newsprint (Hayter 1976; Mackay 1982). The first two companies merged in 1951 to form MacMillan and Bloedel, which then merged with the Powell River Company in 1959, creating the MacMillan, Bloedel and Powell River Company, subsequently changed to MacMillan Bloedel in 1965 (Figure 4.1). All three founding companies had expanded and achieved a limited degree of integration by the 1930s. For example, the MacMillan Export Company had acquired and constructed sawmills, plywood mills, and logging camps to supply its lumber carriers; Powell River had secured a logging base; and Bloedel, Stewart and Welch had expanded into lumber production.

After 1945, as a result of the mergers and large-scale investments, MB became a mass producer of an integrated range of forest-product commodities. In the 1950s, its private timber lands were supplemented by the awards of the highly productive TFL 39 and TFL 44, and the (privately owned) area now called Managed Forest Unit 19 (see Figure 10.1). By 1970, MB had become the leading Canadian producer of lumber and newsprint, the second-largest producer of plywood, and an important producer of market kraft pulp, paperboard, fine papers, and converted paper products, especially corrugated containers, and a number of wood-based products such as shingles and particleboard. The firm maintained and developed worldwide marketing organizations for its products, and owner-operated as well as leased charter vessels. MB concentrated its expansion in the southwestern littoral of BC until the early 1960s, after which it expanded to Europe, the US, and eastern Canada, and to some extent to Southeast Asia.

Integration and Consolidation, 1948-70

The postwar expansion of MB during Fordism can be disaggregated into three broad strategies (Hayter 1976). Between 1948 and 1959, culminating in the Powell River merger, integration (especially vertical integration) strategies within BC, were pursued. Between 1960 and 1970, MB adopted a consolidation or expansion strategy of existing BC facilities. After 1963, an interregional and multinational expansion strategy became the dominant growth direction. Throughout this period, MB's sales were diversified across the major forest-product commodities, about half in wood products (especially lumber and plywood) and about half in pulp, paper, and allied products (Figure 4.2). From 1960 to 1970, a modest trend developed towards an increase in the relative importance of the latter, especially as a result of

Figure 4.1

Location and corporate origins of MacMillan Bloedel's principal operating facilities in British Columbia

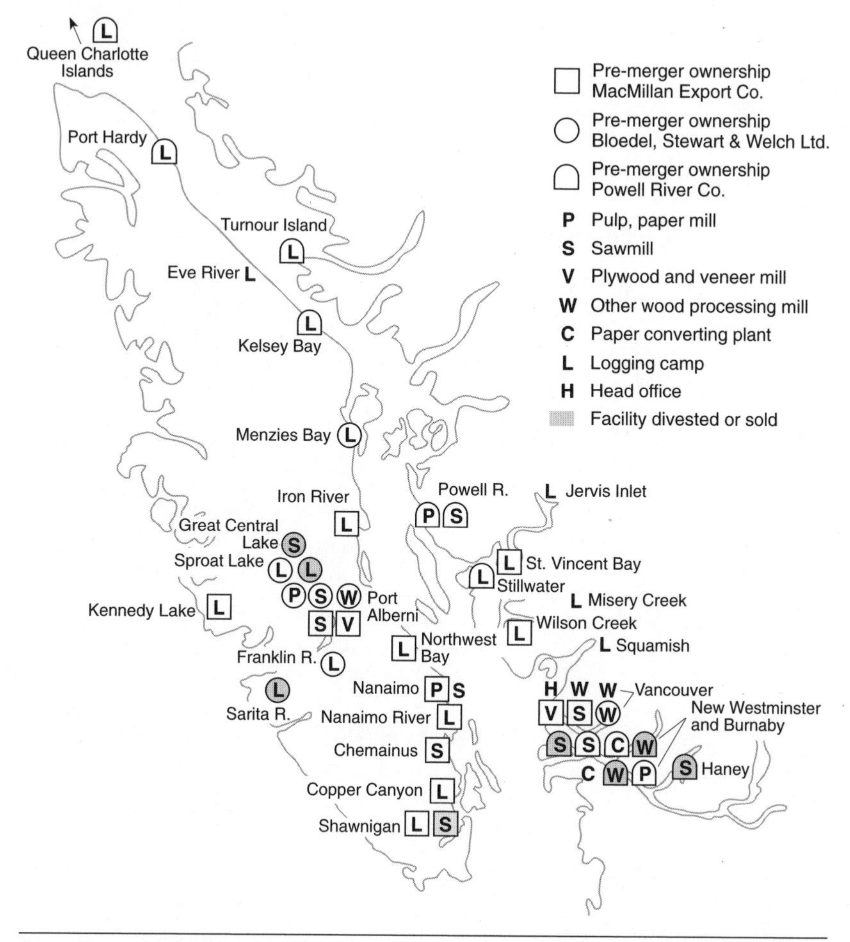

Source: Hayter 1976: 218. Reprinted with permission of the American Geographical Society.

expansion of corrugated containers, folding cartons, and fine papers. In 1970, however, MB's sales were still dominated by primary commodities, namely, lumber, plywood, pulp, paperboard, and newsprint.

The postwar vertical integration policies incorporated the implicit planning of dominant entrepreneurs and were governed by a philosophy of doing "nothing that did not contribute to the utilization of the forest resources," by spatial horizons limited to BC, by aggressive attitudes, and by preferences for tidewater locations (Hayter 1976: 219). All three founding companies independently initiated vertically linked expansions to meet growth objectives, to promote greater stability by reducing the risks of specialized

Figure 4.2

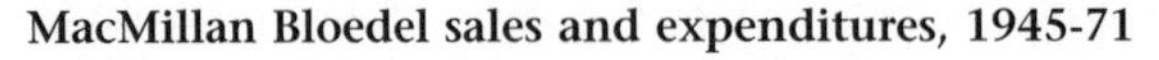

MacMillan Bloedel sales and expenditures, 1945-71

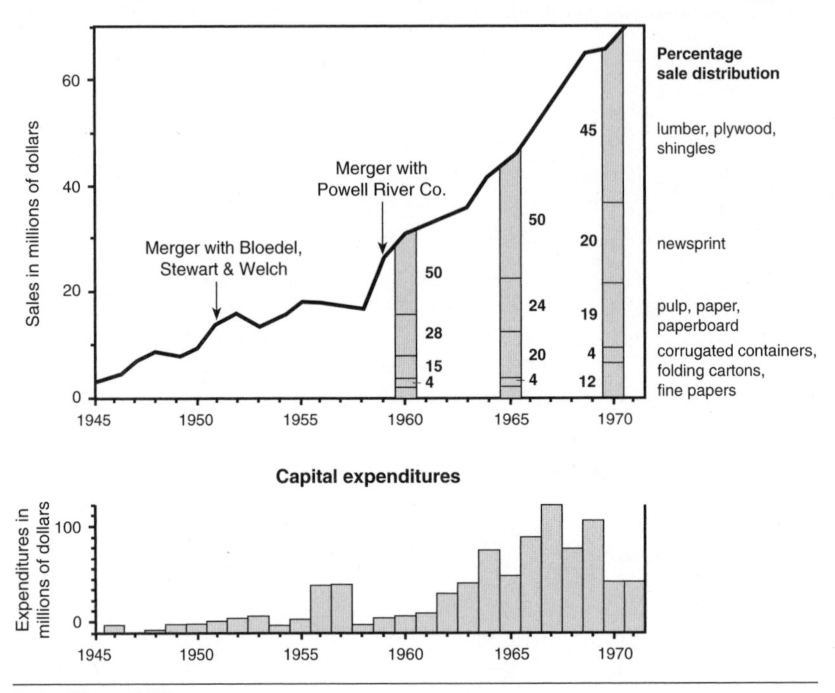

Source: Hayter 1973.

manufacturing activities that supplied widely fluctuating export markets, to achieve greater timber utilization, and to obtain processing economies. Thus, Bloedel, Stewart and Welch and the MacMillan Export Company built kraft pulp mills, and the Powell River Company acquired two lumber companies in New Westminster. Through installation of the newly developed chippers and barkers, the Bloedel firm's 70,000-ton-per-year kraft pulp mill, constructed at Port Alberni in 1948, became the first pulp mill in western Canada to utilize sawmill and plywood waste materials. The MacMillan Company rapidly adopted the same policy, and in 1948, announced plans for a mill of similar size, initially to produce unbleached kraft pulp, to be located at Harmac near Nanaimo, a tidewater location like Port Alberni accessible to company wood supplies. Subsequent expansion in BC was effected principally through the mergers and by successive on-site expansions.

The Rationale for the Mergers

According to the companies, the 1951 and 1959 mergers creating MB were based on the potential contribution of each firm to enhancing growth potential and operating efficiencies as a result of pooling complementary organizational and locational resources (Figure 4.1). The merger decisions were also perceived as increasing competitive strengths in foreign markets while restricting competition among themselves and, with respect to the 1959 merger, preventing threats of takeovers by nonresident firms. Organizational advantages of the mergers were seen in expansion and concentration of research effort, creation of a more effective management team, and particularly, potential marketing synergies to be gained. For example, MB's fledgling newsprint sales organization was collapsed within the established network of Powell River, while the latter's small lumber sales staff was rationalized within that of the former.

The extent to which MB's large-scale, horizontally, and vertically integrated operations resulted in processing efficiencies is questioned by Schwindt (1977), who suggests that MB's major manufacturing plants may have been bigger than could be justified solely by principles of economies of scale. Whether diseconomies of scale existed is another matter; size can provide competitive strengths other than scale economies. Moreover, there was a "spatial" rationale for the mergers because the operations of each company were complementary and the main processing plants occupied tidewater sites accessible, via cheap water transportation, to each other as well as to world markets. The merged operations were soon integrated. Indeed, at Port Alberni, the lumber, plywood, and shingle mills of the MacMillan Export Company were adjacent to the kraft pulp and lumber facilities of Bloedel, Stewart and Welch. With the addition of newsprint and paperboard plants during the 1950s, Port Alberni became the most diversified, perhaps lowest-cost, integrated forest-product site in BC by the early 1960s (Hardwick 1964).

Logging activities and flows were quickly rationalized after the mergers. Powell River, for example, obtained timber from Vancouver Island camps owned by MacMillan and Bloedel, while its high-quality saw and peeler logs were sorted and redirected to the sawmills and plywood plants of MacMillan and Bloedel at Chemainus and Vancouver. In return, these plants supplied Powell River with chips for newsprint manufacture, while Powell River's sulphite pulp was partially replaced by kraft pulp from Harmac as the chemical pulp input to newsprint. In addition, corrugated-container plants in Vancouver, Edmonton, Calgary, and Winnipeg, originally acquired by Powell River, and one constructed in Regina, provided important captive markets for the paperboard plant installed at Port Alberni in 1959.

In the years following the mergers, a few, smaller-scale operations were closed and some new plants built. However, most investment by MB in BC has occurred at nodes in existence before the mergers (Hayter 1976). Increasingly, the integration objectives of the 1950s were modified during the consolidation strategy of the 1960s, which focused on exploiting the growth opportunities of individual manufacturing sites, especially at Port Alberni, Powell River, and Harmac. Geographically, MB preferred to expand manufacturing at existing tidewater locations rather than disperse production to

Figure 4.3

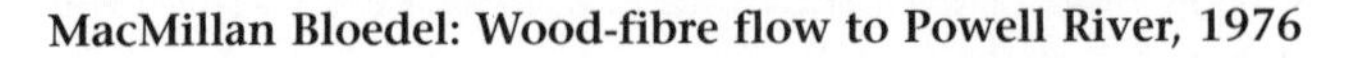

MacMillan Bloedel: Wood-fibre flow to Powell River, 1976

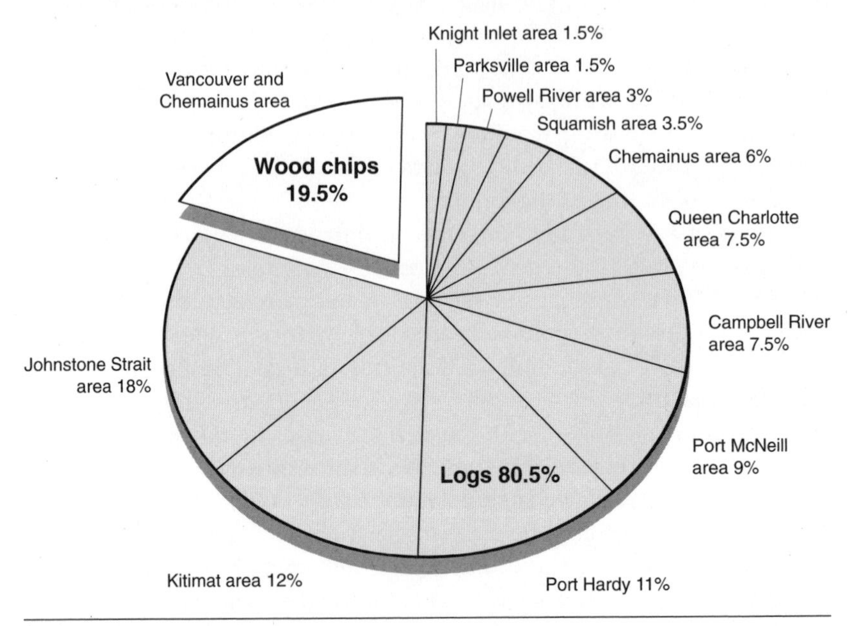

Source: Based on Hayter 1976.

remoter places, a costly and uncertain strategy in relation to infrastructure and labour supply. Logging supply lines became longer, however, as a result of this strategy (Figure 4.3).

In Powell River, the integrated complex in the early 1970s produced over 600,000 tons of newsprint per year, market kraft pulp (about 50,000 tons), and dimension lumber (about 125 million board feet). All employees, then about 2,000 in the mill, were unionized and MB dominated the company town. The chemicals and ground pulp needed for newsprint were manufactured on site by this time, and the entire complex was fed by approximately 67 million cubic feet of wood, 80 percent of which came from MB's logging operations, and by 8 million cubic feet of chips from outside Powell River, the majority (60 percent) originating at MB's Chemainus and Vancouver sawmills. These logs were transported on MB barges directly to Powell River or via sorting grounds. About 18 percent of the timber supply, mainly hemlock, balsam, or spruce, was of saw-log quality, the remainder being pulp logs (although all logs passed through Powell's lumber and chipping facilities). Fibre supplies were purchased from market sources to obtain particular species or as deals with other companies to attain shorter transportation routes. Log supplies from around Georgia Strait, especially the mainland, were extended to include the Queen Charlotte Islands in the 1950s, and

Prince Rupert and Kitimat in the 1960s. In 1972, for the first time, chips were obtained via land transportation from the interior. In fact, MB by 1970 had virtually exhausted its growth opportunities in BC's coastal region. In anticipation, MB initiated a largely multinational diversification strategy.

Multinational Expansion

In 1957, Powell River announced a plan for a pulp-and-paper mill in Oregon, and in 1959 MB indicated an interest in building a newsprint mill in Arizona. Both plans were dropped after the merger. By the early 1960s, however, MB was again interested in geographical diversification, at least partly because of the risks associated with concentration in BC where it considered "opportunities ... were also running out and the best locations gone" (Hayter 1976: 225). With production expertise and marketing connections in the forest-product industries, MB chose to geographically diversify its existing product line (Table 4.2). Thus, international expansion involved an extension of the strategies of horizontal and vertical integration. Both the UK corrugated-container companies and Koninklijke Nederlandse Papierfabrieken (KNP), its principal partner in the construction of fine-paper mills in Belgium, for example, were established customers of pulp and paperboard produced by MB at Port Alberni and Harmac. Virtually all the chosen locations were on tidewater and the great majority in politically stable ("nonsensitive") regions within North America and Europe, a definition that excluded politically sensitive Quebec (Hayter 1976).

In tandem with manufacturing investments, MB expanded its distribution systems, especially in relation to lumber, including a merger with Jardine and Company in 1969 to create MacMillan Jardine, a sales organization for the company's products in Australia and Asia. MB's largest investments were in the US, notably in Alabama where a large-scale mill provided paperboard to the many corrugated-box plants MB acquired in the US. Although options were considered, investments in Latin America and Southeast Asia were limited mainly to hardwood logging operations in Indonesia and Malaysia, primarily to serve the Japanese market.

Location Flexibility

By the 1970s, MB had achieved a considerable degree of horizontal and vertical integration, producing vast quantities of commodities in large-scale, unionized mills. If MB had invested globally, its home region was clearly coastal BC. Moreover, as a quintessential Fordist MNC, MB was not without flexibility. Integration meant access to markets across the forest-product spectrum, an ability to use trees in a variety of end-uses, and bargaining strength with both rivals and customers. Its tidewater locations added to flexibility by providing ready access via cheap water transportation to global markets and in BC to its resource base. MB's trading tentacles were

Table 4.2

Interregional growth of MacMillan Bloedel's manufacturing facilities to 1973

Product and location	Method of growth
Vertically integrated expansion	
Western Canada	
Corrugated-container plants in Winnipeg, Regina, Edmonton, Calgary, and Vancouver	Acquisition in 1954 of the Martin Paper Box Co. by the former Powell River Co.
Europe	
Corrugated-container plants in Hatfield, Nelson, and Southall (United Kingdom); subsequent construction of new plants at West Auckland and Weston-Super-Mare	Acquisition in 1963 of Cook Containers Ltd. and Ily-grade Containers Ltd.
70,000-ton-per-year capacity fine-paper plant, Lanaken (Belgium)	Acquisition of a 36 percent (now 46 percent) interest in Royal Dutch Paper Mills (KNP), a Dutch fine-paper producer, and subsequent $13 million investment in the Lanaken plant completed in 1968
25,000-ton-per-year capacity fine-paper plants, Algeciras (Spain)	Joint venture with KNP and Celupal S.A. (30 percent interest acquired by MacMillan Bloedel) in a $25 million investment completed in 1969
Geographical diversification	
United States	
Corrugated-container plants in New Jersey and in Baltimore	Acquisition in 1966 of the two plants from the St. Regis Paper Co. and the Mead Corporation
Integrated forest-product complex at Pine Hills, Alabama, with capacity for 270,000 tons of linerboard, 50 MFBM of lumber, and 100 million square feet of plywood	Project completed in 1968 for $70 million as a joint venture with the United Fruit Box Co. (which had a 40 percent interest in the linerboard mill); became sole owner in 1970 and announced a $10.5 million particleboard plant in 1973
Corrugated-container plant, Odenton, Maryland	Acquisition in 1971 from the Hoerner Waldorf Co. (Baltimore plant subsequently phased out)
Ten corrugated-container plants in New York, New Jersey, Ohio (two), Illinois, Indiana, Arkansas, Texas, Mississippi, and California	Acquisition in 1971 of the Flintkote Corporation

►

◄ *Table 4.2*

Product and location	Method of growth
Canada	
180,000-ton-per-year capacity newsprint mill in St. John, New Brunswick	Acquisition in 1969 of a 54 percent interest in partnership with Feldmüble A.G. of Germany; the mill (constructed in 1964) subsequently expanded by 180,000 tons per year
Aspenite mill, Hudson Bay, Saskatchewan	Acquisition from the provincial government of Saskatchewan in 1965
Plywood mill, Nipigon, Ontario	Acquisition in 1973 of Multiple Plywood Ltd.
Waferboard plant, Thunder Bay, Ontario	$9.4 million investment announced in 1973
Southeast Asia	
Plywood and blockboard plant, Pekan, Malaysia	Completed in 1973 as 30 percent partner of Mentegon Forest Products Sdn. Bhd.
Europe	
Hardwood pulp plants in France (three) and Belgium (one), with a 600,000-ton annual capacity	Acquisition late in 1973, reportedly securing a 40 percent interest in La Cellulose d'Acquitaine

Source: Hayter 1976: 227. Reprinted with permission of the American Geographical Society.

worldwide and were supported (partially) by its own fleet of ships and barges. MB was also innovative. It had adopted improvements in wood utilization technology, such as chippers, barkers, and sawdust-pulp refiners, pioneered the self-dumping log barge, and in a new R&D laboratory built in the mid-1960s, began to develop new types of pulp, paper, and wood products.

Nevertheless, MB's competitive strengths, rooted in size, integration, resources, and locational flexibility, were vulnerable to the rapidly changing competitive conditions of the 1970s. Its mills were antiquated, not suitable for a changing, lower-quality timber supply, and too rigidly specialized in commodities exposed to price competition. If prices fluctuated, costs rose steadily. For MB, its first (annual) loss in 1975, in part because of losses incurred by long-term, high-priced leases taken out on shipping shortly before shipping rates collapsed worldwide, signalled changing times (Mackay 1982). Within MB, concerns arose regarding recent diversification initiatives beyond the forest sector, the geographical extent of the firm's activities, and the growing cost-price squeeze in coastal BC. These concerns, mitigated by a boom in the late 1970s that encouraged MB to increase capacity and jobs, mainly in support of established commodity markets, suddenly became the source of crisis in the "bust" of the early 1980s.

Recession and Crisis at MB

The recession of the early 1980s hit MB hard. Operating losses in 1982 and 1983 were substantial, reinforced by escalating interest rates on payments for the large-scale capital-investment program initiated in the late 1970s. MB's debt-equity ratio increased substantially. MB had to both create cash and reduce costs immediately. To generate additional funds, in 1983 MB sold its head-office building for $63 million and, after renting part of the space back, rented cheaper premises nearby. Several subsidiaries were sold outright, including its newsprint operation in New Brunswick (for $145 million in 1981) and fine-paper plants in Spain. In 1983, MB created three joint ventures by selling 50 percent of its shares in its UK-based paper-packaging business (to Smurfitt SCA), Canadian-based paper-packaging operations (to Consolidated Bathurst), and its Vancouver-based fine-paper plant (to Fraser Inc.). MB also reduced ownership in its partly owned Dutch pulp-and-paper mills in 1983 and divested properties in Indonesia and Brazil. These sales generated "exceptional" earnings while extensive layoffs and plant closures in BC reduced costs in drastic attempts to limit the extent of losses (see Table 4.1). With all these changes, MB recorded small net earnings in 1983 and 1984.

In terms of organizational structure, MB reduced its head-office staff by over half and devolved decision making to operating regions based at Powell River, Port Alberni, and Vancouver-Nanaimo, more or less based on the operations of the founding companies of MB. While this decentralization was designed to encourage operating divisions to think of themselves as "profit centres," decision-making structures and corporate strategy in practice was complicated by a hostile takeover in 1981 by Noranda, a Toronto-based mining company with forest-product investments in BC's interior. Ironically, this takeover occurred in competition with attempts by the provincial government to increase control of MB, justified by the claim of the province's premier in 1980 that "BC resources are not for sale," through its newly formed British Columbia Resources Investment Corporation. The government, however, could not justify its own attempts at acquisition while denying bids from elsewhere.

The corporate rationale for Noranda's acquisition of MB seemed to reflect a desire for growth and the prestige of conquest. Business synergies were not obvious. Noranda bought a company in crisis and not in its main line of business at a time when even the financial rationale for conglomerate growth strategies was being questioned. For BC, there was no (known) injection of expertise or resources that contributed to MB reestablishing profitability or rethinking its long-term strategy. Indeed, shifts of some managerial functions to Toronto were probably not helpful in this regard. The acquisition in any case proved unsuccessful, and Noranda divested its controlling interest in MB in 1993. Since then, MB's shares have been widely held, although a Texas-based family and an Ontario-based pension fund became noteworthy owners.

Flexibility Imperatives: Moving Back Along the Product Life Cycle
Within the context of financial crisis, cost cutting, and an unwanted takeover, MB sought to develop a new competitive strategy. In particular, MB undertook initiatives to reduce reliance on narrowly defined commodity production and to increase the value and range of its products while maintaining large-scale production. In short, MB's restructuring, implicitly at first but with gathering momentum, represented a shift towards flexible mass production. In Porter's (1985) terminology, MB sought to shift its competitive advantage from a strategy based on cost leadership across a range of industries to a strategy based on differentiation of a wider range of products within fewer industries. If the objectives of this strategic shift were not precisely articulated, at least publicly, in the context of its operations in BC, the dynamics of MB's restructuring from the early 1980s to 1996 first involved a shift in focus from mature products (or commodities) to "younger," faster-growing products. This shift back along the product life cycle linked MB to new markets, most notably in Japan, and required considerable adaptation at existing factories.

Product Cycle Dynamics
In product terms, the central theme of MB's restructuring is to manufacture innovative or new products that are in the early stages of the product life cycle. De-emphasizing standardized or mature products (commodities) in the late stages of product cycle and innovation also helps differentiate mature products and so renews their life cycle. The rationale for such a strategy is that innovative products command a price premium over standardized commodities whose competitiveness derives solely from cost efficiency. Generally, more technologically complex products are harder to copy by lower-cost, less sophisticated rivals, and they potentially provide longer-lasting competitive advantage.

MB has recently sketched its own adaptation of its product innovation strategy, specifically for building products within the familiar context of the so-called product cycle model (Figure 4.4; Hayter 1997a: 100-2). This model represents MB's perceptions of product-growth potentials. Thus, as of the late 1990s, the products with the greatest growth potential within the S-shaped product cycle curve are recently innovated, including engineered wood, medium-density fibreboard, and oriented strandboard. Particleboard is closer to maturity, whereas construction plywood is already in rapid decline. On the other hand, the creation of "baby squares" (that is, lumber cut specifically to Japanese specifications) and the application of kiln drying has created new markets for "differentiated" lumber. Specialty

Figure 4.4

MacMillan Bloedel's building products: Generalized corporate perception of product cycles

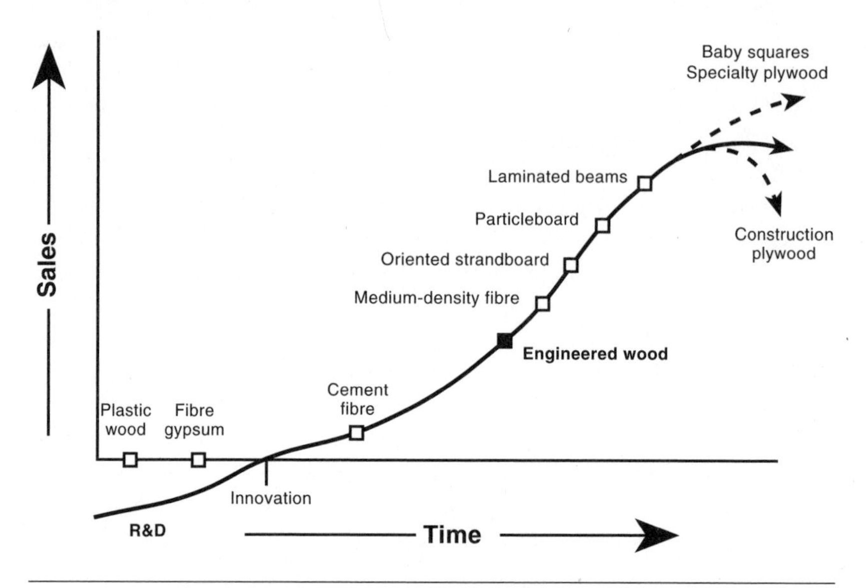

Source: Based on MacMillan Bloedel 1996a: 15. Reprinted with permission of MacMillan Bloedel.

plywoods are also differentiated forms of mature products that serve new markets. Still other building products are in the research and development (R&D) stage, including types of plastic woods and fibre/gypsum-based composite products, while cement-fibre products have been innovated and are on the verge of rapid growth. A similar categorization of products is possible for pulp and paper.

In-house R&D

In shifting back along the product cycle curve towards a more innovative product mix, MB's in-house R&D group has played a vital role. The R&D centre grew out of various technical and research groups located at the pulp mills and was established as a separate laboratory in 1966 in the suburbs of Vancouver. In the late 1980s, the R&D group relocated a short distance to the Discovery Park area adjacent to the British Columbia Institute of Technology. In the early 1970s, a forestry research centre was also established near Nanaimo. Since its beginnings, MB's R&D programs have been among the largest in Canada (Hayter 1988). Until its closure at the end of 1997, the group consistently employed around 100 scientists and engineers and its 1996 budget was approximately $13 million, of which $3 million came from grants and tax credits. In the past, budgets have been higher. MB faced considerable uncertainty in funding its R&D programs; costs are high, the duration for individual projects is over a decade in some instances, and there have doubtless been failures. Even so, MB until 1997 maintained its R&D programs throughout the various crises of the past two or three decades. The evidence suggests that its long-term commitment to R&D paid off.

MB's R&D programs have led to major investments in new technologies and facilities to manufacture new products. Among the larger-scale projects, R&D groups have developed higher-yielding pulping technology and effluent disposal systems, and various new products including engineered wood products, notably Parallam, examples of cement fibre (Cemwood Shake and Permatek), various high-quality groundwood printing papers (such as Electrabrite), lightweight coated papers (notably Pacifica), and Spacekraft®, a disposable and recyclable intermediate bulk container for storing and transporting liquid foods.

For pulping technology, an R&D group spent over five years and $10 million to develop high-yielding CMT processes specifically for the species mix and factory conditions at Powell River and Port Alberni. While this technology has not been transferred or sold elsewhere, it is much higher yielding (more than 90 percent fibre recovery) than kraft pulps (around 50 percent fibre recovery) and greatly increased efficiency at MB's pulp mills while reducing environmental impacts.

MB's R&D group also contributed, along with technical groups at the mill sites, to state-of-the-art environmental systems at Port Alberni and Powell River based on biotechnology, or in vernacular terms, the use of bugs to consume environmental waste. The environmental systems themselves cost around $100 million to significantly reduce environmental pollution (Figure 4.5). At Port Alberni, for example, where environmental regulations are the strictest in the country, biochemical oxygen demand (BOD) is lower than prescribed levels and marine life has increased significantly – local boat owners now have to clean their boats.

Product-based R&D, however, most directly illustrates MB's attempt to move back along the product cycle to create more valuable products. Parallam and Pacifica are two examples from the building-product and paper segments respectively.

Parallam

In the building-product segment, MB's production was dominated by construction lumber and plywood, both mature commodities by the 1970s. The main exception was particleboard, a form of which MB had innovated in the mid-1960s at its then new Vancouver plant, located adjacent to its old Canadian White Pine sawmill. By 1996, MB had innovated a wide range of building products, the best known being Parallam, a brand-name engineered wood. Indeed, Parallam is a good illustration of the nature and value of the in-house R&D process, whereas its manufacture captures the essence of the shift to flexible mass production. Thus, scientists began working on Parallam in the late 1960s, but it was not until 1982 that a prototype plant was built (at MB's Canadian White Pine sawmill in Vancouver). Parallam became a commercial reality and was used in the construction of Expo 86 in Vancouver. Full-scale factory production of Parallam began in 1986 at a new plant on Annacis Island, Delta, in the Vancouver metropolitan area. From R&D to innovation, the Parallam project took over fifteen years and cost over $50 million.

Moreover, as a high-quality, "engineered" product customized over a wide range of sizes at consistent high-quality specifications and produced in volume, Parallam is an excellent example of flexible, value-oriented mass production. Parallam is a structural beam made from long (8-foot or 2.5-metre) veneer strands, mainly hemlock and Douglas-fir but also larch and poplar, that are bonded in a patented microwave pressing process. Beams of a variety of sizes (width and length) can be manufactured and structural properties are consistent and compare favourably with metal beams. Parallam is two-and-one-half times the strength of sawn lumber, and the Annacis Island mill produces beams up to 66 feet long and 2 feet wide, length being restricted only by the size of the factory rather than the physical limitation of the process. Quality control in the process is rigorous and samples are tested several times daily. Markets have grown rapidly in Canada and the US. By 1996, the Vancouver plant manufactured 200 million cubic metres and was shipping four containers of Douglas-fir-based Parallam a week to Japan. In 1996, this plant employed 160 workers organized on teamwork

Figure 4.5

Trends in environmental pollution at Port Alberni in the 1990s

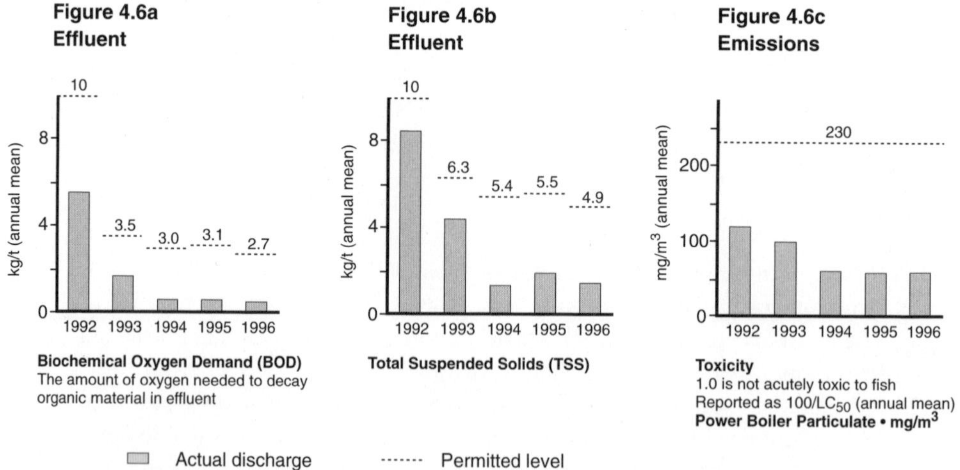

Source: MacMillan Bloedel 1996b: 35. Reprinted with permission of MacMillan Bloedel.

Flexible mass production at Port Alberni. Paper is produced in large volumes but with more varied grades than in the past.

principles and, in contrast to MB's other plants, the workforce is nonunion. In 1997-8, the Annacis Island plant doubled its capacity and increased employment.

In 1991, MB formed a partnership (49 and 51 percent shares) with Trus Joist International, the largest engineered structural lumber producer in the US, which allowed Parallam to be sold within Trus Joist's established marketing network. Within this partnership, Parallam is a consistent source of profit to MB.

Pacifica

MB has long sought to innovate groundwood and lightweight coated papers to reduce emphasis on newsprint. Indeed, in the 1970s, first Powell River and then Port Alberni started to produce such papers developed by MB's R&D group. Telephone directory papers are a major, highly profitable product line at Port Alberni, although first introduced at Powell River. Port Alberni was also the chosen site for the startup of NexGen in 1996, a new ultralightweight coated paper (now known as Pacifica), and used in glossy magazines, catalogues, and advertising flyers. The technology used in the $200 million NexGen project was engineered by MB to allow coating to be applied on both sides of the paper by converting the existing paper machine No. 5. Pacifica is difficult to make, the technology is sophisticated, the paper makers are well trained, and less wood fibre is required in comparison to newsprint. These features make it appropriate for Port Alberni, and local fibre supplies are ideal for a lightweight coated paper. The process uses 125,000 tonnes of recycled paper, mainly from Newstech of Coquitlam, and over 40,000 tonnes of clay (kaolin) from Georgia as a cheap bright filler. The process also uses some kraft pulp for strength, because it initially had a breakage problem. Pacifica also employs a nine-worker team, rather than six employees, the traditional staffing level on paper machines.

Pacifica requires a specialized process. At Port Alberni that process significantly contributes to flexible mass production. Once dominated by newsprint, the mill now produces over forty different grades of paper, including telephone directory paper and Pacifica. The latter is in high demand, commanding premium prices.

The Special Role of In-house R&D

In-house R&D provides MB with technological expertise to resolve firm-specific problems and create specific product market opportunities. MB initially invested in an R&D group because it felt that other forms of R&D would be unable to serve its specific needs at specific times appropriate to the firm (Hayter 1988; Chapter 11). The marketing and profit gains that resulted from product innovations such as Parallam and Spacekraft were only possible because MB had in-house R&D. The fact that both products are manufactured in the US, exclusively so in the case of Spacekraft, does not diminish the value of MB's R&D to the firm or to BC. The profits created by these products came back to MB and helped pay for its R&D program in BC.

It is hard to imagine how MB could have bought product innovations such as Parallam off the shelf or contracted the technology development to some outside group. The reason MB took out patents on Parallam is to hinder such transfer; for the same reason, the company is otherwise extremely secretive about the technology underlying Parallam. MB's R&D has also been valuable by pioneering firm-specific innovations based on secret or quasi-secret knowledge that results in processing efficiencies at specific sites. MB's R&D group, for example, was one of several firms across Canada that innovated high-yielding pulps in the 1970s (Hayter 1988). In this instance, MB,

Table 4.3

MacMillan Bloedel: Global sales distribution by broad business segment, 1965, 1979, and 1996

Business segment	1965	1979	1996	1996 Sales ($M)
Building materials	49%	45%	67%	3,489
Paper and allied	51%	55%	33%	1,710

Note: In 1996, building materials sales principally comprised lumber ($1,975 million), panelboards ($398 million), engineered wood ($519 million), and logs and chips ($380 million). Paper and allied sales principally comprised newsprint ($257 million), groundwood papers ($441 million), corrugated containers ($628 million), and containerboard ($263 million).
Source: MacMillan Bloedel Annual Reports.

Table 4.4

MacMillan Bloedel: Global sales distribution by geographic region, 1965, 1979, and 1996

Geographic region	1965	1979	1996
Canada	24%	19%	26%
United States	41%	47%	53%
Japan	6%[1]	9%[1]	12%[1]
Other[2]	29%	25%	9%

1 Figures include all sales to Asia.
2 The dominant market is the UK (14.3 percent of sales in 1965).
Source: MacMillan Bloedel Annual Reports.

believed that distinct (plant-specific) adaptations were necessary because of regionally specific fibre supplies, site-specific factory conditions, and even specific marketing connections. In other instances, such as Pacifica and the environmental-waste-management system at Port Alberni, MB's R&D personnel cooperated with outside groups as well as with technical staff at the mill site. In the Pacifica project, there was extensive cooperation with a Finnish paper-machine builder (Valmet) that has substantial R&D capability. If MB had not had matching expertise, this project may not have been possible.

In-house R&D is not the sole supplier of technology to MB. Equipment suppliers provide technology in the form of machinery. MB has also obtained innovations from small firms and university-based research, whereas other technical changes have been initiated at mill sites by technical groups and workers. With the closure of its R&D program in 1997, MB's emphasis on innovation relies on these sources, and its activities are now more short term and adaptive rather than long term and pioneering.

The Geography of Markets

In overall corporate sales, in 1996, as in 1970, MB manufactured goods in wood-based building materials, pulp and paper, and packaging products (Table 4.3). In 1996, the building-material segment was relatively more important, accounting for 69.2 percent of MB's sales, compared to 1970. In 1996, lumber remained the dominant building material, whereas engineered woods rapidly emerged as the second sales generator in the building-materials segment. In paper sales, lightweight and (especially) other groundwood papers had replaced newsprint as the main sales generator while packaging was dominated by sales of containerboard (paperboard) and corrugated containers.

By 1996, MB to a considerable degree had shifted from a commodity trader to a marketer of a more refined range of products. In tandem with this product-market shift, the broad geographic distribution of sales of MB's overall sales (and operations) was more integrated within North America

Table 4.5

MacMillan Bloedel: Sales distribution by geographic region from British Columbia, 1996

Geographic region	Lumber	Paper
Canada	18%	11%
United States	25%	67%
Japan	42%	4%
Other	15%	18%

Source: MacMillan Bloedel 1996a.

than had been the case in the 1970s (Table 4.4). This market shift reflected expansions of MB investments in the US and the rest of Canada, especially in Ontario. In terms of productive facilities, MB had become less multinational and more continental. Most of the sales in the packaging segment, and the engineered wood sales in the building-product segment, are generated in the US and, secondarily, in central Canada.

For MB's Canadian-based operations, the domestic market was the most important in 1996, accounting for 41.3 percent of the $3.1 billion sales generated. MB's facilities in Ontario are strongly connected to the Canadian market. The US accounted for a further 30.6 percent of Canadian-generated sales and the Japanese market for 17.3 percent. In comparison to MB's sales as whole, Canadian-generated sales are more oriented towards Canada and Japan. BC is MB's most important regional centre of production, especially for lumber and pulp-and-paper production. Many of the sales to Japan are from MB's operations in BC, especially its sawmills (Table 4.5). In fact, Japan in 1996 was the most important market for MB's sawmills in BC, over twice as important as the domestic market and substantially more so than the US. Moreover, the value-added trend among established sawmills in BC has been predicated on accessing the Japanese market (Hayter and Edgington 1997). On the other hand, the US remained the dominant market for pulp and paper.

Getting Inside the Japanese House

Although MB's exports of lumber can be traced to the Kanto earthquake of 1923, when a lumber shipment destined to China was diverted to Tokyo to help in the reconstruction, MB's lumber sales to Japan did not become important until 1960, when the Japanese government permitted softwood lumber imports. Until the early 1980s, MB only sold logs and large cants (large sections of timber, one step along the value-added chain from logs) to Japan and relied completely on its trading company subsidiary, MacMillan Jardine, to arrange sales with the *sogo shosha* (trading companies). For MB, however, the recession of the early 1980s revealed the weakness of dependence on the *sogo shosha* for Japanese sales. Because the trading companies simply take a commission on volumes traded, MB argued that the *sogo shosha* have little interest in maintaining prices or even price stability. MB had also become frustrated by its lack of understanding of the Japanese market, claiming that the *sogo shosha* either would not or could not provide the necessary information. MB's frustration is reflected in the view of a sales manager:

> At this time [1982], I went over on one trip and I was pretty much down on my hands and knees saying "what are we going to do," and basically the answer [from the *sogo shosha*] that I got was "Well we can sell your volume for you if you drop your prices a little more." I said "We are shutting our mills down now due to lack of price, I can't give you more volume at a lower price. We have to get some value added." ... We found out that the *sogo shosha* didn't know [what MB needed to know about Japanese lumber specifications] or they didn't want to tell us ... (Hayter and Edgington 1997: 202-3).

MB's reactions were aggressive (Figure 4.6). In 1983, the company bought out the Jardine interests and, to gain control over prices and develop a better understanding of the Japanese market, took the unprecedented step of establishing its own distribution company in Japan, MBKK. In 1985, MBKK began to compete directly with the *sogo shosha,* with the mandate to develop (as a principal) its own customer base within Japan, for which it absorbs the risk of buying and selling, and to serve as an agent for sales negotiated by MB's sawmills in BC to the *sogo shosha,* for which it receives a commission. In support of MBKK, MB in 1984, led by a Japanese-speaking manager, transported two loads of logs to Japanese sawmills, requesting that they saw the logs and explain exactly what they were doing. MB found that Japanese specifications converted readily into metric, that a Japanese house comprised a much larger range of lumber specifications than in North American construction, and that Japanese lumber demands were extremely quality conscious. In addition, MB began in the late 1980s an annual lumber auction at one of its Vancouver Island sawmills, obtained permission from Japanese authorities to apply the Japanese Agricultural Standard, and completely rebuilt its "whitewood" (hemlock, Douglas-fir) mills at Port Alberni, Chemainus, and Nanaimo during the 1980s to manufacture lumber to Japanese specifications.

In 1978, the decision to rebuild the Alberni Pacific Division (APD) at Port Alberni for $20 million assured continuation of traditional four-by-four lumber manufacture. However, operating ideas were modified to allow the cutting of "baby squares" for the Japanese market. The mill began by cutting rough 105-by-105 mm squares in 1980; later, it began to finish these squares, and since then, it has introduced a variety of lumber sizes used in Japanese houses. At Chemainus, the sawmill was completely rebuilt in 1985 for $22 million, and designed to cut the largest "clears" or highest-quality species

Figure 4.6

MacMillan Bloedel's timber exports to Japan: Pre-1983 and 1993

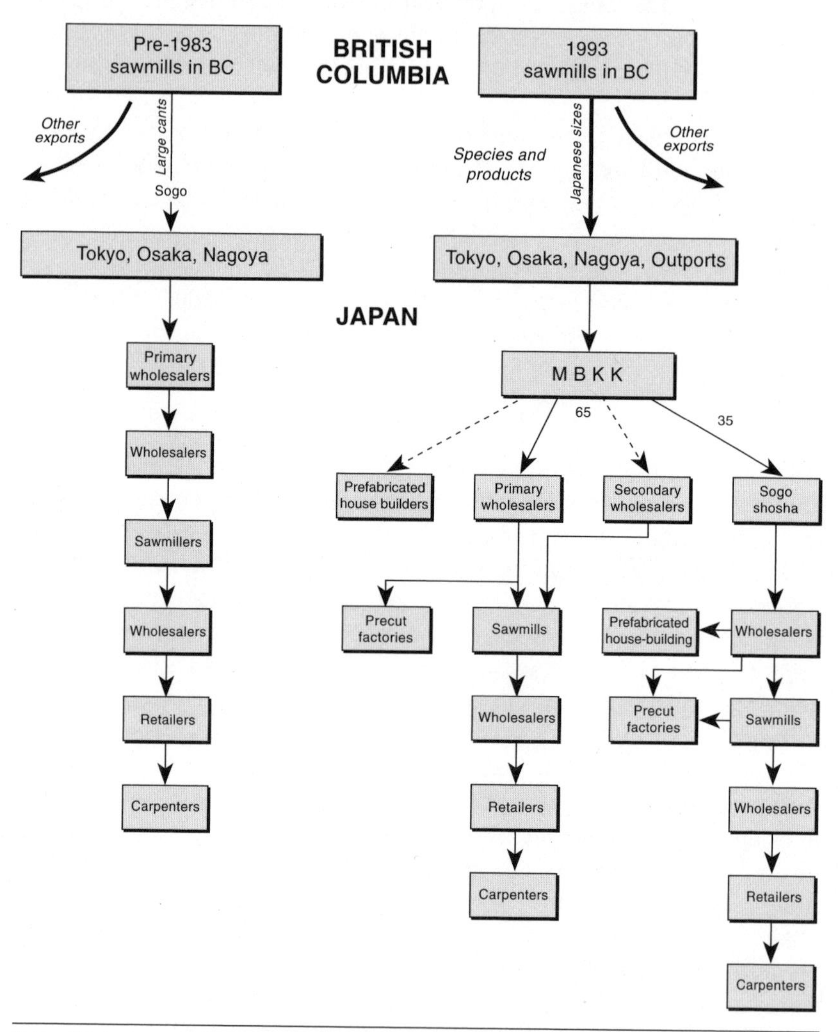

Source: Edgington and Hayter 1997. Reprinted with permission of *Economic Geography.*

of Douglas-fir and hemlock, specifically for export markets such as Japan where remanufacturing for door and window components occurs. The new Harmac sawmill (Island Phoenix) which opened in 1989 at a cost of $70 million, also cuts to Japanese specifications, using lower-quality hemlock for construction purposes. These three whitewood mills, which complement each other in terms of timber species and quality and cooperate in serving the Japanese market, provide MB with half its lumber sold in Japan, the rest originating with MB's Vancouver area sawmills, its cedar sawmill at Port Alberni, and a subcontractor.

Figure 4.7

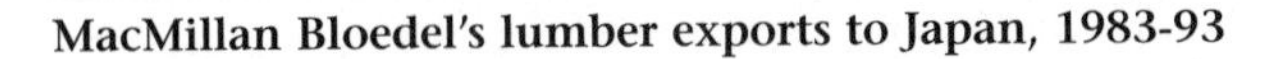

MacMillan Bloedel's lumber exports to Japan, 1983-93

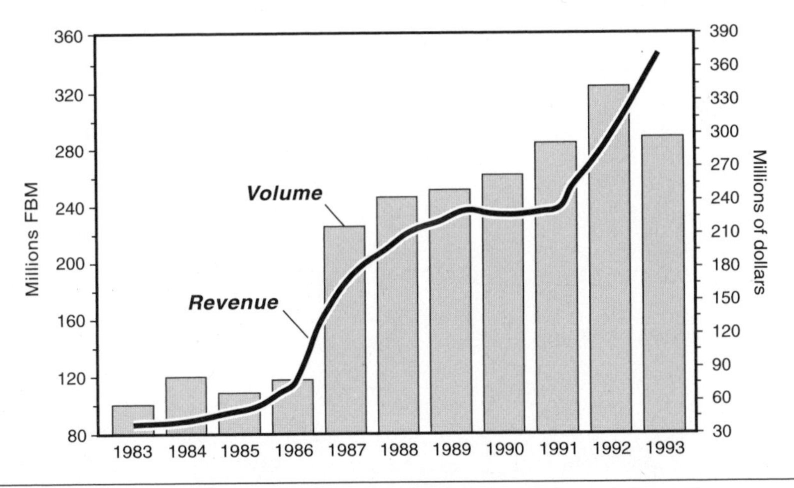

Source: Hayter and Edgington 1997. Reprinted with permission of *Economic Geography.*

With the creation of MBKK, MB is unique among BC firms in becoming part of the internal Japanese distribution system. MBKK clearly ruffled the feathers of the *sogo shosha* who had previously bought MB's lumber, and other BC firms continue to rely for sales on the *sogo shosha* and other Japanese organizations (Hayter and Edgington 1997: 201; Chapter 7). MB, however, is the leading BC-based firm in the Japanese lumber market. By 1994, 65 percent of MBKK's sales stemmed from its own deals, mainly with primary and secondary wholesalers, and it served as MB's agent with the *sogo shosha* for 35 percent of its sales. MB's lumber exports to Japan have grown rapidly (Figure 4.7). In 1994, they comprised almost half of MB's lumber exports from BC, or about one-quarter of the province's lumber exports to Japan. MBKK has established connections with prefabricated house builders in Japan and, from offices in Tokyo and Osaka, arranges for timber shipments to these cities and Nagoya, and to a variety of smaller outports around Japan's coast. MBKK's original plan to supply small, niche products in Japan has been replaced by an emphasis on flexible mass production in which large volumes combine with product differentiation and higher value.

In Situ Adaptation of Manufacturing Operations in BC, 1980-96

In 1980, MB's facilities in BC were generally at least forty years old and in Powell River's case, seventy years. Most facilities required modernization; some were obsolete. During Fordism, in situ change was dominated by expansion, integration, and commodity diversification. Since then, in situ change has featured modernization, rationalization, and product diversification in support of flexible mass production. By 1996, the sites that MB

Table 4.6

MacMillan Bloedel: Production change at main BC facilities, 1980 and 1996

	1980	1996
Lumber (MFBM)[1]		
Port Alberni (APD)	372[2]	161
Port Alberni (Somass)	–	75
Nanaimo	256	146
Chemainus	167	97
Powell River	75	28
New Westminster	68	87
Vancouver (White Pine)	176	118
Vancouver (custom cut)	–	72
Panelboards (M.Sq.ft.)[3]		
Vancouver, plywood	100	closed
Port Alberni, plywood	150	closed
Vancouver, particleboard	106	104
Pulp and paper (tonnes)		
Port Alberni, paper	371	318
Port Alberni, pulp	56	closed
Powell River, paper	471	400
Powell River, pulp	47	34
Nanaimo, pulp	334	sold
Port Alberni, linerboard	96	closed
Vancouver, fine paper	37	sold

1 MFBM refers to million board feet.
2 The production totals for APD and Somass are combined for the year 1980. (APD was rebuilt in that year, resulting in a lower than average production level.)
3 M.Sq.ft. refers to million square feet, 3/8" base.
Note: Corrugated-box and bag plants and Parallam plant excluded.
Source: MacMillan Bloedel Annual Reports, 1980 and 1996a.

abandoned included most notably the Nanaimo pulp mill (sold for $254 million in 1994) and the Vancouver plywood mill (which was closed) (Table 4.6).

At Port Alberni, MB's most integrated forest-product site, three commodity lines (market pulp, paperboard, and shingles and shakes) were closed. The lumber and paper operations were diversified in product mix while maintaining large volumes. As such, Port Alberni reflects the thrust of corporate restructuring since 1980. Also, at Port Alberni and in MB's other BC-based facilities, employment downsizing occurred everywhere, in association with more flexible work practices (Chapter 8). Among surviving mills, the realization of flexible mass production has not been easy. In just three plants (the so-called whitewood sawmills at Chemainus, Somass Cedar at Port Alberni, and Nanaimo), modernization was achieved by a comprehensive or virtually comprehensive rebuilding of operations to establish state-of-the-art facilities. At Chemainus, modernization was also accompanied by an innovative labour-relations agreement (Barnes, Hayter, and Grass 1990). In the other operations, most notably at the Port Alberni and Powell River pulp-and-paper complexes, modernization and rationalization have been implemented as piecemeal processes; innovative change has thus been fit into preexisting building designs and layouts. In a few facilities, notably the White Pine sawmill in Vancouver, built in 1928, and the nearby particleboard plant, few changes have been made.

Chemainus: An Exemplar of In Situ Industrial Adjustment

The modernization of the Chemainus sawmill in the mid-1980s was designed to renew a mature product by shifting to principles of flexible mass production in which high volumes combine with product variety, teamwork, and a commitment to value maximization (Barnes, Hayter, and Grass 1990; Hayter, Grass, and Barnes 1994). In 1980, the existing sawmill was a coastal dinosaur: it employed around 650 people to produce 167 million board feet, and it was unprofitable. In the 1982 recession, MB closed down the mill, intimating the decision was permanent (Table 4.7). The community and workers reacted strongly to this decision, and representations were made to MB to reopen the mill. Encouraged by the mill's location, which offers tidewater access to high-quality resources and markets, skilled labour, and infrastructure, and the fact that MB owned the land on which the mill stood, MB built, for $22 million, a new mill with state-of-the-art equipment. The mill opened in 1985 and soon after was fully computerized.

Whether or not MB had always planned to rebuild the mill, it is clear that the closure of the old mill gave the firm a bargaining advantage over the union. Indeed, in accordance within the union contract, by not opening the new mill until two years after the old mill's closure, MB had no obligation to rehire former workers on the basis of seniority (or any other criteria). Moreover, the mill received 2,500 applications for the 125 positions available at startup. In the event, MB agreed to stay within the union, whereas

Table 4.7

MacMillan Bloedel's Chemainus sawmill, selected characteristics for selected years, 1980-96

			Sales to		
	Employment	Production (MFBM)	North America	Japan	Other
1980	650	167	45%	40%	15%
1981	550	135			
1983	0	0			
1985	125	69	30%	35%	35%
1989	140	101	15%	49%	36%
1996	150	97	12%	55%	33%

Note: MFBM refers to million board feet.
Source: Fieldwork (information provided by various managers).

the union (specifically, IWA-80) accepted substantially changed, more flexible work conditions in the new mill, notably the principle of teamwork.

At Chemainus, the new mill's production is smaller than the old one, but it remains a large-volume producer with an annual capacity of 106 million board feet, on the basis of a five-day, three-shift working week. Employment levels are much lower. Moreover, in terms of production philosophy, work organization, and marketing, the new mill represents a strong shift to flexible mass production. Thus, in the old mill, sawing principles were based on maximizing throughput (volume) and the first cuts by the head sawyer (headrig saw) progressed from one edge of the log to the other. In the new mill, the headrig team "turns" logs to ensure that outside cuts, which contain the largest amount of clear, most valuable wood, are maximized. These initial cuts take into account the fact that trees "taper." Subsequent sawing by the edgers and trimmers also searches to maximize "clears," according to sizes determined by market information on specifications and price.

In the new mill, logs are cut to order and, on site, packaged and computer-coded for specific customers. Moreover, in contrast to the previous product mix, the new mill cuts "baby squares" for the Japanese market and kiln dries much of its output at nearby subcontractors. These trends typify how mature products are differentiated and renewed (see Figure 4.4). The mill has moved out of low value commodity markets anchored in the US to high-value markets based in Japan. In 1998, the top price (in Japan) of 1,000 board feet of sawn wood was $8,500 (compared to commodity wood that sold at less than $1,000 per 1,000 board feet). Since its opening, apart from during a provincewide strike, the mill has consistently been profitable.

The new mill is more quality conscious and far less wasteful. While the old Chemainus mill recovered about 35 percent of the timber cut (as sawn wood), the new mill recovers 70 percent (and the waste chips are sold to pulp mills, sawdust to farmers, and bark to the boiler rooms of Powell River and Port Alberni). Teamwork underlies the new mill's greater efficiency and commitment to value maximization. Although the present management group is much smaller than in 1980, this group has more marketing and production autonomy than before. Among the workforce, ongoing training is the basis for teamwork, the worker incentives for which are an hourly wage rate that depends on the number of skill levels passed – regardless of which task the worker is performing at a particular time – and relief from boredom. The plant maintains an apprenticeship program for tradespeople and has a small group of on-call temporary workers who also have to pass entry-level tests.

At Chemainus, there is an ongoing search to improve value recovery and efficiency. MB has experimented with X-ray technology to assess the inside of logs to inform cutting decisions. In 1996, a computerized system was introduced at Chemainus to monitor (and reduce) saw-blade vibration to reduce waste and maintain accuracy. This system cost $850,000 and paid for itself within a year. In addition, the plant continues to innovate with respect to labour relations. In 1992, for example, a gain-sharing system was introduced whereby employees were awarded a year-end bonus based on a formula-based share of production value. In 1996, tradespeople agreed to spend two weeks a year learning each other's jobs. Although difficult to implement in mills with established labour contracts, principles of teamwork and bonus payments are gradually spreading to other MB operations in support of trends towards flexible mass production.

Somass Cedar: Decorative Wood and Flexible Mass Production

At the factory level, the Somass cedar mill is perhaps MB's most comprehensive attempt to combine scale and scope economies. The Somass mill was originally built in 1935 (by Bloedel, Stewart and Welch, which also established its head office there) and cut fir and pine as well as yellow and red cedar to manufacture commodities, initially mainly for Commonwealth markets. By 1980, however, the Somass Division, which included a shingle mill and two sawmills, was unprofitable, with antiquated equipment and poor labour relations. Following the 1981 recession, the 1,130 employees were reduced to 320, the shingle mill closed, and the sawmills reduced to one shift per day. Since then, the division has increasingly specialized in western red cedar, which serves a wider range of markets and emphasizes value maximization as well as efficiency. In 1986, the division was comprehensively modernized and, following subsequent expansions, employment in 1996 reached 450 and production 75 million board feet, about two-thirds of its former capacity (but still large).

Cedar is a soft, easily damaged wood and requires more careful handling than other species. Thus, in contrast to the whitewood sawmills, the Somass Division is relatively labour intensive and the products are primarily decorative (rather than structural). Products are also remanufactured to a greater degree; that is, more value is added. The manufacturing core of the Division comprises four closely integrated mills. Both the A-mill, which uses knotty and low-quality "mirch" wood, and the Hival sawmill, which uses the best-quality clears, have a trimmer and resaw as well as headrig saws and edgers. More than 80 percent of the output from both mills is remanufactured in the Planermill. In addition, the Speciality Mill recovers and remanufactures clear wood from the Hival sawmill, previously converted to wood chips for pulp mills, into various garden, industrial, and finger-jointed products.

The Somass Division can potentially produce 300 different red cedar products. In practice, the number of products is fewer, partly because of the need to achieve scale economies and to fully use specialized machinery, and partly because of the need to emphasize products in the early stages of the product life cycle, or at least newly differentiated products. In this latter respect, for example, a manager estimated that the life cycles of cedar products at Somass before 1980 were about ten to fifteen years but in the 1990s only two to five years.

To remain viable, products at Somass must command a price premium. In 1996, the cost of wood at Somass before conversion was approximately $500 per 1,000 board feet, and the division needed an average price of $1,150 per 1,000 board feet to break even. Consequently, production emphasizes high-priced mouldings, beveled sidings, and panelling, and the full use of wood fibre, including making low-priced products such as sawdust bricks. Given this focus on value, efforts are then made to minimize costs. In each part of the mill, cost and revenue data are posted for all to see and to promote awareness of the importance of cost reduction and revenue generation. By 1996, about 50 percent of the division's workforce were part of job rotation schemes in an effort to enhance the goals by job flexibility.

A Corporate Perspective on Flexible Mass Production of Lumber

Flexible value-oriented mass production has implied complementary specializations among MB's wood-processing mills. Thus, the three whitewood sawmills on Vancouver Island focus on certain sizes of particular species: Chemainus cuts large (more than 24 inches in diameter) Douglas-fir or hemlock up to 24 feet, whereas Nanaimo and APD cut smaller, lower-quality fir and hemlock (15 to 24 inches in diameter). Canadian White Pine in Vancouver has not been modernized to the same extent as these mills – indeed, the pressures of operating a site within a dynamic urban environment may encourage closure and replacement by a different kind of land use. In the meantime, this mill remains profitable by using machinery long since depreciated and by its ability to cut hemlock and Douglas-fir logs of greater length than is possible at Chemainus. The four MB whitewood sawmills are further complemented by cedar mills in Port Alberni (Somass) and New Westminster, the former cutting larger cedar logs. In addition, MB contracts out high-value yellow cedar and Sitka spruce for custom cutting by plants, mainly in the Vancouver area.

Within BC, MB is a diversified producer of a range of lumber products used for structural and decorative purposes, while at the factory level, individual sites are more specialized. Consequently, factories are able to focus on and become familiar with specific market segments and the processing problems of particular species and sizes of timber, which in turn reduces the problems of cleaning and modifying machinery that arise from the use of a wider species mix. In other words, from both factory and firm perspectives, this system of related specializations enhances economies of scope; that is, the ability to efficiently utilize the same resources for different, best-value purposes, as well as for economies of scale.

Powell River and Port Alberni:
Piecemeal Modernization in Pulp and Paper

MB's search for flexible, value-oriented mass production in lumber and building products is complemented by pulp and paper. However, in this industry, MB followed Canadian industry practice in pursuing piecemeal modernization at Powell River and Port Alberni until their sale in 1997, and at the pulp mill at Harmac until its sale in 1994.

The Powell River and Port Alberni pulp-and-paper complexes were primarily built before 1970, in Powell River's case beginning in 1912 (Hayter and Holmes 1993). Indeed, over the decades, eleven paper machines were built at Powell River, and in the late 1960s, ten were in operation. At Port Alberni, five paper machines were built before 1970. During the Fordist years, these mills produced vast tonnages of newsprint and smaller amounts of market pulp, while Port Alberni established a paperboard line (another commodity) to serve MB's corrugated-box plants. Newsprint was almost exclusively exported to the US, whereas paperboard supplied MB plants in the UK and the US.

At both Port Alberni and Powell River, restructuring has involved a shift from commodity production to the flexible mass production of a wider range of higher-value papers (Hayter 1997b). By 1996, newsprint had been de-emphasized in favour of a wider range of specialty papers. Thus, in 1996, the two mills produced 718,000 tonnes of paper of which 42 percent was standard newsprint, 8 percent comprised lightweight coated papers, and 50 percent comprised telephone directory and other groundwood printing papers. Port Alberni produces forty, and Powell River about fifty, different grades of paper, most developed by MB's in-house R&D group.

In contrast to APD or Chemainus, flexible mass production at Powell River and Port Alberni occurred in a series of steps in which investments to modernize combined with investments to combat pollution and with rationalization. Thus, both mills have diversified product mix on fewer paper machines; each mill in 1996 operated just three paper machines. In both mills, the shift to flexible mass production has occurred by converting existing machines. In association with these changes, both mills have shifted from chemical and groundwood pulps to high-yielding pulps, a shift that was initiated at Powell River in the 1970s. They have invested in recycling technology and, for around $100 million each, in new environmental systems. At both sites, existing layouts imposed physical problems on modernization. The Pacifica (NexGen) project at Port Alberni, for example, could not install the specialized drying systems (super calenders) in the conventional manner at the end of the paper machine because of insufficient space.

The combined shift towards flexible mass production has increasingly differentiated paper making at Powell River and Port Alberni. Thus, the transition has gone farthest at Port Alberni, as two-thirds of its output was in noncommodity grades by the mid-1990s compared to just one-third at Powell River, even though MB first manufactured specialty papers at Powell River in the 1970s. Indeed, after its innovation at Powell River, telephone-directory-paper production was shifted to Port Alberni, where, MB claimed, fibre conditions were more appropriate to high-quality paper making. Rationalization

Table 4.8

Powell River: Geographic distribution of sales, 1993

	Standard newsprint	Speciality grades	Total
Canada	22.4	30.0	52.4
United States	144.8	89.2	234.0
Asia-Pacific	90.0	17.9	107.9
Latin America	47.3	2.2	49.5
Europe	39.0	5.0	44.0

Note: Data are in metric tonnes.
Source: Hayter and Holmes 1993: 31.

at Powell River also involved closure of an old, small paper machine that had been the first to convert to specialty papers; the new paper machine (No. 11) installed at Powell River in 1980 is highly efficient but is designed to manufacture newsprint. As a result, Port Alberni produces large quantities of lightweight coated Pacifica grades on one completely rebuilt machine and telephone directory on another, whereas Powell River uses two smaller machines to manufacture magazine paper grades as well as newsprint. Even so, given that Powell River's entire production of newsprint historically supplied (western) North America until the 1970s, a significant geographic diversification of sales has occurred in recent years (Table 4.8).

Questions about the efficacy of seeking flexible mass production at two giant complexes can be raised, however. As of 1996, the extent of the shift from newsprint to specialty papers roughly amounted to the production levels of one mill. Alternatively put, in restructuring both mills, MB was unable by 1996 to fully meet the principles of flexible mass production at either site. Would it have been better to have sold Powell River in the mid-1980s – and to have sold Harmac – and concentrated product diversification and capital expenditure in the Port Alberni complex? MB may also have contemplated keeping its paperboard line at Port Alberni, which supplies market-oriented box plants that are typically less cyclical than other forest-product businesses. Alternatively, MB might have undertaken its 1997 restructuring plan a decade earlier.

Adding Value: How Far to Go?

In shifting from a commodity-trading bias to a product-marketing bias, MB faces significant questions about just how much of the value-added chain it should internalize and make in-house. The most complex end-markets for building products are houses, comprising an array of wood-based structural and decorative components, subassemblies or subunits (such as cabinets, doors, and windows), and distinct products such as furniture. For MB, the rationale underlying flexible mass production is to permit production of a range of closely related components that have greater quality and design requirements than highly standardized construction materials. Consumer demands for housing, and related products, however, are incredibly diverse. At some point, product differentiation limits mass production. Huge companies manufacture housing, especially in Japan, but this business is very different from forest-product manufacturing.

For MB, the value-added challenge is to cut wood economically into sizes and qualities as close as possible to the final use. For example, should MB build houses and compete with Mitsui Home (or the vast number of small house contractors), or make furniture and compete with IKEA? Given the specialized and complex nature of these individual consumer-oriented product markets, such a strategy is unlikely to make sense in relation to supplying industrial consumers with components that they assemble and remanufacture. Whether or not MB has realized its potential in this regard is hard to say. While the shift towards a more value-added product mix has been strong, MB still exports wood in large dimensions that is resawn elsewhere. The dilemmas facing MB in moving up the value-added ladder in BC are partly revealed by its 1995 acquisition of Plenk's, a small door and tilt window manufacturer located in Chemainus (Chapter 6).

Plenk's Wood Centre employed sixty-seven people in 1997 and is an excellent example of a high-value "reman" operation that manufactures and retails top-of-the-line products. Its products are expensive and are purchased for architecturally designed homes as far away as California. The long-run implications of this acquisition for MB, however, are not clear. Plenk's is a tiny operation that developed a market niche as a result of its innovations in manufacturing and marketing high-quality, high-priced products to individual consumers. It may be neither easy nor appropriate for MB to expand in this direction. Successful innovation of high-quality, wood-based consumer products is an uncertain process, and the acquisition possibilities in BC for firms like Plenk's are few. Moreover, the advantages of entrepreneurial firms like Plenk's relate to their sharp focus on highly differentiated market niches and an ability to make quick decisions; neither of these attributes can be assumed in a large corporate culture.

On the other hand, important implications of a value-added emphasis are flexibility and differentiation. Possibly, MB will widen its interests in doors and windows to include medium-value market segments. Moreover, if MB wants to refine its markets, wood-remanufacturing operations need financing and a wood supply. MB also has options other than total control, such as joint ventures and long-term sales contracts, in moving towards value-added markets. In the meantime, the Plenk's acquisition intimates a further dimension of MB's restructuring, namely, how to reinject entrepreneurialism into a large corporation.

Restructuring Beyond BC: An Overview

In addition to in situ modernization and rationalization of wholly owned facilities in BC, MB between 1980 and 1996 restructured in other ways, notably by severing connections in some joint ventures while creating others and by initiatives outside BC (see Table 4.9). Thus, in 1993, MB sold its 50 percent share of its fine-paper plant (to E.B. Eddy Forest Products, a subsidiary of George Weston Ltd.) and its 50 percent share of corrugated-box interests in the UK. These plants had been MB's first international acquisitions (see Table 4.2). During the 1980s and early 1990s, MB's shares in the Dutch-based KNP increased and decreased as shares were bought and sold

Table 4.9

Restructuring beyond British Columbia: Major changes in MacMillan Bloedel's manufacturing empire, 1980-96

Company/region	Year	Type of change	Comment
Canada			
MacMillan Rothesay, New Brunswick	1981	Divestment from joint venture	Newsprint mill, sale to the Irving family for $145M
MacMillan Bathurst, C. and W. Canada	1983	Sale of assets to form joint venture	MB sold 50 percent of its packaging interests across Canada to Consolidated Bathurst
Green Forest Lumber, Ontario	1995	Acquisition	$122M purchase of lumber mill, distribution outlets
Saskfor MacMillan Partnership	1995	Joint venture with SFPC (Saskatchewan Forest Products)	MB transferred its OSB mill to SFPC, Limited and SFPC transferred its sawmill and plywood mill
Jager Strandboard, Ontario	1996	Acquisition	$84M purchase of 340,000 cubic metre board mill plus harvesting licence
United States and Mexico			
Trus Joist MacMillan	1991	Joint venture with TJ International	Produce composite lumber in Idaho (MB has 49 percent control); two new mills built in Kentucky and West Virginia in 1995
American Cemwood, California	1993	Acquisition	$40M purchase, cement-fibre roofing tile company
MacMillan Bloedel, Mexico	1994	Acquisition	Particleboard plant, with lamination
MacMillan Bloedel, Kentucky	1995	New plant	100 percent recycled liner-board mill for $105M

►

◀ *Table 4.9*

Company/region	Year	Type of change	Comment
Europe			
UK Corrugated, UK	1983	Joint venture with Smurfit SCA	Merger of MB's eight container plants in UK with those of Smurfit
UK Corrugated	1993	Divestment	Sale of equity share for $39M
KNP (Koninklijke Nederlandse Papierfabrieken)	1983 -93	Changes of equity in pulp/paper	MB's share declined from 48 percent to 40.7 percent, to 38.9 percent in 1985 and to 25 percent in 1986 for $88M; equity increased to 28.5 percent in 1986 and to 30.6 percent in 1989 for $36M following KNP acquisition and declined to 17 percent in 1993 following KNP merger
KNP	1996	Divestment	Sale of interests for $258M

Source: Financial Post Company, 1998.

and KNP itself merged with other European distributors of fine papers in 1993, leaving MB with a 17 percent interest. This interest was then sold in 1996 for $258 million (and an after-tax profit of $32 million).

MB also expanded its interests. In particular, in 1991, MB created its partnership with Trus Joist, and in 1995, this joint venture completed a recycled linerboard mill in Kentucky (for $100 million) as well as two new engineered lumber plants in Kentucky and West Virginia ($49 million). In 1994, MB acquired a particleboard plant in Mexico; in 1995, a lumber company in Ontario (for $122 million); and in 1996, substantial oriented strandboard (OSB) capacity in Ontario (for $84 million); and a 50 percent interest in a new OSB mill in New Brunswick ($90 million). In 1999, MB took full control of Saskfor MacMillan, with a view to building a $180 million OSB plant near its existing OSB mill at Hudson Bay, Saskatchewan. MB also invested around $250 million to build two new medium-density fibreboard (MDF) mills in Pennsylvania and Ontario in 1996 and 1997, in which it has majority control.

Reassessing Restructuring: Towards a Leaner Corporation

The extent to which MB restructured itself between 1980 and 1996, within and beyond BC, was considerable and in some senses desirable. Movement back along the product cycle occurred and new initiatives were profitable. A shift to flexible mass production is clearly discernible. In addition, MB's 1996 sales (albeit in nominal terms) were twice the 1979 levels, even if jobs had been almost halved (see Table 4.1). Yet, MB's operations remained highly volatile, and this volatility in turn affected strategy. Thus, substantial losses in the first years of the 1990s encouraged the sale of interests in joint ventures, while the boom of the mid-1990s encouraged acquisitions and the building of new mills outside BC. By 1996, earnings were down again, and in 1997, MB incurred a substantial loss, which was followed by another in the first half of 1998. Moreover, these losses were not simply a result of the Asian economic crisis – MB's building products that are strongly connected to Japan performed profitably while other segments lost money. After a decade of restructuring, MB again had be restructured.

The urgency of the situation was underlined by the hiring of Tom Stephens as chief executive officer; a man with a reputation for restructuring corporations in the interests of shareholders, but with no prior connections to MB. The new CEO wasted little time in assessing MB's strategies and initiating another round of restructuring (Hamilton 1998a, 1998b). In early 1998, MB announced 2,700 jobs would be terminated, including in BC where head-office jobs were cut by 100 and R&D operations closed. Pulp-and-paper operations in BC have been spun off as a separate company to private investors, and Canadian Transport Company, a shipping subsidiary of MB, has been sold to a US-based MNC. Increased demands to be profitable have been made on all operations and for labour to agree to flexible work rules or face loss of jobs (Chapter 8). Changes have extended beyond BC to include sale of the two new MDF mills for $160 million at a loss. Moreover, Mr. Stephens has turned conventional wisdom on its head by asserting that MB will stop clear-cutting (supposedly the most economical and safe way of logging in BC) within five years. He has also called for the "privatization" of BC's industrial forests, and most recently, he offered some of MB's land to the provincial government in exchange for tenure reform (Chapter 10).

The immediate question is why does MB have to restructure in the late 1990s after a decade of restructuring? What went wrong? Two perspectives, not mutually exclusive, are offered here to shed light on MB's dilemmas. These dilemmas have to be seen as "structural" (deep seated and long term) rather than temporary, occurring only because of the Asian economic crisis. The perspectives focus on MB's strategic choices and turbulence in MB's business environment.

Internal Problems in Strategic Mission

Porter (1985) claims that corporations achieve sustainable success in an industry by following one of four generic strategies: cost leadership across a range of goods, differentiation across a range of goods, focused cost-leadership on a limited set of goods, and focused differentiation on a limited set of goods. In his view, problems arise when firms do not clarify their fundamental competitive base, whether rooted primarily in cost minimization or value maximization, and whether their activities should range across many or a few product markets. The underlying problems stem from the fact that different strategies require different managerial competencies, forms of investment, corporate structures, and work skills, and imply different marketing production and marketing economies of scope. In a given region, only one strategy will likely be appropriate to local conditions.

In Porter's terminology, MB aspired during Fordism, with considerable success, to be a leader based on low costs derived from economies of scale and size across the full spectrum of forest-product industries. In its restructuring of the 1980s and 1990s, MB sought to shift the basis of its leadership to value maximization and a more focused range of higher-value industries. However, MB failed to shift fast enough towards this goal, especially in pulp and paper. Thus, in 1996 MB still manufactured large quantities of newsprint and pulp, which is strongly price sensitive and whose competitive base primarily rests on low costs. Moreover, while MB's extensive US-based operations remain oriented towards domestic paper-packaging markets amenable to product differentiation, MB in BC (and across Canada) sold 50 percent of its interests in these markets. Possibly, MB should have divested market pulp and newsprint while retaining packaging.

If some commodities were dropped completely or in large measure (plywood, shingles and shakes, and market pulp), rationalization was piecemeal, sometimes driven by short-term crises, and MB still sought to become a flexible mass producer in *both* paper and building products. Yet, such a shift meant that MB faced much more complex business environments than its traditional commodity markets, requiring new technologies, new marketing initiatives, and new forms of labour relations centred on flexibility (Chapter 8). Possibly, MB overestimated its abilities in trying to push value-added initiatives across such a wide spectrum of product types. Finally, Noranda admitted its acquisition of MB had not been successful when, after three years of net losses, it sold its controlling interest in MB in 1993. Indeed, during a time of extraordinary ferment in BC's forest economy, it is unlikely that control by a Toronto-based mining company, itself taken over by an even bigger conglomerate, helped MB to sort out an effective long-term strategy. Yet, confusion in MB's strategic vision remained after Noranda had divested control; thus, the recently built MDF mills in Kentucky and Ontario, which cost $270 million, have been sold for $160 million.

MB took important steps towards flexible mass production. By 1996, those steps remained incomplete. With the arrival of Tom Stephens as CEO, however, the restructuring of MB, featuring the sale of its pulp-and-paper mills in 1997, became much more focused. But similar decisions could have been implemented a decade earlier.

MB as a Victim of Circumstances

In the early 1980s, MB's restructuring plans sought to redefine corporate strategy in response to changing fibre-supply conditions, technological developments, and market dynamics. As volatile as they were, these trends represent conventional market forces. In BC, however, MB faced remarkable shifts in political economy that could scarcely have been foreseen in the nadir of the early 1980s recession. Environmental protest, Aboriginal land claims, American protectionism, and government (especially provincial) legislation rapidly and unpredictably altered the environment of BC industrial forestry. The collapse of the Canadian dollar and the Asian economic crisis, not unrelated events, have further added to the uncertainty of BC's forest economy.

These changing circumstances, which have clearly not yet worked themselves out, have had pervasive effects on forest-product firms. For several reasons, however, MB may be considered a special "victim." First, MB is the largest, most diverse forest-product corporation in BC; its business environment is the most complex. Second, MB is concentrated in coastal BC, where conflicts are most intense and where costs in the province are highest. Third, MB has been targeted by environmental groups that have blockaded MB's logging operations and attempted boycotts in Europe and North America of MB's products, more so than other corporations. Fourth, recent government policies have specific implications for MB because of its size and location. Thus, the federal government imposed the country's strongest environmental regulations on MB's Port Alberni operations. The provincial government has reduced the AAC in its TFLs to provide timber for the SBFEP and land for parks, the corporate impact of which has been greatest on MB. Thus, the biggest parks and ecological reserves transferred from MB licences are at Schoen Lake (8,170 hectares), Carmanah Valley (3,600 hectares), Tsitika watershed (164 hectares), and Moresby (964 hectares). Although MB agreed to participate in a broadly based committee to deal with logging in Clayoquot Sound, where it has logging rights, the provincial government modified the initial agreement to cater to demands by environmental groups that had pulled out of the decision-making process (Chapter 10). The provincial government also increased taxes on MB's private lands.

The reduction in MB's AAC is substantial, in gross terms from over 8 million cubic metres in the 1970s to 6.2 million cubic metres in 1995. In TFL 44, an AAC of 3.5 million cubic metres in the 1970s became 2.2 million cubic metres in 1995 and is now 1.8 million cubic metres, a 50 percent reduction. MB's initiative to shift from clear-cutting to partial retention schemes, and its own proposals to exchange Crown tenures, will further reduce these levels.

Calling MB a victim – for a corporation as large and as important as MB – may seem far-fetched. Yet, it is equally surprising that MB, as a BC-based corporation, should have been so targeted. From this perspective, it is ironic that a recent report should advocate the idea of a "corporate champion" for BC's forest economy, because MB has the most appropriate credentials for such a role: size, local control, and innovativeness (Ernst and Young 1998).

Instead, the ties that have so closely bound MB and BC in the past appear to be loosening, a trend to be formally confirmed by Weyerhaeuser's pending (29 October 1999) acquisition of MB.

MB and BC: Unravelling Ties?

Prior to Weyerhaeuser's acquisition of MB, questions about MB's commitment to BC had long been raised. Typically, these questions were rooted in the multinational expansion of MB, often expressed as a search for lower wood costs (Hamilton 1996). In this view, MB's global expansions equate to fewer benefits for BC and reduced loyalty. In such a view, MB's takeover by Weyerhaeuser scarcely makes a difference; it is simply part of the same "globalizing" process that is limiting the ability of BC's forest economy to shape its own future. The truth about MB's "unravelling ties" with BC is more prosaic than is commonly supposed, however, and with the realization of Weyerhaeuser's takeover, MB's relations to BC will take on fundamentally new dimensions.

Clearly, a sea change in MB's role in the province began in the depths of the early 1980s recession when the firm extensively rationalized facilities and employment. This became a tidal wave with the arrival of Mr. Stephens as CEO. MB's head office was cut back to absolute bare bones, the cherished and successful R&D program was closed down, the pulp-and-paper operations were sold, and a long-standing transportation subsidiary was sold. In addition, MB withdrew its membership from the Council of Forest Industries (COFI) and the Forestry Alliance, signalling that it was no longer willing to contribute to, let alone lead, an "industry voice" on matters of general concern.

In assessing these "unravelling ties" between MB and BC, prior to the implications of Weyerhaeuser's takeover, several points are worthy of note. First, many aspects of MB's restructuring reflect the imperatives of technology and the search for flexible mass production. In 1980 MB was a dinosaur and it needed to adapt its operations and head-office functions more comprehensively to an electronic age in order to enhance productivity. It also needed to shift away from commodity production. Probably, it did not push changes in these directions fast enough in the late 1980s; otherwise the general direction of change was in the best interests of BC as well as MB.

Second, the argument that MB's multinational expansions were a "threat" to its commitments in BC has always been bogus. Thus, from the 1960s onwards, multinational expansion did broaden MB's options for investment and reduce its dependence on BC as a source of profit. But multinational growth did not add up to geographic abandonment of BC. In the 1960s, with a substantial fixed commitment in BC, geographic diversification made corporate sense in terms of growth opportunities and the reduction of the uncertainties associated with concentration in one jurisdiction. At the same time, BC realized many benefits from MB's multinationalism that was as much about a search for markets as it was for lower wood costs. Foreign investments were closely integrated with BC operations. For example, they helped fund MB's investments in R&D and head-office activities in BC while manufacturing operations in the province often gained favoured access to affiliated markets in foreign countries. Moreover, in the early 1980s, MB's restructuring affected its foreign investments as well as those in BC. In fact, MB was able to divest some of its foreign investments to generate cash flows to help save its core facilities in BC.

Third, during Fordism and since, MB's core control, manufacturing, and logging functions were focused in coastal BC; the largest share of its forest resources, capital equipment, and employees was always concentrated in this region. It has been the leading corporate innovator, including with respect to flexible mass production. Even following the implementation of Mr. Stephens's restructuring plans in the late 1990s, MB remained firmly "embedded" in the forest economy of BC. In 1998, MB's asset base in BC was considerable, and it has skills and resources in BC, including 6.2 percent of the provincial harvesting rights, that cannot be readily relocated elsewhere. MB's product mix within BC is distinctive compared to what it does in other regions, and it has developed specific and valuable trade connections, including with Japan. In the 1990s, major environmental expenditures, tree planting programs, and other measures improved its fibre supply, in turn reflecting a long-term commitment to BC. Indeed, it has made strenuous efforts to meet environmental concerns and in connecting with Aboriginal (and other local forest owners) in re-organizing fibre supplies (see Chapter 10). Its "forest project" of 1998 aims to establish MB as a global leader in environmentally sensitive forest management, and its North Island woodland division qualified for the ISO 14001 environmental management system and has passed its first audit on the way to becoming the first Canadian operation certified to the Canadian Standards Association's sustainable forest management standard.

Fourth, it may be argued that ties between MB and BC were increasingly undermined by local forces, that BC felt it did not need MB. As noted, as the province's largest corporation, MB has long been the target of corporate concentration criticisms (Schwindt 1977). With its operations concentrated in the coastal region, MB has been the centre of environmental criticism. Significantly, the close, mutually supporting alliance between MB (and other large corporations) and the provincial government that was such a central plank of Fordism was no longer the case by the 1990s. Indeed, the high-stumpage regime is a critique of this alliance, and provincial policies have been particularly severe on MB (Gunton 1997). Escalating costs, new regulations, less wood fibre, taxes on private lands, new parks, incentives to SMEs, support for land claims, support for community forests, and the Clayoquot Sound Compromise very much put MB on the defensive.

MB was more than a little bruised by these developments. As a result of higher stumpage, the Forest Code, timber lost to parks, and the SBFEP program, its fibre costs have escalated in BC and its AAC declined. Thus, MB's harvesting costs on Crown land in BC have increased from $72 per cubic metre in 1993 to $111 per cubic metre in 1996, and total forest costs increased from $75 million to $215 million in the same period. At the same time, its AAC from its TFLs has been drastically reduced, by more than half since the 1970s (see Chapter 10). These developments reflected local choices, not globally imposed imperatives.

Mr. Stephens's restructuring strategies were therefore formulated in the context of a provincial framework that had become highly hostile to MB. Indeed, the appointment of Mr. Stephens, a man with a reputation for "rescuing" large corporations in trouble and with no prior affiliation or attachment to BC (or MB), doubtless reflected the growing disquiet of key shareholders over MB's declining fortunes in BC. Certainly, Mr. Stephens was transparent about his objectives: return MB to profitability in a way that maximizes values to shareholders, apparently the most influential being the Bass family in Texas, an Ontario-based pension fund, and, on his appointment, Mr. Stephens himself.

In terms of his stated goals, Mr. Stephens has been successful. In two years MB was returned to profitability and share values increased. Moreover, with great skill, Mr. Stephens turned the tables on local opponents to MB, including the provincial government. Since 1996, as its product strategy has become more focused, MB's publicly announced shift from clear-cut logging, its public promotion of the idea of privatization, its public proposals to receive private land (worth $83 million) as compensation for lost AAC and then to exchange part of its AAC in return for tenure reform, and its reduction of logging on Crown land while increasing cutting on its private lands, has put the provincial government on the defensive. In this latter regard, for example, MB doubled its harvest from its private lands from 800,000 cubic metres in 1997 to 1.5 million cubic metres in 1998 (Hamilton 1999e). This increased harvesting of private lands has reduced costs, because stumpage is not charged and regulations are fewer, including export restrictions, and has further increased returns because many of the logs are exported to foreign markets willing to pay premium prices. If jobs are lost in BC because of these exports, thus thwarting the government's Jobs Accord, they have helped MB return to profitability. Moreover, the high-stumpage regime wanted higher value for the forest resource, but it did not offer a comprehensive plan for tenure or plans to eliminate clear-cutting, and it most certainly did not include privatization. These are now active policies or serious proposals. MB revealed the unworkability of the government-ratified Clayoquot Sound Compromise by stopping logging and then laying off workers. MB's closure of its R&D laboratory counters government hopes for more R&D and innovation. If the government was abreast of environmental thinking regarding forestry in 1990, MB became the leading innovator by 1997. Finally, MB's public reference to the BC "discount factor," which implies shareholder return is significantly enhanced by investing elsewhere, places further pressure on the government to improve the BC business climate. The government was now responding to an agenda set by MB.

In the restructuring of MB between 1996 and 1998, the closure of its R&D program appeared to make the least sense. The R&D group had successfully contributed to MB's shift to higher-value production and its closure is not consistent with further shifts in this direction. Indeed, as the only major forest product in-house R&D program in BC, its demise was a blow to BC as a whole. Mr. Stephens simply stated that there was no longer need for an in-house group and responsibility for innovation would now be devolved to individual factories. Just how factories would become "smarter" was not indicated (see Chapter 11). On the other hand, the R&D divestment did provide an immediate cost savings (with no immediate loss of revenues), and help enhance shareholder values in the short run. The R&D divestment also did nothing to reduce the attractiveness of MB as a take-over target since most giant MNCs, such as Weyerhaeuser, already have their own R&D and would be unlikely to want additional programs.

In any event, when publicly announced, Weyerhauser's proposal to acquire MB was endorsed by Mr. Stephens, even though it countered the logic of his own restructuring of MB, which was to make the firm more globally efficient by becoming leaner and more focused. In his view, acquisitions of Canadian firms by US MNCs are to be expected given the low value of the Canadian dollar. Yet, it should also be noted that once it was proposed, the takeover increased shareholder value, at least in the short run, and that the Bass family, an Ontario pension fund, and Mr. Stephens gained much from this increased value. One newspaper story has claimed that they soon realized some of this increased value, prior to a recent reduction in share value (Hamilton 1999e). Whether or not MB's most recent restructuring was designed to facilitate an acquisition may never be known. But such an acquisition is consistent with the immediate goal of enhancing shareholder value (and the divestment of R&D makes more sense from this point of view).

The acquisition of MB by Weyerhaeuser, will, of course, further significantly unravel ties between MB and BC. BC's flagship corporation will now be controlled from Washington State by an MNC that will enjoy enormous flexibility in choosing the nature, size, and location of investments on either side of the border. The costs to BC are likely to be considerable, and the benefits negligible (see Chapter 5). Weyerhaeuser will also almost certainly reinforce Mr. Stephens's plea for privatization of log markets in BC – a plea that has already led to increased log exports from BC in 1998 – to increase its flexibility in directing logs to its facilities on either side of the US-Canada border (or to serve Weyerhaeuser's Japanese customers). It does seem that corporate interpretation of privatization means the flexibility to export logs.

Conclusion: A Requiem for MB

MB's restructuring towards flexible mass production is broadly consistent with changes sweeping through the global economy and with BC's changing competitive advantages. Inevitably, the transformation from a Fordist to a flexible corporation driven by know-how has been problematical. The reasons are various. Changes have been implemented in situ, rationalization and downsizing have paralleled modernization and product diversification, new value-added markets are not self-evident, new products sooner or later become standardized and subject to cost-based competition, and the capital and human investment requirements implied by flexible, value-adding strategies are substantial. Simultaneously, new environmental codes of conduct have increasingly shaped all aspects of MB's behaviour.

The sea change in MB's strategy has become especially pronounced in the last years of the 1990s, affecting its attitudes towards wood-fibre supply as well as the product cycle dynamics documented in this chapter. Thus, as a Fordist corporation in the 1950s and 1960s, a secure supply of timber was the basis for MB's vertical and horizontal integration strategies. Control of its timber supplies, specifically from TFL 39 and TFL 44, as well as from private lands, was considered essential to the competitive strength of the company (see Figure 10.1). In 1999, by contrast, MB is willing to forfeit tenure rights, and to rely on a smaller land base that is preferably privatized. The fact is that MB lost considerable money in recent years harvesting Crown timber because of changes in government policy and because large-volume logging is less appropriate for contemporary times (Chapter 10). In present, more dynamic times, MB was apparently seeking to reestablish a more entrepreneurial style of leadership and become a more innovative, leaner, and more focused corporation. MB may still be international in scope, but it has taken on the strategies and structure of a flexible, large firm rather than those of a giant. Yet, innovation and flexibility in turn celebrate R&D. The closure of MB's R&D group is undoubtedly a loss for BC; it may have become a problem for MB, too. However, following the Weyerhaeuser acquisition, MB will be reduced to the status of a foreign-owned subsidiary, primarily responsible only for operational matters within BC. Decisions about strategy and structure will be made elsewhere. The forest economy of BC has lost its corporate champion.

5
Foreign Direct Investment: Help or Hindrance?

In BC, public controversy over foreign direct investment (FDI) in the local development process is perhaps less evident than in the rest of Canada. In the Sloan and Pearse Royal Commissions, the foreignness of firms was not much of a consideration. Pearse (1976: 322) referred to FDI but only in passing, even though in the early 1970s economic nationalism in Canada was on the rise and the federal government passed some mild legislation to control FDI (Britton and Gilmour 1978). Criticisms of FDI, however, were perfunctorily dismissed. In Pearse's view, foreign firms behaved no differently from local firms, and there were no important policy implications to be drawn. Schwindt's (1979) overall favourable critique of Pearse (1976) did question the neglect of issues related to corporate concentration but without reference to FDI. Recently, Schwindt and Heaps's (1996) economic accounting of BC's forest industry lists the presence of foreign-owned firms but no implications are drawn or intimated. As throughout Canada, conventional economic wisdom in BC has always been a staunch supporter of an open door policy for FDI; from this perspective, the key issue is investment, not its ownership or control.

Indeed, in a province noted for its polarized political parties, changes in government, specifically the changes from free enterprise to socialist governments that occurred in the early 1970s and 1990s, have not been associated with changes in philosophy towards FDI. Bill Bennett, a free enterprise premier, did once publicly proclaim in the early 1980s that "BC's resources are not for sale," but he was targeting Canadian Pacific, a Montreal-based Canadian MNC in its attempted acquisition of MacMillan Bloedel (MB). His proclamation was not intended to signal any change in the government's long-standing policies of attracting FDI. These policies remained intact with the election of left-wing NDP governments, and the latter's concern over corporate concentration has not implied a hidden agenda towards FDI. As the province's largest corporation, the BC-based MB has long been the centre of these criticisms (Schwindt 1977). In contrast, an announcement in early 1999 by Louisiana Pacific, a large US-based MNC, of intentions to invest further in northeastern BC has been publicly applauded – and subsidized – by the province's NDP premier. Even more startling, the US-based Weyerhaeuser's proposal to acquire MB has occasioned no apparent political concern within the NDP (or opposition parties).

From both the left and the right of the political spectrum, there is concern over FDI by Canadian firms outside the province, and about foreign firms pulling out of BC in favour of other places (Marchak 1983). In this view, investment is the bellwether of economic health and such "flights of capital" are automatically assumed to be negative for BC. But if FDI is only beneficial to BC as a host or receiving economy, why do the major metropolitan powers of the world urge MNCs to invest outside their borders? Do not host countries and regions pay for FDI? Questions can also be asked about how the linkages that foreign subsidiaries have with their international parent companies affect the structure of the forest industry in BC.

In practice, FDI is a mixed blessing for host economies. In BC, FDI has served both to stimulate the growth of the forest industries and to limit potentials for diversification. The precise balance of costs and benefits is hard to summarize, but the general contours of the debate can be traced. This chapter draws such a map. In particular, the chapter examines the role and impact of foreign-controlled firms in the evolution and structure of BC's forest economy (Hayter 1981, 1982a, 1985). The first part of the chapter examines the patterns of entry and postentry behaviour of foreign firms until the early 1970s, especially during the Fordist long boom. The second part of the chapter examines FDI in the most recent period of corporate restructuring. Finally, some of the long-term implications of high levels of FDI in BC's forest economy are identified. This discussion notes that if FDI fuelled the Fordist boom, it also institutionalized a commodity structure and limited industry's ability to create technology.

This chapter is part of a bigger debate on the role of FDI in Canadian economic development. Conventional economic wisdom insists that FDI fosters growth and efficiency by providing scarce technical, marketing, managerial, and financial resources as well as by stimulating competition and innovation (Daly 1979; Safarian 1979). Critics argue that FDI "truncates" economic development by reinforcing the staple trap and limiting opportunities to diversify (Innis 1956; Watkins 1963; Rosenbluth 1977; Britton and Gilmour 1978). As a policy contest, the debate is over. In contemporary times of globalization and free trade, conventional wisdom regarding FDI is taken for granted. The critics are lonely voices, ignored by public policy. Indeed, since the National Policy of the 1880s, federal and provincial governments have pursued open door policies for FDI. The marginalization of the critics is unfortunate, however. Their concerns are valid. If a change in policy towards FDI is remote, the inheritance of high levels of FDI has consequences for policy that should not be ignored.

As a final note, it is interesting to speculate about why the provincial government in BC during the 1950s and 1960s, which was so strongly committed to free enterprise, apparently never seriously entertained the privatization of the forests (Drushka 1993). Was it concerned about foreign control? If not, foreign control may well be a concern facing privatization now.

Foreign Firms in the Long Boom from the 1940s to the 1970s

In BC, and for Canada as a whole, it is useful to distinguish five broad phases in the evolution of the forest-product sector during which the nature, motivation, and impact of foreign participation has varied (Hayter 1985). In the first phase, before the 1860s, British control and demands exercised the most pervasive influence in the commercial exploitation of the Canadian forests. In the latter part of the nineteenth century, individual American entrepreneurs and family-based concerns became an increasingly important influence on logging and lumber activities. The third phase, between about 1910 and 1945, was marked by an extension of American corporate interests, most notably in pulp and paper. After 1950, all aspects of the forest-product industries became increasingly influenced by foreign-based MNCs largely, but by no means exclusively, based in the US. In the last and present phase, which began in the mid-1970s, foreign firms have been involved in industrywide restructuring that has seen some long-established foreign firms leave Canada, and others, including several Japanese-based MNCs, enter.

BC was only marginally involved in the first of these stages, as British Admiralty ships captained by James Cook and John Meares made contact with BC, and some logs were shipped to China (Taylor 1975: 2; Chapter 2). With the onset of large-scale industrialization in the 1880s, BC became fully engaged with foreign capital along the lines that occurred elsewhere in Canada. This engagement meant American involvement. In fact, American logging interests began to expand very rapidly in the 1870s in central and eastern Canada, while in BC the first reference to an American company acquiring timber limits was in 1887 when the Minneapolis and Ontario Lumber Company bought 1.5 billion board feet of standing timber. To force American lumber companies to establish sawmills in Canada, the Canadian government increased its export duty on logs to $2 per 1,000 board feet.

Until 1910, however, foreign investments were primarily made by family-owned companies or small groups of businessmen, and decision-making autonomy often shifted to BC. For example, the Powell River Company, a founding company of MB, was set up in 1908 by the Scanlon family of Minneapolis, where control was maintained until the 1930s when it was shifted to BC (Schwindt, 1977: 14). In other instances, as demonstrated by the Bentley Brothers and Koerner Brothers, who left Austria in the 1930s to respectively establish Canadian Forest Products (Canfor) and Alaska Pine and Cellulose (acquired by Rayonier in 1954), the transfer of family wealth and control to BC occurred simultaneously.

In general, American enterprise occupied an important though hardly dominant position in the BC (and Canadian) lumber industry from 1850 to 1890 (Marshall et al. 1936: 6). By the end of the nineteenth century, however, a wood-based pulp-and-paper industry was established in Canada. This development eventually was far more strongly influenced by American capital and directed through corporate rather than family control. Indeed, the great northward migration of the newsprint industry occurred after 1910. Given the declining raw material base in the northern US, this migration was stimulated by, first, the prohibition of pulp wood exports by provincial governments (and the federal government on behalf of federally owned Crown lands) between 1900 and 1910, and second, the removal of US tariff on newsprint in 1911. BC was very much a part of this trend: newsprint mills at Powell River (1912) and Ocean Falls (1917) were some of the earliest built in Canada.

Following the Second World War, however, the characteristics of FDI in the province's forest sector changed radically. In particular, in a manner consistent with global trends, foreign participation overwhelmingly involved equity capital controlled by parent companies whose head offices remained outside BC. Inevitably, FDI meant big commodity mills, typically for export.

Entry Characteristics of Foreign Firms in the Fordist Period

Between 1950 and the mid-1970s, foreign companies entered BC's forest sector by establishing both majority-owned subsidiaries (Table 5.1) and joint ventures (Table 5.2). One important subsidiary (namely, British Columbia Forest Products or BCFP) is excluded from these tables. Originally set up by eastern Canadian financing (involving acquisitions) in 1949, foreign companies held a majority interest in BCFP between 1950 and 1969, after which the US-based Mead Corporation (29 percent ownership) became its principal owner along with the Toronto-based Noranda Mines (29 percent ownership). The ownership characteristics of BCFP are therefore complex, although it can be said that ultimate control lies outside BC. Although the reasons are not obvious, attempts by the long-established Toronto- and Montreal-based forest-product giants to participate in BC's forest sector at this time were small scale or unsuccessful; Noranda Mines, which entered the forest industry in 1961 by acquiring several lumber producers in northcentral BC where it operated several mines, was the most important central Canadian firm to control large subsidiaries in the provincial forest-product economy.

Foreign firms entering BC's forest sector invariably pursued horizontal and vertical integration strategies and typically were already important manufacturers of forest products. Thus, foreign participation was dominated by the forest-product giants of the US (notably, Crown Zellerbach,

Table 5.1

Entry characteristics of principal foreign subsidiaries operating in British Columbia's forest-product sector, 1950-73[1]

Company	Locus of control office (parent)	Year	Method, activity, and location(s)
Crown Willamette Paper (Crown Zellerbach Canada since 1954)	San Francisco	1914	Acquisition of the pulp-and-paper mill at Ocean Falls, then under receivership, to form Pacific Mills; expanded and restarted in 1917
Columbia Cellulose[2]	New York (Celenese Cor.)	1949	Investment in a 200-ton-per-day sulphite mill, Prince Rupert
Scott Paper	Philadelphia	1954	Acquisition of New Westminster Paper's tissue-paper converting plants, New Westminster
Rayonier Canada	New York[3]	1954	Acquisition of Alaska Pine and Cellulose's sulphite mills at Woodfibre and Port Alice, and three lumber mills in New Westminster and Vancouver
The Pas	Minneapolis (Winton Family)[4]	1954	Relocation of Manitoba mill to Prince George
West Fraser Timber	Seattle (Ketchum Bros.)[4]	1955	Acquisition of Two Mile Planer's sawmill, Quesnel
Weldwood Canada	New York (US Plywood)[5]	1961	Acquisition of Western Plywood's plywood mills, Vancouver and Quesnel
Weyerhaeuser Canada	Tacoma	1965	Joint venture in a 250-ton-per-day pulp mill at Kamloops; acquired full control in 1970
Triangle Pacific	New York	1968	Acquisition of Pacific Pine's sawmill, New Westminster
Evans Forest Products	Portland	1969	Acquisition of the Kicking Horse and Commercial Lumber sawmills, Golden and Lillooet
Pope and Talbot	Portland	1969	Acquisition of Boundary Sawmills and C and F Lumber Mills, Midway and Grand Forks

►

◄ *Table 5.1*

Company	Locus of control office (parent)	Year	Method, activity, and location(s)
Balfour Guthrie	London (Dalgeti)	1970	Acquisition of Netherlands Overseas's lumber mills, Prince George

1 Foreign-owned companies that controlled more than 1 percent of the province's harvested timber as reported by Pearse (1976) are included. Itochu of Japan's CIPA sawmill, established in 1968 in a joint venture with Pacific Logging, is therefore excluded.
2 In 1972 sold to the BC government to form Canadian Cellulose.
3 In 1968, IT&T of New York acquired Rayonier.
4 Family-owned companies.
5 Champion Paper of New York subsequently became the parent company.
Source: Hayter 1981: 104. Reprinted with permission of Blackwell.

Weyerhaeuser, Rayonier, Scott, International Plywood, International Paper, Mead, and Pope and Talbot), Europe (notably, Reed, Feldmühle, Enso-Gutzeit, and Svenska Aktiebolaget), and Japan (notably, Honsho, Jujo, and Daishowa). The Japanese firms, however, arrived late in the period. For Scott and International Plywood, horizontal motivations related to participating in the Canadian tissue and plywood markets respectively clearly dominated their entry decisions. For the others, entry into BC characteristically reflected desires to (vertically) secure new resources from which to supply inputs expanding affiliated and nonaffiliated downstream plants around the globe *and* to (horizontally) exploit accumulated know-how in the forest industries. The entry decision by forest firms also involved large-scale operations. Indeed, several subsidiaries (notably, Scott Paper, Rayonier, and Weldwood) were formed by the takeover of substantial and well-known provincially owned enterprises, whereas all the joint-venture agreements were negotiated to build expensive, capital-intensive, export-oriented pulp-and-paper mills.

Two qualifications to the generalization made here concerning corporate entry characteristics may be noted. First, in two instances, entry was accomplished by small, family-owned enterprises (West Fraser Timber and The Pas); that is, in a manner more consistent with pre-1950 than post-1950 trends. Second, Columbia Cellulose's (Cocel's) entry into BC's forest sector was made without previous experience in the forest industry. This decision, however, was part of a vertical integration strategy, and in particular it was designed to supply dissolving pulp grades to textile plants owned by Cocel's New York-based parent company.

In the method and timing of entry, some interesting differences are related to the national origins of foreign firms. In particular, American companies dominated FDI in BC's forest sector until the mid-1960s, and they preferred to establish majority-owned or wholly owned subsidiaries rather than negotiate joint ventures (or some other arrangement for technological

Table 5.2

Foreign firms in joint ventures established in the pulp-and-paper industry of British Columbia, 1960-73

Company	Location of mill	Startup		Partners: Locus of control	
		Year	Size[1]	Local firms[2]	Incoming firms
Kamloops Pulp & Paper	Kamloops	1965	250	Four sawmill firms (49%)	Weyerhaeuser, Tacoma (US)[3]
Prince George Pulp & Paper	Prince George	1966	750	Canadian Forest Products (50%)[4]	Reed, London (UK)
Intercontinental Pulp & Paper	Prince George	1968	750	Canadian Forest Products (37.5%)[4]	Reed and Feldmühle A.G., Dusseldorf (Germany)
Northwood Pulp & Timber	Prince George	1967	625	Northwood Mills, subsidiary of Noranda Mines, Toronto (50%)	Mead Corporation, Dayton (US)
Tahsis	Gold River	1967	750	East Asiatic Co., Copenhagen, Denmark (50%)	Canadian International Paper of Montreal, a subsidiary of International Paper, New York (US)
Skeena Kraft	Prince Rupert	1967	750	Columbia Cellulose, subsidiary of Celanese Co., New York (60%)[5]	Svenska Cellulose, Aktiebolaget, Sandsvall (Sweden)
Crestbrook Forest Industries	Skookumchuck	1969	400	Crestbrook Timber (49.9%)	Mitsubishi and Honshu Paper Mfg. Co., Tokyo (Japan)
Eurocan	Kitimat	1970	915	(Local Entrepreneur: 3%)[6]	Enso-Gutzeit Oy, Kymin Oy-Kymmene Ab. and Oy Tampella Ab, Helsinki (Finland)

Finlay Forest Products	Mackenzie	1970	160	Cattermole Timber (75%)	Jujo Paper Mfg. and Sumitomo, Tokyo (Japan)
Bulkley Valley Forest Industries	Houston	Pulp mill not started (no local partners)			Consolidated Bathurst of Montreal and Bowater Canadian, a subsidiary of Bowaters, London (UK)
Cariboo Pulp & Paper	Quesnel	1972	750	Weldwood, subsidiary of U.S. Plywood-Champion Papers, New York (50%)	Daishowa Paper Mfg. & Marubeni, Tokyo (Japan)

1 Tons per day.
2 Local firms are defined as those already operating in BC.
3 Weyerhaeuser acquired full control in 1970.
4 Reed's interests acquired by Canadian Forest Products in 1979.
5 Obtained full control in 1970 and in 1973 was purchased by British Columbia government.
6 This interest purchased by Eurocan in 1971.

Source: Hayter 1981: 105. Reprinted with permission of Blackwell.

transfer). Indeed, only one such subsidiary (Balfour Guthrie) was not American, whereas just three American firms entered the BC forest sector by participation in a joint venture, and one of these (Weyerhaeuser) subsequently bought out its local partners. The larger American subsidiaries (specifically, Cocel, Crown, Scott, Rayonier – until 1968 – and Weldwood) have permitted some local ownership and traded shares at the Vancouver Stock Exchange. Even in these cases, however, parent companies maintained at least 60 percent ownership.

The most striking characteristic of the way in which foreign firms entered BC's forest sector is the overwhelming dominance of acquisitions and joint ventures rather than by majority-controlled direct investments in new facilities. This pattern of entry clearly reflects corporate desires to reduce as far as possible the cost and uncertainties ("spatial entry barriers") associated with establishing operations in the unfamiliar environment of BC (see Hayter 1981). The subsidiary companies, for example, with only one exception (Cocel), were formed by the acquisition of existing firms. Similarly, in all but one (Eurocan) of the successful joint ventures, incoming foreign firms combined with partners based in BC, and the local firms were typically responsible for much of the planning of the mill projects, including assessment and negotiation of timber leases and selection of regions, communities, and sites (Hayter 1978c). The significance attached to previous operating experience within the provincial environment by foreign firms was not misplaced. Indeed, the extremely costly problems faced by Eurocan after startup and the failure of the Bulkley Valley forest-product complex, another joint venture lacking local partners, both resulted from decisions made without adequate understanding of local conditions.

In addition to providing valuable geographical know-how, local partners were variously responsible for marketing studies, equipment selection, and hiring employees while at Prince George; Canfor supplied the technological expertise for two projects. For their part, the foreign partners contributed to the joint ventures by sharing financial costs and marketing responsibilities, both roughly in accordance with equity positions (Barr and Fairbairn 1974). As expected, much of the pulp marketed by the incoming firms has supplied affiliated plants overseas, although important exceptions (specifically, the projects at Gold River and Kamloops) were planned to compete on the open market. In some instances (notably, at Kamloops, Gold River, and Skookumchuck), foreign partners supplied technological know-how in the construction and operation of the mills, whereas at Quesnel Weldwood's American parent provided similar expertise (Hayter 1973: 210-20). In fact, the contributions by foreign firms were welcomed, actively sought by the local partners and constituted a necessary condition for an investment go-ahead decision.

As of the mid-1970s, the ownership structure of the joint ventures had remained relatively stable. In one case, a foreign firm (Weyerhaeuser) acquired its local partner, whereas Cocel obtained full control at Prince Rupert (and later sold its entire operations to the provincial government). In terms of post-startup investment activity, the joint-venture companies have largely emphasized in-plant modifications, although in some cases wood-processing operations were established. By 1977, however, none of the joint ventures had expanded interregionally, and of course, any such commitment would require agreement among all principal partners. Subsidiary companies, on the other hand, enjoy more flexibility in plotting strategies of growth.

Investment Strategies of Subsidiary Companies

Subsequent to their formation in BC, forest-product subsidiary companies have to varying degrees expanded, modernized, and even entirely replaced acquired facilities, while a few pursued active policies of spatial and/or industrial diversification. Consequently there are considerable variations in the extent to which foreign (and externally controlled) companies became horizontally and vertically integrated within BC in the Fordist period. Several firms remained specialized in one of the main forest-product commodities. These firms include West Fraser, The Pas, Triangle Pacific, Evans Products, Pope and Talbot, and Balfour Guthrie (all lumber producers), and Scott (a

Table 5.3

Production levels of the principal integrated companies in the forest-product sector of British Columbia, 1977

Company	Lumber (MFBM)	Plywood (M.sq.ft.)	Pulp	Newsprint (000 tons)	Paperboard
Foreign owned					
Crown	499	315	216	270	88
Rayonier	282	–	310e	–	–
Weldwood	470	637	129	–	–
Weyerhaeuser	413	–	400e	–	–
Externally owned					
BCFP[1]	812	210	423	270	–
Northwood	735	–	130e	–	–
BC owned					
MB[2]	1,150	350	440	920	126
Canfor	787	184	500e	–	107
Cancel[3]	250	–	350	–	–

1 BCFP's ownership structure is complex and could justifiably be classified as foreign because Mead owns 29 percent and Scott a further 14 percent.
2 Data refer only to BC operations.
3 Canadian Cellulose (Columbia Cellulose until 1973).
Note: MFBM refers to million board feet; M.sq.ft. refers to million square feet, 3/8" base; e = estimated.
Source: Hayter 1981: 107. Reprinted with permission of Blackwell.

tissue-paper manufacturer). Except for Scott, which acquired a large tissue-paper plant in Quebec in 1964 (and so captured about one-third of the Canadian market for tissue-paper products), these firms have also not expanded interregionally.

A few subsidiaries, on or since entry, established integrated operations in the wood-processing and pulp-and-paper sectors (as well as logging). As such, they have been among the largest firms in BC's forest economy. These subsidiaries are Crown, Rayonier, Cocel, Weldwood, and Weyerhaeuser; BCFP and Northwood Mills also belong to this group. In fact, in the mid-1970s, only two BC-controlled enterprises rivalled them in size and diversity of their forest-product operations: MB and Canfor. Together, these integrated companies accounted for 100 percent of the newsprint, 60 percent of the chemical pulp, 61 percent of the paperboard, 45 percent of the lumber, and 72 percent of the plywood, as well as substantial proportions of various converted-paper products manufactured in BC in 1977 (Table 5.3).

Crown Zellerbach Canada

Among the integrated foreign companies, Crown was the largest in the Fordist period and manufactured the most balanced range of forest products, including lumber, plywood, pulp, newsprint, paperboard, and converted-paper products. In the retail sector, it operated several large home-building stores. Consequently, Crown's development path, as represented by production and investment profiles between 1950 and 1977, provides a distinctive and useful example by which to summarize the investment behaviour of subsidiaries in BC (Figure 5.1). Several observations are particularly worthy of note:

- Crown's entry and subsequent geographical expansion *within* BC was accomplished by joint venture or acquisition. Thus, Crown (as Crown Willamette) began operating in the province in 1917 at Ocean Falls following the acquisition of Pacific Mills, which had been unable to bring on stream its planned newsprint mill. Consequently, the company contacted Crown, the only newsprint manufacturer on the west coast, to rescue its venture. Since 1950, Crown's geographical expansions were predicated on a joint venture with Canadian Western Lumber to build a newsprint mill at Elk Falls (Campbell River), the acquisition of Canadian Western Lumber (and its Vancouver area wood mill), and the acquisition of lumber and plywood firms in the southern interior of the province.
- Crown's expansion comprised horizontal and vertical integration strategies involving locationally integrated in-site, new-site, and inter-site adjustments in the corporate product mix designed to maximize growth potentials, enhance stability, and increase efficiency in converting timber to the most profitable end-products. On the coast, Elk Falls was the focus for much of this activity. Thus, the newsprint mill, originally based

Figure 5.1

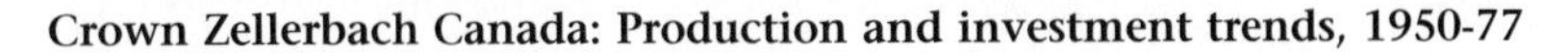

Crown Zellerbach Canada: Production and investment trends, 1950-77

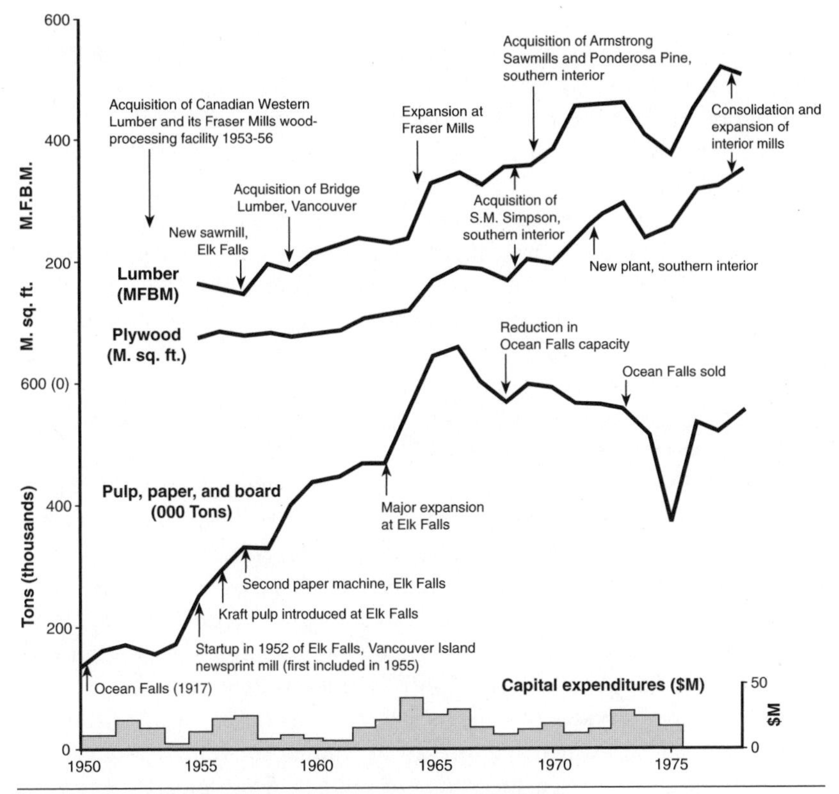

Source: Hayter 1981: 108. Reprinted with permission of Blackwell.

on converting salvage logs from company timber leases and sulphite pulp from Ocean Falls, was soon expanded to include kraft pulp (used in part to replace sulphite from Ocean Falls as the chemical input for newsprint), a machine to jointly produce newsprint and kraft papers (the latter to supply new converting facilities in the Vancouver area), and a small-log sawmill (which, with Vancouver area wood-processing plants, supplied the pulp-and-paper mill with wood chips, sawdust, and bark). Subsequent expansions at Elk Falls during the 1960s were similarly backwardly and forwardly linked with expansions of the firm's wood-processing and paper-converting facilities as well as with intermittent divestments of pulp-and-paper capacity at Ocean Falls. This latter facility, facing increasingly high operating costs, was finally sold in 1973 to the provincial government.

- Crown's investments in pulp and paperboard were closely coordinated within the global integration plans of the parent company. In particular, the successive expansions at Elk Falls during the 1950s and 1960s were phased to supply new parent company paper-converting plants in California, the Netherlands, and South Africa. In fact, by the mid-1970s approximately three-quarters of Crown's pulp and paper served "captive" markets. In this industry, such internal linkages are extensive, but they vary in degree; for example, Weyerhaeuser (1975: 6) claimed little of its BC production was so tied. Expansions of wood products were more oriented towards open markets with exports distributed by means of a cooperative sales agency.

- Crown's investments were "bunched" and typically occurred as the result of similar decisions by rivals. Investment interdependencies are an inherent characteristic of the forest sector, and they are intimately associated with cycles of overcapacity and the ensuing employment and production cutbacks that have so marred the sector's evolution (Hayter 1976; Schwindt 1979). In Crown's case, as with MB, a diversified forest-product base and access to affiliated markets provided some corporate flexibility in meeting depressed market conditions.
- Since entry, Crown has financed its investments mainly by internally generated profits. Thus, between 1955 and 1977, internal savings (less dividends) amounted to $347 million or 77 percent of the total additions to properties. Such behaviour was not unusual. Thus, a federal government study (Canada 1974: 42, 86) reported that between 1969 and 1971, 63 percent of the funding of investments by foreign subsidiaries in Canada's forest sector was generated internally by the subsidiaries.
- Subsidiary status typically imposes spatial limits on growth horizons. In Crown's case, investment planning was limited to western Canada but potentially encompassed the whole of Canada. Scott and ITT Rayonier, however, are parent companies that operated subsidiaries in different parts of Canada as separate entities, thus substituting north-south international lines of communication and authority for east-west national linkages. It is true that BCFP, despite its subsidiary status, acquired a large paper-converting firm in Minnesota in the late 1970s. Whether this acquisition represented an independent decision is unlikely, a view shared by US anti-trust investigators who accused BCFP of acting on behalf of the Mead Corporation, BCFP's American parent, in attempting to monopolize the paper-converting industry in the Midwest after earlier court rulings had restricted Mead from so doing.
- Crown's investments, apart from acquisition of a plastics paper manufacturer and construction of home-building retail outlets, remained almost exclusively within traditional industry lines. Given that forest-product companies as a whole have long experienced difficulties in diversifying outside their specialized area of expertise (Schwindt 1979), diversification of subsidiary companies is even less likely.

Crown's expansions throughout the Fordist period served to maintain its position as one of the province's forestry giants and more generally to strengthen foreign control over BC's forest sector. Admittedly, some subsidiaries have been less active in modernizing existing operations (for example, Rayonier) while Cocel's experience offers a particularly sharp contrast to that of Crown. In Cocel's case, an overly ambitious program of growth, technological problems, and failure to establish fully integrated operations led to losses exceeding $100 million between 1966 and 1971, and ultimately to its purchase by the provincial government, resulting in the formation of Canadian Cellulose. Foreign ownership, therefore, has not constituted an irreversible process; other examples of Canadianization have occurred as foreign firms have restructured. As well, other foreign firms have entered BC.

Extent of Foreign Control in the 1970s

In terms of aggregate levels of foreign control, Statistics Canada (1977) indicated that in 1970 foreign companies (defined as companies that are at least 50 percent foreign owned) controlled 37.9 percent of the value of shipments generated by the wood-processing industries (sawmilling/plywood) in BC and 24.7 percent in Canada as a whole as well as almost 50 percent of the paper and allied industries at both provincial and national levels. More disaggregated estimates of foreign (and external) equity ownership of the forest sector for BC in 1979, on the basis of the conventional 50 percent equity criterion and the 25 percent equity criterion, confirm the high level of foreign participation (Table 5.4). The conventional estimates, based on the more than 50 percent equity criterion, are conservative because joint ventures with 50 percent Canadian ownership are assigned as Canadian, as

Table 5.4

Percentage of foreign and external control of selected forest products in British Columbia in 1979

Commodity	Capacity	Foreign controlled (> 50% equity)	Externally controlled (> 50% equity)	Foreign Control (> 25% equity)
Newsprint	19,050 tons	17 (39)	35	34
Chemical pulps	4,460 tons	45 (50)	63	
Lumber	29,373 MFBM	35 (28)	42	46
Softwood-plywood	2,728 M.sq.ft	47 (46)	56	52
Tissue papers		100 (78)	100	100
Fine papers		0 (-)	0	0

Notes: Figures in parentheses refer to level of foreign control in Canada as a whole. MFBM refers to million board feet. M.sq.ft refers to million square feet, 3/8" base.
Source: Based on Hayter 1981 and Hayter 1985: 445-5.

Figure 5.2

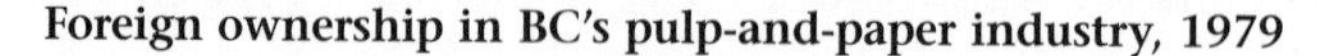

Foreign ownership in BC's pulp-and-paper industry, 1979

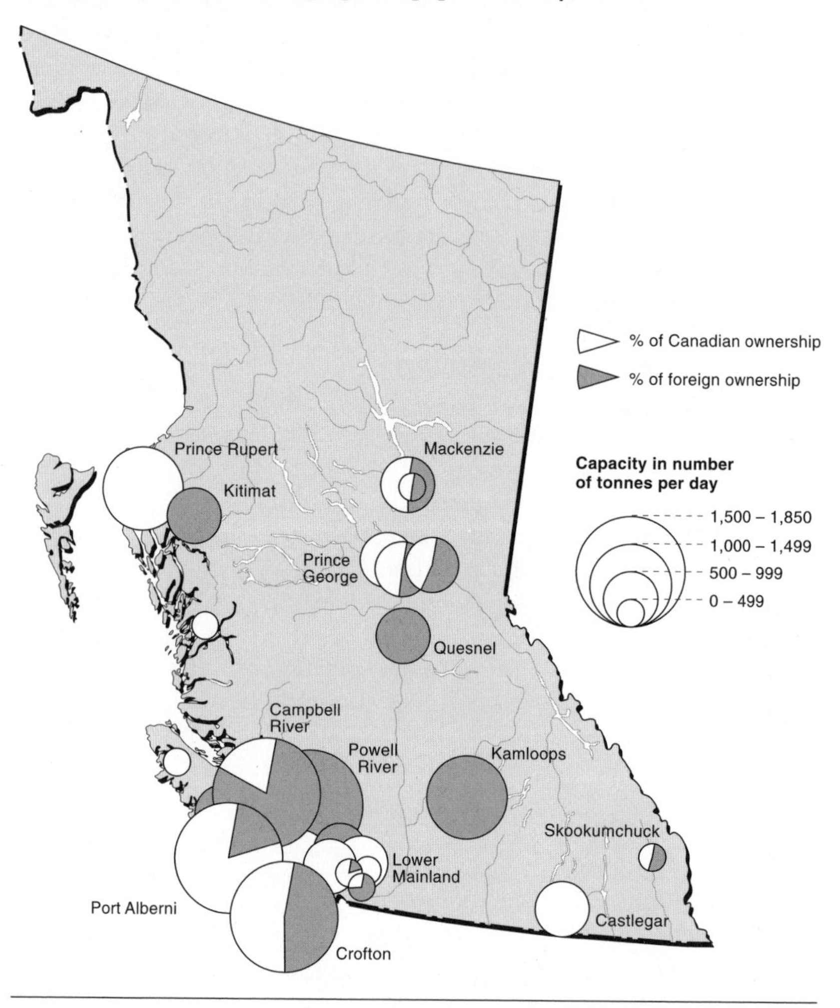

is BCFP, whose Canadian parent's headquarters were in Toronto. Even so, levels of foreign participation are high, and they are considerably higher, especially in the case of lumber and newsprint, when foreign control is defined in terms of the more than 25 percent equity criterion.

Industry variations in degrees of foreign control are in the expected direction. In general, levels of foreign ownership are higher in pulp-and-paper than in wood-processing activities. Thus, the tissue-paper industry, which is characterized by marked product differentiation and firm-level economies of scale in terms of R&D and advertising, exhibits an extremely high degree of foreign ownership; it is a good example of powerful American

Figure 5.3

Foreign ownership in BC's sawmill industry, 1979

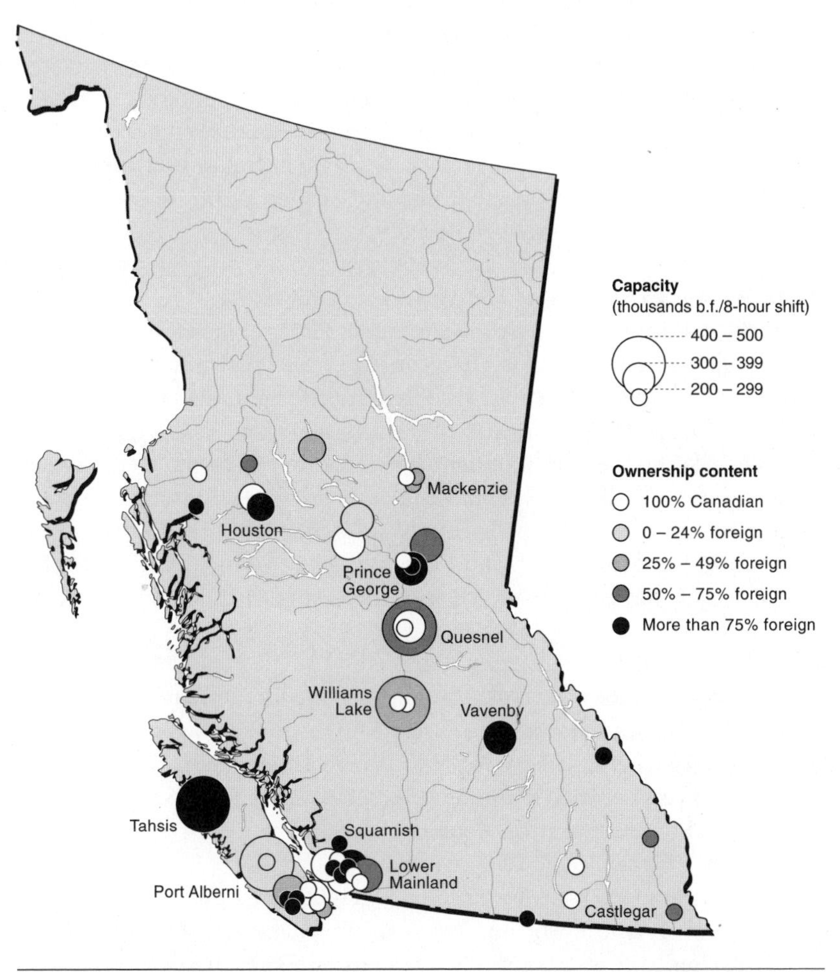

Note: Only sawmills with a capacity greater than 200,000 board feet per eight-hour shift are shown.

oligopolists extending their rivalry into the smaller Canadian market to establish so-called miniature replicas of parent operations. In contrast, the competitive advantages of international firms in the lumber industry in comparison with local firms are not so marked. Accordingly, the degree of foreign ownership is lower.

Maps of pulp-and-paper mills and the largest sawmills in 1979, classified by size and degrees of foreign ownership, graphically reveal the importance of FDI for the BC forest economy (Figures 5.2 and 5.3). In some instances, Canadian ownership implies a central Canadian-based parent. BCFP, with its pulp-and-paper mills at Crofton and Mackenzie, along with sawmills at

Mackenzie and in the southwestern littoral, is a case in point. Moreover, among the largest sawmills, foreign ownership is as important as in pulp and paper. For many forest communities in BC, the locus of control is not only outside the province but outside the country.

As a summary indicator of foreign ownership, Pearse (1976) estimated conservatively that 34 percent of the leased timber supply was controlled by foreign firms and all externally controlled firms controlled just over half. Another indicator of the vulnerability of local control at that time is the fact that six of the nine largest firms, and sixteen of the twenty largest, forest-product firms were foreign or externally controlled.

Patterns and Behaviour of FDI Since the Mid-1970s

Generally, Fordism meant a massive influx of FDI into BC. By 1970, foreign firms were well established in the economy. The origins of FDI generally buttressed trade patterns. Thus, the dominant foreign firms were American, reflecting BC's primary continentalist trade orientation, while Japanese and European FDI pointed to important but secondary markets. Over the last twenty-five years, the entry, behaviour, and exit of FDI into BC have become more volatile and generalizations are harder to make. Nevertheless, the relationship between FDI and trade has remained intact as growing Pacific Rim trade has been associated with increasing Pacific Rim investment.

Entry Characteristics of Recent FDI

In terms of geographic origins, a significant trend in recent times is the increase in FDI from around the Pacific Rim, especially by Japanese firms but also by Fletcher Challenge of New Zealand (Table 5.5). The first Japanese investments occurred in the late 1960s, beginning in 1968 when Itochu, a *sogo shosha* (trading company), participated in CIPA Lumber, a joint venture with BC-owned Pacific Logging, the first sawmill to cut lumber to Japanese dimensions (see Table 5.1, note 1). Much larger investments were then made in 1969 by Mitsubishi, another *sogo shosha*, and Honshu Paper Manufacturing Company as part of a joint venture in Crestbrook Forest Industries, whose big new pulp mill started in 1969; by Jujo Paper and Sumitomo, which provided 25 percent of the equity in a relatively small pulp mill that opened in 1970 in Mackenzie; and Daishowa Paper and Marubeni (a trading company) as part of joint-venture, much larger-scale pulp mill that began operations in Quesnel in 1972. Virtually all important Japanese trading companies or *sogo shosha* established offices in Vancouver during the Fordist years, beginning with Mitsui in 1956, followed by Okura (1957), Nissho Iwai (1959), Marubeni (1960), Mitsubishi (1960), Sumitomo (1961), Itochu (1969), Kanematsu (1972), Nichimen Canada (1974), and Tomen (1974) (Edgington and Hayter 1997: 154).

After the completion of Cariboo Pulp and Paper's mill in Quesnel in 1972, further investments by Japanese companies in forest-product manufacturing

Table 5.5

Entry characteristics of principal new foreign subsidiaries in British Columbia's forest-product sector since the mid-1970s

Company	Locus of control office (parent)	Year	Method, activity, and location(s)
Mayo Forest Products	Japan (Mitsubishi)	1978	Acquired 40 percent interest in sawmill in Nanaimo; increased to 60 percent in 1980
Quesnel River Pulp	Japan (Daishowa Paper and Daishowa Ashitaka)	1981	50-50 joint venture with West Fraser Timber, 310,000-tonne-per-annum new pulp mill in Quesnel
Fletcher Challenge Canada	New Zealand (Fletcher Challenge)	1982	Acquisition of Crown Zellerbach Canada (see Figure 5.1)
Louisiana Pacific	US	1985	New OSB mill at Dawson Creek; new pulp mill at Taylor in 1991
Elk Wood Company Speciality Sawmill	Japan (Emachu)	1988	$2 million new samill, Maple Ridge
Howe Sound Pulp and Paper	Japan (Oji Paper)	1988	50-50 joint venture with Canfor to modernize Port Mellon pulp-and-paper plant for $600 million
Sanyo-Kokusaku Pulp Chip Mill	Japan (Sanyo-Kokusaku)	1989	50-50 joint venture with Primex, $1 million chip mill
Stone Container	US	1989	Acquisition of Celgar pulp mill
Canadian Chopstick Manufacturing	Japan (Mitsubishi Canada and Chugoku Pearl Co.)	1990	$9 million new plant in Fort Nelson
Anderson Sawmill	Japan (Mitsui and Mitsui Home)	1994	Acquisition of small custom-cut sawmill
Kimberly Clark	US	1995	Acquisition of Scott Paper
Bowaters	US	1998	Acquisition of the Gold River pulp mill owned by Avenor of Montreal

Note: Several central Canadian firms acquired significant interests in BC's forest economy in this period, including Noranda (Toronto) which acquired MB in 1982, Repap (Montreal) which acquired Cancel's pulp-and-paper mill in Prince Rupert, Avenor (Montreal) which acquired the Gold River mill, and Kruger (Montreal) which acquired Kimberly Clark's pulp mill plant (the former Scott Paper holdings). All these firms, apart from Kruger, have sold their interests in BC.

did not occur until 1978 (Table 5.5). Since then, a number of new companies involving Japanese FDI have been formed. Typically, Japanese FDI has continued to favour joint ventures with companies already operating in BC. The largest and most expensive of these joint ventures have involved Japanese forest-product companies in the construction of large pulp-and-paper mills whose activities are closely integrated with Japanese facilities (Edgington 1992). In the case of the Quesnel River Company, Daishowa Paper, a partner in the existing pulp mill in Quesnel, was joined by a related company, Daishowa Ashitaka, to partner West Fraser Timber. The Daishowa group now has substantial pulp capacity across Canada, reflecting the limitations of its supply base in Japan. Oji Paper, with larger timber holdings in Japan than Daishowa, chose BC for its major North American investment, providing $300 million for its 50 percent share of the financing of Howe Sound Pulp and Paper in 1988 (Edgington 1992: 27). All output is planned for shipment to Japan, with Oji having prime responsibility for newsprint marketing.

A number of other, smaller-scale investments have been made by various kinds of Japanese organizations in BC's forest economy, extending and cementing BC's timber trade with Japan (Chapter 6). In the case of the *sogo shosha*, two others finally followed Itochu's (1968) lead of establishing wood-processing capacity in BC, albeit with variations. Thus, in 1978 Mitsubishi acquired a 40 percent interest in Mayo Forest Products Ltd., a sawmill located in Nanaimo and 60 percent owned by Pacific Forest. Mitsubishi's rationale was to secure access to the supply of timber that its partner had on Vancouver Island (Stoffman 1990) and to cut this timber according to Japanese specifications (that is, for the traditional Japanese house). In 1994, Mitsui acquired the Anderson Sawmill, a small custom-cut sawmill. Partly to reduce dependence on the *sogo shosha* and to increase control and awareness of building-product supply, in 1988 Emachu, a large ("tier 1") wholesaling company, also acquired a small custom-cut sawmill, the Elk Wood Company in Maple Ridge; Sanyo-Kokusaku, a Japanese house builder, formed a joint venture with Primex, a small (and innovative) custom manufacturer, to build a wood-chip mill; while Mitsui Home, another Japanese home builder, established a log-sorting operation in Langley. In addition, Mitsubishi and the Chugoku Pearl Company (a Japanese catering firm) invested $9 million in 1989 in a chopstick factory in Fort Nelson (Canadian Chopstick Manufacturing Company).

A dozen or so of Japanese timber wholesalers and leading building-supply companies also followed the earlier lead of the *sogo shosha* in establishing offices in Vancouver (Edgington and Hayter 1997: 154). Increasing wood-fibre uncertainty and cost in BC and market downturns in Japan undermined further investments in forest-product manufacturing in BC by Japanese firms. A varied range of Japanese organizations, however, having gained important direct knowledge of the BC forest economy, are in a position to quickly make further investments. In contrast to foreign firms from elsewhere, Japanese MNCs have invested in small, value-added operations, including house-building firms (Reiffenstein 1999). Admittedly, outside of pulp-and-paper mills, Japanese investments have not been substantial, and rapid withdrawal from BC is also possible. Most Japanese MNCs have options outside BC, including (and increasingly) in Alberta.

Outside Japan, sources of FDI in BC's forest economy have been limited. However, Fletcher Challenge of New Zealand, by acquiring Crown Zellerbach Canada, became a major presence in BC on entry. It has also been unusually active since. Cocel's pulp mill in Celgar, which became part of Cancel in the early 1970s, was subsequently acquired by Chinese as well as by US and Venezuelan interests. The latter owners also contributed to some new continental links.

New Continental Links

US-based MNCs dominated FDI in the BC forest economy during Fordism. Since then, there have been few new corporate entrants from the US. The most significant is Louisiana Pacific (LP), which built an OSB mill at Dawson Creek in 1985 and a pulp mill in Taylor in 1991, the first BC mills to use aspen. Kimberly Clark, headquartered in Wisconsin, temporarily acquired Scott Paper's pulp-and-paper mill in New Westminster (1995-7), and Bowaters Inc., a spinoff of a UK company of the same name, became the last, short-lived owner of the Gold River pulp mill in 1998, as it closed the facility down in early 1999. Bowaters's divestment completed a bewildering series of corporate changes that began in 1981 when Tahsis and its original American owner, International Paper of New York, sold the operations to Canadian Pacific (Montreal) who then spun off the mill to Pacific Logging, which in turn sold it to Avenor of Quebec.

Central Canadian companies became unusually active in BC's forest economy in the 1980s and 1990s. Thus, the Bronfmans of Toronto, who own a diverse range of businesses, acquired a 21 percent interest in Scott Paper of Philadelphia in the late 1980s, whereas MB was controlled by Noranda of Toronto from 1981 to 1993. In turn, Noranda was acquired by Brascade, a company owned by the Bronfman family. More recently, several central Canada-based forest companies have entered the BC industry. Repap (New Brunswick) acquired Cancel's pulp mill in Prince Rupert, adding to the saga of ownership changes of that complex, MB sold its fine-paper plant to E.B. Eddy (Toronto) in 1993, Kruger Inc. (Montreal) bought the New Westminster pulp-and-paper mill of Kimberly Clark in 1997, and Donohue (Montreal) acquired 40 percent control of Finlay Forest Products' pulp, paper, and sawmilling operations in Mackenzie in 1994, acquiring full control in early 1999, and Avenor's (short-lived) acquisition of the Gold River pulp-and-paper mill has been mentioned. In early 1999, Ontario-based Tembec announced plans to acquire the Japanese-controlled Crestbrook Forest

Industries (whose main asset is the pulp mill at Skookumchuck) for $270 million. The mill underwent a $291-million modernization program in 1991, and it is reportedly the most efficient mill in BC. Investments and over-expansion in Alberta, however, created a significant debt load for the main parent companies (Mitsubishi and Oji Paper).

The Restructuring of Fletcher Challenge Canada

The behaviour of foreign subsidiaries within BC has continued to be highly varied, as might be anticipated in a period of volatility. As a group, Japanese firms have been relatively stable and a few have expanded. CIPA, for example, has become more vertically integrated by acquiring the higher-value-added operations of Delta Plywood Mill in 1993 on Annacis Island in Vancouver (Hamilton 1994). Even in the late 1990s, which found Japan and BC in recession, Japanese firms have not hinted that they may downsize in BC. Firms such as Bowaters and Repap, on the other hand, entered and exited BC after a relatively short period. Indeed, Repap invested virtually nothing in modernization in BC and has been accused of profit skimming before leaving. The most active, high-profile foreign subsidiary of the contemporary period of restructuring is Fletcher Challenge. Its nine-month-long strike/lockout in 1997-8 kept it in the public eye.

One of Fletcher Challenge's first international investments was the purchase of Crown Zellerbach Canada in 1982 (LeHeron 1990). In 1987, Fletcher Challenge then acquired a controlling interest in BCFP to become one of the biggest companies in BC, employing 12,000 people. BCFP was also a major partner in Western Forest Products, which controlled the former assets of Rayonier, and had pulp-and-paper facilities in Quebec and the US. Within BC, Fletcher Challenge had become a major company in the coastal region, with an integrated forest-product complex at Elk Falls, Campbell River, a major pulp-and-paper plant at Crofton (as well as a nearby sawmill at Youbou), and various large sawmills and plywood mills in the Vancouver region. In the interior, Fletcher acquired sawmills and other facilities in the Okanagan (from Crown) and an integrated forest-product complex in Mackenzie (from BCFP).

Yet, soon after its acquisition of BCFP, an acquisition that implied an ambition to be a highly integrated producer of all forest-product commodities, Fletcher Challenge radically changed its strategy. This strategic shift complemented, in some ways foreshadowed, the shift at MB (Chapter 4) in that Fletcher chose to be a more specialized company and one that has especially strong connections around the Pacific Rim. Both firms have also given priority to developing a more flexible labour force (Chapter 8). Fletcher's strategy differs from MB's in that while the latter focused on building products, Fletcher chose the paper segment as its priority. Its sea change in strategy occurred shortly after the introduction of the new stumpage formula and increase in stumpage levels in 1988.

Fletcher Challenge Canada (as it then became) began its restructuring ("disintegration") towards a smaller, more specialized company when the former BCFP cedar mill at Hammond, and its timber supply, was sold. Then several sawmills were auctioned and a number of operations closed. Various packaging plants in the Vancouver area and in the Okanagan that had belonged to Crown were sold, followed by more sawmills and timber rights. In addition, Fletcher sold its interests in Western Forest Products and Finlay Forest Products, and its Quebec operations. According to Wilson (1997: A13), Fletcher, from 1990 to 1994, received $156.2 million from the sale of fixed assets.

More recently, the company placed its remaining sawmills and timber investments (and logging activities) in a company called Timberwest and sold 49 percent of the shares. Fletcher then sold the remaining shares to TAL Acquisition Ltd., an investment group with no particular background in the forest industry. This radical downsizing left Fletcher Challenge Canada with the pulp-and-paper complexes at Elk Falls, Mackenzie, and Crofton in BC, and its pulp-and-paper operations in Blandin, Minnesota, were sold in 1998 for an after-tax gain of $390 million, a sum that helped offset the losses created by the lengthy strike. By 1997, Fletcher reduced its employment in BC to 3,000 and sought to become a company of just 1,000 employees who run paper machines. Thus, Fletcher cut 100 jobs from its head office and divided its three BC mills into two companies: Pulpco and Paperco. The mills remain integrated, but they have separate management and the pulp business is for sale.

According to a union leader, Fletcher's strategy represents "a bizarre corporate reorganization [and it is] difficult to see how FCC's sorry history could be any guide to a better future for the BC forest industry" (Wilson 1997: A13). For the union, Fletcher's restructuring is controversial, because it involved substantial employment downsizing, the breakup of affiliated operations in 1997, comprehensive demands for employment flexibility, and a higher priority for maximizing shareholder value. These demands provoked the nine-month strike/lockout. This battle was clearly won by Fletcher Challenge; employment is now lower and flexibility is the principle governing labour relations. Management claims it will save $30 million every year, given the new labour agreement (Hamilton 1998h: D1, D2).

From management's point of view, restructuring towards a leaner, differentiated manufacturer of a more specialized product line is not bizarre, but rather makes sense in contemporary global and local realities. Thus, two main forces pushed its restructuring to enhance shareholder value. First, the changes in stumpage formula and levels in 1988, reinforced by the onset of the high-stumpage regime in the 1990s, dramatically increased the cost of logging. Fletcher's response has been to externalize or contract out this activity. It also expects its suppliers to be eco-certified. Second, the increasingly complicated business dynamics of the global forest industry encouraged Fletcher, like MB, to become a more focused manufacturer. Fletcher is also seeking to become a flexible mass producer of a more differentiated range of paper products as it reduces its reliance on newsprint.

Fletcher's ten-year restructuring is considered successful by management: shareholder value has increased, and the company is cash rich, with $750 million in cash reserves (Hamilton 1998h). Moreover, the company acquired a newsprint producer in the Philippines in 1998, and announced in 1999 a plan to make Vancouver (and Fletcher Challenge Canada) the control centre for its paper-related activities (Mertyl 1999: D1). This deal, if struck, would provide the parent in New Zealand with cash and the Vancouver-based subsidiary with the responsibility of organizing Fletcher's global (that is, Pacific) paper activities. Within BC, a key long-term issue facing Fletcher Challenge Canada, similar to that facing MB, is whether its restructuring will result in the greater potential for innovation that is essential to remaining competitive, and how much of its cash will be invested locally.

Exit of Foreign Firms

FDI is not irreversible, and foreign (and central Canadian) firms have exited the BC forest economy. Even during the Fordist boom in the BC forest economy, when there was a tremendous influx of FDI, not all anticipated FDI occurred and some initiatives failed. Thus, a giant forest-product complex planned for the Houston area of northwestern BC by Bowaters of the UK and Consolidated Bathurst of Montreal failed after major losses (reportedly in excess of $60 million in 1970 dollars) had been incurred in the first phase of the project, a highly automated sawmill built in 1968. At Prince Rupert, the joint venture between Cocel and Svenska Cellulose (see Table 5.2) never reached profitability, and its problems led both firms to pull out of the project and BC in the early 1970s. For Cocel, this withdrawal meant the sale of sawmills and its Celgar pulp mill. Indeed, Cocel's operations were saved only by government intervention and the creation of Canadian Cellulose (Cancel) in 1973.

These exits were relatively few and were occasioned by failures of specific projects that, it could be argued, had not been adequately planned. With increasing volatility in BC's forest economy, however, exits of foreign (and externally controlled) firms appear to have become frequent. Recent exits are also voluntary and strategic, in a way they were not in the past, and they have consequently taken on more alarming connotations (Table 5.6). Thus, in 1979, as part of a global restructuring, Reed (UK) sold its entire Canadian holdings, including equity shares in two pulp-mill joint ventures to Canfor, and Triangle Pacific (US) sold its sawmill to Slocan, a fast-growing BC-controlled firm that began operations in 1978. Bigger changes were stimulated by the recession of the early 1980s, and included two major integrated US-controlled subsidiaries, specifically Rayonier Canada, owned by ITT since 1968, and Crown Zellerbach Canada. While Crown sold out to Fletcher

Table 5.6

Exit of foreign firms from British Columbia's forest-product sector since the mid-1970s

Company	Origins of parent	Year	Method, activity and location(s)
Reed	UK	1979/85	Sale of interests in two joint-venture pulp mills in Prince George to Canfor
Rayonier	US	1980/1	Acquisition by local consortium
International Paper	US	1981	Sale of Gold River pulp mill to Pacific Logging
Crown Zellerbach Canada	US	1982	Acquisition by Fletcher Challenge of New Zealand
Feldmühle	Germany	1985	Sale of interests in Prince George pulp-and-paper mill to Canfor
Mead Corporation	US	1987	Sale of equity interest in BCFP to Fletcher Challenge of New Zealand.
Balfour Forest Products	UK	1989	Acquired by Canfor
Eurocan	Finland	1990/5	Acquisition by West Fraser (US), 50 percent in 1981 and 50 percent in 1993
Scott Paper	US	1995	Acquisition by Kimberly Clark (US)
Kimberly Clark	US	1997	Acquisition by Kruger Inc. (Montreal)
Bowaters	US	1999	Closure of Gold River pulp mill
Crestbrook Forest Industries	Japan	1999	Proposed acquisition by Tembec of Ontario
Stone Container	US	1999	Withdrawal from Celgar pulp mill

Challenge in 1982, a year earlier Rayonier had sold its operations in BC, mainly concentrated in the coastal region (as well as other interests in Quebec), to a consortium of three firms, two of which were Canadian owned. In addition, the Bronfmans of Toronto acquired 21 percent of Scott Paper of Philadelphia, although this acquisition did not represent a controlling interest and Scott's BC-based pulp-and-paper operation continued to operate with little if any change. In the early 1980s, International Paper also sold its Gold River pulp mill, as well as its much more substantial facilities in eastern Canada, to Canadian Pacific.

According to Marchak (1983), the withdrawal of American firms at this time reflected the declining competitive position of the BC forest-product industry. In this view, American firms consolidated their holdings in the US because of higher rates of return on investment while selling their subsidiaries to firms with less flexibility and perhaps prescience. While this view has merit, it needs qualification. Thus, the American subsidiaries that pulled out of BC (and Reed, for that matter) also sold their operations throughout Canada, if they had such holdings. In addition, corporate restructuring is not confined to American firms nor to Canada (Hayter 1985). In fact, an important factor underlying the sale of subsidiaries in BC was to generate cash flow at a time of recession.

Indeed, the forest-product sector is highly capital intensive and the 1980s recession created enormous profitability and cash-flow problems for forest-product firms. Many large forest-product firms needed large cash injections to prevent losses and to facilitate restructuring; the sale of foreign subsidiaries served this purpose. Much is suggested by the reasoning of the president of Crown Zellerbach for the sale of its Canadian operations: "Crown Zellerbach has had a long and satisfactory investment ... Now we believe we can best increase shareholder value in the company by monetizing and redeploying this investment into improving our US core facilities and product lines" (Forest Industries 1982:11).

MB pursued a similar strategy, except that many of its divestments were outside BC (Chapter 4). In Crown's case, the parent claimed that its Canadian subsidiary had always been profitable, and its sale to Fletcher Challenge realized a gain of $600 million. It is also questionable to assume that Crown, in comparison to Fletcher Challenge, was more rational, profit oriented, and less mistake-prone. The same point can be made in relation to the other divestment-acquisitions (Hayter 1985).

The BC Forest Economy as a Pacific Rim Branch Plant

Since the mid-1970s, if the entry, post-entry behaviour, and exit of foreign (and externally) controlled investments have been bewildering in detail, patterns can be discerned. Two clear trends need to be underlined. First, despite the (mistaken) "fears" expressed over divestments by foreign, especially American, firms in the 1980s, the level of foreign (and external) control remains significant (and has likely recently risen). Second, the BC forest economy has increasingly come under the control of MNCs based elsewhere around the Pacific Rim.

In the 1990s, the level of foreign control of the BC forest economy remains important. According to Schwindt and Heaps (1996: 81), in 1993 firms with 50 percent or more foreign ownership accounted for 27.8 percent of the AAC, 31.9 percent of lumber, 55.9 percent of market pulp, and 54.1 percent of paper production. These figures are not that different from those calculated by Hayter (1981; 1985) for the late 1970s according to the same criterion (Table 5.4). Foreign firms have continued to enter BC, while the acquisition of MB by Weyerhaeuser will almost certainly increase foreign control of BC's forest economy. Admittedly, ownership share need not correlate exactly with control. However, the 50 percent ownership criterion provides a commonly used comparative yardstick to indicate levels of control and, on balance, this criterion probably understates effective control.

If levels of foreign control have remained high at a more or less constant level, the geography of this control has evolved from the Fordist period to the present time. During Fordism, US MNCs were dominant while European MNCs were also present. During the present period of flexibility, European MNCs (based in the UK, Germany, Sweden, and Finland) have all but disappeared. Furthermore, while US MNCs are also still a significant force, the head office locations of these firms have shifted from the centres of the old "manufacturing belt" to Pacific Rim States. Thus, International Paper and ITT (which owned Rayonier), both of New York, Scott Paper of Philadelphia, and others, such as Kimberly Clark of Neenah, Wisconsin and Stone Container of Chicago have left BC. On the other hand, Louisiana Pacific and Pope and Talbot of Portland, Oregon, and Weyerhaeuser of Tacoma, Washington are now more important, and other firms based in the Pacific Northwest, such as West Fraser, Evans Forest Products, and Boise Cascade, are also present. In addition, from across the Pacific, Japanese MNCs have become a significant presence in BC's forest economy while the California-based CZC was acquired by Fletcher Challenge of New Zealand. As of 1999, two Quebec-based MNCs have important operations in BC, but the role of Noranda (Ontario), Repap (New Brunswick), and Avenor (Quebec) proved short-lived (Table 5.6).

Thus, industrial forestry in BC forests is in danger of becoming a narrowly specialized hinterland servicing Pacific Rim markets and controlled by organizations based elsewhere in the Pacific Rim, notably the Pacific Northwest and Japan. It is hard to think of a BC-controlled firm, especially if MB is acquired by Weyerhaeuser, that has significant investments outside the province. Decision makers in BC's forest economy have very little impact elsewhere, and that which was evident is now in decline. It is also interesting to note that in the present age of flexibility, the locus of foreign control of BC's forest economy has become much more regionally defined around the Pacific Rim than it was during Fordism when control lines were more globally diffuse. For the BC forest economy, geography (qua proximity) seems to have become more, not less, important in this regard.

High levels of foreign control have been important to the BC forest economy for a long period of time, since at least the early 1950s, and these levels are much higher than experienced by most forest economies elsewhere in the industrialized world. The BC forest economy, therefore, is a good case by which to explore the rival claims regarding the role of FDI in development, although such explorations have not been thought worth much of an effort (for exceptions see Hayter 1981; 1982b; 1985).

Developmental Implications of FDI

It is not straightforward to empirically assess the contributions of FDI to the local development process in the BC forest economy (or elsewhere for that matter). In effect, any such evaluation faces, at least implicitly, a counter-factual situation defined by the economic structures that "would have been" created in the absence of FDI or at least high levels of FDI (Hayter 1982b). This counterfactual structure cannot be defined with certainty. Moreover, there is no a priori reason why FDI should not impose both benefits and costs on a local economy; that is, the benefits and costs are not mutually exclusive. Yet, there is no acknowledged way of balancing these impacts. In practice, conventional economic wisdom in BC has assumed the benefits of FDI, and expressed little or no concern for its costs. As a result, the benefits may have been overrated and the costs underestimated.

The Foreign Firm as Catalyst

As a catalyst to local development, FDI potentially provides a bundle of benefits. In particular, foreign firms can provide locally scarce financial and managerial inputs, facilitate access to international markets, and lessen the uncertainties of investments in new facilities. In addition, foreign firms may add leading-edge know-how, in terms of management, marketing, and supply systems, as well as new product and process technology, to local economies.

These potential contributions have not been subject to systematic analysis in the BC forest economy. It is highly likely, however, that FDI (and externally controlled investments) helped fuel extremely rapid rates of growth in BC's forest sector, especially in the pulp-and-paper industry during the Fordist period. Thus, the explosion of kraft pulp mills that occurred in the 1950s, 1960s, and early 1970s featured joint ventures involving extensive foreign participation (see Table 5.2). Typically, the foreign firms provided their share of financing and marketing responsibilities according to their equity share. And typically, markets were their affiliated plants elsewhere, and such guaranteed exports added to the stability of the expansion process. Frequently, foreign firms provided vital production and engineering know-how, including managers and supervisors, to ensure the successful building and operation of pulp mills. The importance of relevant production experience to the success of joint-venture pulp mills is underlined by the deliberate searches undertaken by local firms for a suitable foreign partner. In Tahsis, for example, the local firm commissioned a study by the Stanford Research Institute to undertake a market analysis and to find a list of partners; this study took eighteen months to complete (Hayter 1973: 221).

Moreover, the activities of foreign firms encouraged the growth mentality evident in the sector by their competition for resources and willingness to invest. The presence of so many firms willing to compete for BC's timber provided the provincial government with bargaining power. In practice, the government chose to play off rival firms, not to promote value-added activities or increase stumpage, but to stimulate as fast and large-scale growth as possible. The ownership of firms was simply not a concern. In 1965, for example, MB rejected a timber award (TFL 41) because it was deemed insufficient for major industrial development (Hayter 1973: 198). The same TFL, however, was accepted by Eurocan which outbid Crown Zellerbach Canada because of its willingness to proceed with a pulp-and-paper mill at an earlier date. Crown opted out of plans for this region because, as a senior executive of the firm claimed, the government "wanted too fast a development ... timber was poor ... and not that many good sites ... Crown's idea was start small and build a sawmill and maybe in the long run build a pulp mill" (Hayter 1973: 201).

Encouraged by the presence of so many foreign firms wishing to participate in BC's forest economy in the Fordist period, the government used its considerable powers not only to aggressively promote industrialization of the forests, but to help limit the options to large-scale commodity-based developments. From the provincial government's perspective, the impressive expansions of forest products in the 1950s and 1960s at least justified the massive infrastructure investments that were made in roads, railways, and power facilities throughout the province.

Since the late 1970s, foreign firms, especially Japanese firms, have continued to show interest in developing BC's forest resources. Some of the same kinds of benefits are evident. Thus, Oji Paper and Daishowa Paper have contributed financially to major expansions of pulp capacity and have helped market the output. Although the rate of growth of kraft pulp expansions declined after the early 1970s, absolute increases in capacity have been substantial (see Figure 3.1); new FDI has been a factor underlying this increase. The arrival of Fletcher Challenge in the 1980s also helped shift pulp-and-paper output to Pacific Rim markets where the firm was already a presence.

The extent to which foreign firms have contributed to state-of-the-art expertise, in the form of novel technology, silvicultural practices, labour relations, and production know-how, for the BC forest economy is problematical. Anecdotal evidence suggests positive impacts have occurred but that they are modest. Thus, in the pulp mills built at Crestbrook in 1969, the Japanese partners reputedly introduced new machinery designs (Hayter 1973). More recently, Edgington (1992: 28) notes that Canfor chose Oji Paper for its partner in the Howe Sound pulp-and-paper expansion, in part because of its exceptional technology. In wood processing, CIPA renovated its plant by installing Japanese technology that improved recovery and labour efficiency in turning second-growth Douglas-fir logs into smooth veneer for plywood mills in Japan and the US. More generally, the *sogo shosha* and Japanese wholesalers and building companies have stimulated greater awareness of quality control as well as information about Japanese-market requirements to an increasing number of BC lumber suppliers, small and large. In this regard, Japanese practice is very much hands on, and mills that fail to comply with requirements are not offered business.

In northeastern BC, Louisiana Pacific has pioneered the use of aspen in OSB and pulping, and its pulp mill at Taylor has a novel closed-loop environmental system. But much technology is designed outside of BC, and much of the machinery is imported, especially in the pulp-and-paper industry. For those who argue that FDI promotes local capability in technological know-how, the BC forest economy is a disappointing case.

Limitations to the Catalytic Effect

The contributions of FDI to forest-product expansions in BC should not be overstated. Many foreign firms have entered BC's forest economy by acquiring existing facilities. In such cases, there is no reason to assume injections of more capital or various forms of know-how. The social merits associated with entry by acquisition are particularly questionable when FDI occurred not to obtain inputs for export but to penetrate domestic markets – as was the case in the tissue and plywood industries. To the contrary, acquired firms have provided foreign firms with valuable resources and know-how.

Similarly, the argument that foreign firms provide access to captive markets and therefore enhance stability needs qualification. Throughout the Fordist period, the forest industries experienced cycles of excess capacity as rival firms often chose to expand in lockstep with one another, investment go-ahead decisions occurring during business upturns, and new capacity coming on stream two years later helping to put pressure on prices. These cycles were inevitably reinforced by FDI that had to invest to gain timber resources, especially as the exports of foreign firms were often only partially to affiliated markets. Moreover, the high levels of FDI that characterized the sector by 1970 have not apparently offset the extremely volatile market fluctuations of the last two decades.

Foreign firms have not been immune from planning errors either. In northwestern BC, for example, Bowaters and Consolidated Bathurst experienced expensive failure, and Eurocan's mill at Kitimat suffered numerous expensive problems. Within two years of its 1970 startup, for example, the firm had to introduce a new water-supply system at a cost of $25 million, because water levels had been more variable than anticipated. Even though Eurocan survived, an internal "post-mortem" indicated a lack of understanding of operating in a foreign environment, including failure to anticipate the lack of local experience in handling small interior wood, overreliance on local consulting opinion, and a gap between design and application of technology. Indeed, these problems have had a long-lasting effect (ultimately contributing to Eurocan's decision to exit BC).

Moreover, FDI in the BC forest economy has done little to improve innovation in the sector. Generally, foreign firms in BC have favoured the use of proven equipment to manufacture standard commodities. Although MNCs are often seen as technologically sophisticated, foreign firms in BC have not contributed to a spirit of technological dynamism. Indeed, as recently as 1998, a major report (Ernst and Young 1998) concluded that the industry was still locked in to a traditional commodity, a criticism that has been made for some time (Marchak 1983; Woodbridge, Reid 1984; Hayter 1987). Apparently, FDI has not helped the industry restructure towards flexible mass production and higher value. In fact, the critics argue that this failure addresses the most significant structural weakness associated with high levels of FDI.

Foreign Firms as Truncated Subsidiaries

The term "truncation" was popularized by the Gray Report (Canada 1972), itself a result of broadly based concern for the political, social, and economic impacts arising from high levels of foreign ownership in Canada. In that report, foreign subsidiaries that rely on their parent companies for various services and functions, and whose autonomy is circumscribed by head-office dictates, were declared to be truncated firms (Canada 1972: 405). The report recognized that the type and nature of branch-plant and parent company connections vary. Given these variations, the basic idea of truncation is that because branch plants rely on their parents for various functions and services, they do not, indeed cannot, invest in these same activities in host economies such as BC. If levels of foreign ownership are high, truncation exercises pervasive effects on industry structures. Generally, truncation is defined as a situation in which FDI preempts or replaces economically viable indigenous development (Hayter 1982b: 277; see also Britton and Gilmour 1978).

In the BC forest economy, foreign firms illustrate the idea of truncation by limitations on their decision-making scope and mandate. Subsidiary status almost invariably implies that investment decisions are subject to approval by the parent company. Only rarely, and in special circumstances, as was the case in BCFP's acquisition of a Minnesota paper mill, do subsidiaries invest outside their host economy. Many subsidiaries have no mandate or decision capability to search out growth opportunities. Most fifty-fifty joint ventures fit into this category; for example, they are not growth-oriented companies in the sense of enjoying independent authority.

Indeed, many subsidiaries, including most joint ventures, have limited marketing connections, especially subsidiaries that largely serve affiliated markets in international corporate systems. Such links are not conducive to the accumulation of marketing know-how and typically have precluded thoughts about market diversification. Pulp mill branch plants, for example, may well benefit for many years from assured access to parent company markets, but should such links be reduced or terminated, alternative markets and connections are not readily obvious. Even if these links are maintained, subsidiary status has reduced the motivation within BC to produce higher-value goods, especially where the motivation of FDI is to secure supplies of commodities for value to be added elsewhere. Moreover, affiliated trade links (of all kinds) provide the parent company with some ability to affect transfer prices.

Ultimately, foreign subsidiaries in BC's forest economy all depend on their parent companies for various high-level corporate functions, notably head-office decision making and R&D. Indeed, reliance on parent company R&D is a significant cost of high levels of foreign ownership in the BC forest economy (see also Chapter 11).

Technological Truncation

Virtually every MNC that has operated in BC's forest economy has invested in R&D centres. Overwhelmingly, however, foreign forest-product companies operating in BC have centralized their R&D investments in their home countries. US-based MNCs that have invested in BC, for example, have concentrated their R&D centres in the US, including several in Washington State (Table 5.7). Rayonier and Scott Paper actually closed laboratories in BC after they, respectively, acquired Alaska Pine and Cellulose Company and New Westminster Paper Company in the early 1950s (Hayter 1980: 61). In only one case has an American subsidiary in BC invested in substantial R&D facilities for any length of time over the last fifty years. Thus, Cocel, which provided the *only* internal source of dissolving pulp to its parent company, employed about ninety people in the mid-1960s at its then new Vancouver laboratory where R&D on pulping processes were closely integrated with the parent's paper-converting R&D in New York. Following spiralling corporate losses, however, this facility was reduced considerably, and since 1972, it has been the basis of a small, private company (see Hayter 1980: 30). No Japanese or European MNC (or Fletcher Challenge or central Canada-based firm) has ever created an R&D laboratory of any note in BC.

The general policy, then, is for subsidiaries to rely on parent companies for R&D. From the corporate perspective, such a policy is sensible. Concentration of R&D activities allows for economies of scale and scope, brainstorming, and insurance that at least some innovations will be developed. Geographic concentration eliminates duplication of costly facilities, providing a form of integration and control for the entire corporation, while subsidiaries that function as commodity suppliers have little direct need for R&D. But such corporate preferences do not mean that corporate R&D is not a viable activity in BC. The existence of several corporate R&D centres, most notably the highly successful laboratory operated by MB for over thirty years, empirically demonstrates that BC has the location conditions

Table 5.7

The R&D locations of US forest-product firms with subsidiaries in British Columbia since the 1970s

Company	Main R&D centres by state	Size, c. mid-1980s (Number of professionals)
Pope and Talbot	Oregon	50
Weyerhaeuser	Washington	500
West Fraser	none	
International Paper	New York	130
Crown Zellerbach	Washington	120
Rayonier	Washington	100
Scott Paper	Pennsylvania	300
Kimberly Clark	Wisconsin	300
Mead Corporation	Ohio	200

to support corporate R&D in the forest sector. Given the size of that sector, and the implications for local demand for technology, BC offers a sophisticated educational infrastructure, a supply of skilled labour and professionals, a wide range of cultural and environmental amenities, sophisticated social and economic infrastructure, and good international communications. There are also long-standing government-controlled forest-sector R&D centres, and more recently, some cooperative R&D centres that contribute to the pool of technological expertise in BC's forest economy and that could potentially interact with corporate R&D (Chapter 11).

Given the size, rate of growth, and relative importance of the forest-product sector in the BC economy, and the evident success of existing laboratories, there are legitimate reasons for believing more investments in corporate R&D could have been made in BC under different institutional circumstances, most notably less foreign equity ownership. Certainly, per capita corporate R&D spending is less in BC (and Canada) than in other forest-product regions (Chapter 11). The lack of corporate R&D cannot simply be reduced to the effects of foreign ownership; the commodity structure of the industry, ingrained attitudes, and government policy over the decades are also relevant factors. But the effects of foreign ownership cannot be dismissed either. The few major corporate R&D laboratories that have existed in BC – Alaska Pine and New Westminster Paper's activities in the late 1940s and early 1950s, and more recently, MB's and Canfor's R&D centres – were all owned by locally controlled companies. Cocel's R&D activities were unusual, but the company's American parent was not itself engaged in the forest sector.

The fact that foreign companies, apart from the special case of Cocel, all conduct their R&D in their home economies, constitutes a powerful institutional deterrent to similar R&D investments in BC. In effect, foreign ownership institutionalizes imported technology for local activity. Because levels of foreign ownership are so high in BC, possibilities for BC-based companies to grow to a size that would justify R&D investments are accordingly reduced. The numerous joint ventures found throughout the BC forest-product economy that involve foreign partners in both minority and majority positions, not only lack the mandate to undertake risky investments in R&D, but also typically enjoy access to a parent-company laboratory. In the terminology of the Gray Report (Canada 1972: 405), subsidiaries are technologically truncated. Moreover, government R&D incentives have failed to change this behaviour.

The Costs of Technological Truncation

The most obvious social costs of technological dependency in the forest-product sector relate to lost job opportunities for science graduates in industry and net losses on balance of payments for R&D services. To cite an individual case, Scott Paper of Canada, a medium-sized subsidiary, paid its parent $850,000 for R&D services in 1980. While the total loss for BC as a whole cannot be estimated precisely, there is little doubt that BC incurs trade deficits as a result of payments for R&D services. Direct job losses from reduced R&D are also difficult to estimate. But MB's R&D centre employed around 100 professionals for over thirty years, and Weyerhaeuser's Tacoma complex at one time employed over 500 professionals. Employment of over 1,000 professionals in corporate R&D in BC would not have been an unreasonable expectation.

In more subtle, indirect ways, technological truncation may serve to weaken and dislocate technological capability throughout the sector. The R&D departments of forest-product firms, for example, are frequently required to assess possible equipment purchases on the company's behalf. For BC, this role means that subsidiary companies act to channel decision-making responsibility out of the country, placing BC equipment manufacturers at a disadvantage. This pattern of behaviour, in association with such other factors as foreign ownership and a lack of tariff protection, also militates against R&D investments by equipment manufacturers and limits their export abilities (see Hayter 1980: 32-52). Further, a lack of corporate R&D weakens the development of technological liaisons with equipment manufacturers and with government laboratories and universities.

Similarly, limited industrial R&D functions reduce the demand for small, specialized R&D suppliers that provide complementary services to the major laboratories (Hayter 1982b: 30-1) and for intersectoral R&D groupings involving forest products. An example of the latter is provided by R&D focusing on gypsum, asbestos, cement, paper, wood, and other building products, which, despite BC's raw material and manufacturing base, has also been largely concentrated in the United States by firms such as Johns Manville. Admittedly, the evidence regarding the nature of interestablishment technological liaisons involving forest products in both the public and the private sectors is only fragmentary. Nevertheless, the indications (and logic of the situation) are that underrepresentation in industrial R&D implies losses in terms of spinoffs in related activities.

Finally, corporate R&D is vital to enabling the BC forest economy to diversify its commodity bias and move towards higher-value production. MB's creation of Parallam and Pacifica attest to this potential (Chapter 4). But foreign firms do not conduct R&D, and their diversification must conform to parent company priorities, regardless of BC's aspirations.

Reflections on Weyerhaeuser's Acquisition of MB

The takeover of MB by the Weyerhaeuser Company illuminates the strange juxtaposition of public policy disinterest and profound implications of FDI for BC's forest economy, especially in regards to innovation. Neither the NDP provincial government nor the right-wing opposition voiced any concerns about this loss of control over BC's largest firm. The only public criticisms were by environmentalists who feared that this particular loss of sovereignty would undermine the recent deals brokered with MB regarding environmentally sensitive forest practices (Chapter 10). Yet, Weyerhaeuser's takeover offered few benefits to BC while imposing profound costs. Some reflection on these costs and benefits is instructive.

While Weyerhaeuser has gained greatly in size and flexibility associated with control over extensive forests (and productive capacity) in the Pacific Northwest on either side of the US-Canadian border, the benefits to BC are minuscule. Local shareholders may benefit from increases in share value but it seems that the most important shareholders are based elsewhere. Other tangible benefits are hard to discern. Weyerhaeuser is not offering production know-how, global marketing connections, technological capability, or forest management expertise. MB was already well proven in these respects. Rather, MB's accumulated human and capital resources, centred in BC, will now contribute to the competitive strengths of Weyerhaeuser.

On the other hand, the Weyerhaeuser deal does mean more foreign ownership, less competition, restrictions on innovation potentials, and leakages of revenues from BC's already troubled forest economy. After the acquisition, there will be sustained costs in the form of regular payments from MB's operations to Weyerhaeuser. Undoubtedly, these payments will be substantial and will help compensate Weyerhaeuser for the cost of MB's acquisition. This transfer of funds, in the form of profit payments, payments for head-office services, and payments for R&D services, will be internal transactions at the discretion of Weyerhaeuser. Meanwhile, the former foreign operations of MB will now be controlled by Weyerhaeuser and payments formerly made to MB (and BC) re-routed accordingly.

The acquisition will significantly impair BC's ability to generate innovative value-added activities. MB was BC's leading corporate innovator, especially with respect to product innovations, and it pioneered the vitally important high-value Japanese market. Now, MB's surviving facilities operate as branch plants as directed from Weyerhaeuser's head office in Tacoma. MB's pioneering Japanese subsidiary, a vital source of learning about Japanese markets to the benefit of mills in BC, will almost certainly be closed down and absorbed within Weyerhaeuser's Japanese office.

MB had been competing with Weyerhaeuser for Japanese (and other) high-value markets. Weyerhaeuser will now control this competition. The most likely scenario is that Weyerhaeuser's Washington State mills will pioneer new markets, while its newly acquired mills in BC, like its existing mills in BC, will focus on standard, large volume commodities. And there will be nothing that the BC mills will be able to do to autonomously change their product mix.

Special mention must be made of MB's R&D program, which was unexpectedly closed in 1997 (Chapter 4). Whether or not this closure was motivated by the desire to make MB an attractive acquisition target to firms, once Weyerhaeuser has acquired MB there is no chance for a re-commitment to MB's R&D. Rather, profits from MB will now support Weyerhaeuser's R&D complex in Washington State.

Weyerhaeuser's acquisition of MB will also reinforce recent pleas for "privatization" of BC's forests, led by MB, that will likely imply the sale of logs to the "highest bidder," even if this means log exports. Needless to say, log exports can scarcely help the development of technological expertise in BC, or employment in wood processing.

There is also no evidence that Weyerhaeuser's acquisition will create greater efficiency based on the exploitation of economies of scale, as is often intimated. Three points may be made in this regard. First the recent restructuring of MB was acclaimed for making MB a smaller (if still large), more efficient enterprise. Second, available evidence suggests that MB was plenty big enough to fully realize economies of scale, including funding an R&D department. Third, any benefits of increased size and power realized by the acquisition will be absorbed by the parent company. The argument that being part of a huge MNC will create efficiencies is meaningless from BC's point of view and a totally inadequate justification for permitting even greater foreign control over BC's forests.

The president of Weyerhaeuser Canada was recently quoted as saying that one of the reasons that Canadian firms do not invest in R&D is that they are not big enough. They are often not big enough because firms such as Weyerhaeuser acquire them and refuse to do R&D in Canada (Chapter 11).

Environmental opposition to the takeover on the basis of loss of sovereignty is to be applauded. Admittedly, their criticisms are doubly ironic because, first, they have deliberately targeted MB with blockades and boycotts, and, second, in cooperation with international environmentalist groups based in New York, London, etc. – that is, the same locations where the big MNCs are often centred – they have frequently unhinged local democratic processes (Chapter 10). Indeed, environmental opposition, along with hostile provincial government policies, helped bring MB to its knees, leading to its most recent restructuring (and the present acquisition). Nevertheless, environmental fears that Weyerhaeuser may limit provincial discretion in forest policy may well prove valid (Gibbon 1999b).

There was no reason to oppose the acquisition on the basis of xenophobia. FDI has long been important in the BC forest economy. But why should BC have allowed another major foreign-based MNC to control so much of its forests simply by acquisition? What does BC get out of this deal? The record indicates that high levels of foreign ownership, and an organizational structure biased towards large firms, have not helped the BC forest economy become more innovative. In fact, foreign ownership has helped arrest innovation because it institutionalizes a division of labour in which BC manufactures commodities for value to be added elsewhere.

The fact that Canada (and BC) continues to welcome American control in the forest sector in a period when the US has adopted protectionist policies towards Canadian, especially BC, wood exports only underlines the confused nature of forest policy making in Canada.

Industrial forest policy in BC must be developed to encourage innovation, because it is only through innovation that the economic and environmental goals of industrial forestry can be fully realized (Chapter 11). BC's forest economy needs big corporations, including foreign ones. But we already have too much of both, and not enough community control and BC ownership. Judged from an innovation-based forest policy for BC, Weyerhaeuser's acquisition of MB should have been rejected as costs to BC exceed benefits and, most especially, it will reduce the potential of BC's forest economy to be innovative.

Conclusion

Foreign firms became a powerful presence in the BC forest economy during Fordism, contributing to the province's growth and shaping its structure ever since. FDI has been a mixed blessing for BC; it has helped the provincial forest economy grow faster than it would otherwise have done in the past but curtailed its structure to commodity production that now limits the growth of the sector while adding to its volatility. Foreign countries did not force BC to open its doors to FDI. Governments, federal and provincial, chose this path. Indeed, in the crucial period of Fordism, the provincial government not only encouraged FDI, without any thought of limits or constraints, but used its powers to encourage large-scale and narrowly based industrialization based on commodity exports. Such a policy dovetailed neatly with the strategies of MNCs.

FDI has been so important to the BC forest economy for such a long time that foreign firms are not really foreign anymore. The established foreign firm's understanding of BC's business culture is as deep as any local firm's and its rights the same. Indeed, the idea of the *foreign* firm is foreign to contemporary debates on forest policy in BC; it is a taboo subject. FDI is simply direct investment whose adjective is for descriptive purposes only. Foreign control is deeply embedded in the BC forest economy, and its legacy, good and bad, lives on at the heart of the structure of BC's forest economy.

As well, the legacy of foreign control is at the heart of the debate between (some form of) industrial forestry and alternative, "ecological" models. In the former, local control and ownership are nonissues; in the latter, local control and ownership are the keys to a sustainable future. Free trade, continentalism, and (unregulated) privatization are powerful forces insisting on open door policies to FDI and maintaining access to BC's forest resources for MNCs. Provincial government hopes for immediate investment and job creation decisively reinforce these forces. Yet, the provincial government also advocates small firms, local control, and value-added activities. These contrasting, and in some ways conflicting, ideas, will not be easy to sort out. Yet, a powerful case for promoting community forests in BC rests in large part on the unfortunate legacy of foreign ownership for local innovation and development.

6
Small Firms: Towards Flexible Specialization in BC's Forest Economy

In theory, small and medium-sized enterprises (SMEs) play a defining role in the shift from the Fordist to the information and communication techno-economic paradigm (ICT) (see Table 1.1), or equally in the shift towards flexible specialization (see Figure 1.1). In these models, increasing uncertainty driven by dynamic and highly differentiated markets, more rapid flows of information, and escalating technological change that is widening production possibilities are demanding more flexible, fragmented production structures. In terms of industrial organization, the imperatives of flexibility value SMEs and an entrepreneurial culture in which economies of scope and external economies of scale are vitally important. Indeed, in the flexible specialization model, the ideal form of industrial organization is an industrial district comprising geographic concentrations of flexibly specialized populations of interacting SMEs. Individually, SMEs are specialized, focusing on realizing specific market and production opportunities, while collectively the many material and informational linkages among SMEs are a source of flexibility in response to market changes (Piore and Sabel 1984).

Whether SMEs have been revitalized, and the extent to which flexible specialization has occurred, has been contested in Europe and North America. Giant MNCs dominate many industries, for some observers increasingly so (Amin 1993), and flexible production is not inconsistent with mass production (Chapter 4). The apparent ambiguity in contemporary global assessments of the relative importance of the SME and MNC sectors is evident in the BC forest industries. Thus, corporate concentration is still a feature of the organization of these industries, especially when measured by share of production and harvest levels *within* BC (Schwindt and Heaps 1996). Yet, concentration in terms of global market share is not strong, and few large forest firms can be considered giant MNCs. Moreover, since the 1980s, in terms of number of firms and employment, the SME sector has grown, and flexible specialization is evident in some industry segments, especially the so-called value-added industries (Rees 1993). Several studies have further noted the *potential* of SMEs to add value (see Drushka et al. 1993). Indeed, this potential is important to the ill-fated Jobs Accord initiative of 1997 that proclaimed 20,000 new direct jobs would be created in the forest industries (Ministry of Forests n.d.).

This chapter first outlines the role of SMEs in the BC forest sector, noting the factors that suppressed entrepreneurialism during Fordism and recent policy initiatives that seek to counter this trend. Second, the chapter provides an overview of the so-called value-added sector where the growth of SMEs is strongest. Finally, it offers an assessment of the extent to which flexible specialization is occurring in the wood-remanufacturing industries (Rees and Hayter 1996).

Two prefatory observations need to be noted. First, it is common in BC, including in reports commissioned for the government, to classify secondary wood-processing activities (such as door manufacturers) as value added, to distinguish them from primary manufacturing (such as sawmills). Value added, however, is the difference between the cost of material input purchases at the factory gate and the price of the finished goods before delivery charges. Value added, in other words, is created by labour, management, and machinery on a set of inputs after they arrive at the factory to the point of distribution from the factory. All manufacturing is thus value adding. Indeed, primary manufacturers may add more value than secondary manufacturers. The matter is further clouded by a tendency, in some reports, to include all SMEs as part of the value-added sector. There are small firms, however, whose operations are unsophisticated, pay low wages, and add little value.

The second prefatory point relates to the strong association often made, even if implicitly, between SMEs and entrepreneurialism (and highly competitive markets). Entrepreneurialism is a complex concept, often thought of in terms of a range of characteristics such as owner-management, independence, small absolute size, limited market power, and willingness to innovate and take risks. Empirical estimates of these characteristics invoke judgment, and no matter how defined, SMEs vary in the extent to which they reveal these and other attributes of entrepreneurialism. Nevertheless, the social rationale for SMEs is conventionally rooted in their entrepreneurialism and what this orientation implies for competition and community development.

SMEs and Entrepreneurialism in BC's Forest Economy

Industrialization of BC's forests was led by the lumber and logging industries organized by entrepreneurs, many of them new to the province (Chapter 2). In sawmilling and especially logging, new businesses could be created with limited capital and formal education – the barriers to entry were low. In the early 1900s, an individual could lay claim to a square mile of forest around a planted stake and begin logging. Schooling was not necessarily considered helpful. In well-known views, Gordon Gibson, a famous "bull of the woods," noted that he dropped out of elementary school, and he claimed even those five years were wasted: "I learned so little I have no trouble remembering at all" (Gould 1975: 161). He knew enough, however, to become a millionaire, a noted entrepreneur in logging (as well as other businesses), and a parliamentarian.

In practice, the nature of business organizations established among the forest industries from the 1880s to the 1930s varied considerably. If entrepreneurship often literally meant one-person operations or tiny SMEs employing a handful of workers, in other cases the size of operations was more substantial, with ownership connections to organizations elsewhere. For example, the Fraser Mills sawmill in Coquitlam, built in 1890 by a Vancouver-based company, employed 877 in 1913 (many from Quebec), by which time it had been acquired and sold by a Winnipeg-based firm. After 1900, a flood of American immigrants, many of whom had previous business experience and established connections, fuelled investments in BC's forests. In turn, these investments stimulated related activities (backward and forward linkages) that were typically entrepreneurial in nature. A coincidence of entrepreneurial and societal self-interests, however, should not be assumed. Hardwick's (1963) chronology of the coastal forest industry, in which an "era of development and speculation" (1885-1909) gave way to "eras of liquidation" (1909-45), underlies this point. Indeed, this record of speculation and exploitation partly encouraged Sloan to prefer large corporations in the formulation of a forest policy for BC in the 1940s.

The actual number and size distribution of firms in BC's forest industries is difficult to state precisely for the first half of the twentieth century. Gould (1975: 66), without specifying the period, refers to an estimate of about 4,000 hand-loggers that once populated BC's coastal industry, and more were present in the interior. Published estimates for sawmills indicate that 334 existed in 1914 and 461 in 1939, roughly split between the interior and coastal regions, the latter being much bigger (see Table 2.1). Other types of mills also existed then, including shingle-and-shake mills. Pulp-and-paper mills are relatively capital intensive and fewer in number, as are the related paper-converting activities.

After 1945, during Fordism, the number of sawmills greatly expanded until around 1960 when over 2,500 sawmills operated in BC, all but 350 in the interior (Farley 1979: 64). The majority of the mills were entrepreneurially run – in the vernacular, *gyppo* operations – and in the interior, the majority were diesel-powered, small-scale operations, including mills that could be moved. On the coast, mills remained fewer and larger, and after 1960, the number of mills declined rapidly throughout the province, including the interior. As production increased substantially, sawmills declined to 442 by 1979 (see Table 3.1). Moreover, fewer mills after 1950 were increasingly controlled by multi-plant integrated firms and the new growth industry of the period, kraft pulp, was corporately owned. In logging, Statistics Canada broadened its definition of a logging operation so that clear trends became hard to discern, although SMEs remained important. In forest-product manufacturing, however, the relative decline of the entrepreneur was evident.

Fordism and the (Relative) Decline of the Entrepreneur

The chronological beginnings of Fordism did not immediately imply a downward trend in the role of the small firm. Corporate integration, however, was a growing force and constituted the new "ideal" type of organization, in the sense of being the model preferred by public policy. By the early 1970s, the entrepreneurial firm was of less importance, both practically and as an "ideal." Several interrelated factors may account for this trend.

First, the Forest Policy of 1947, following the Sloan recommendations, favoured large firms on the basis of arguments related to their expertise, greater economic impacts, and (supposedly) greater stability and ability to ensure sustained-yield forest management (Chapter 2). The creation of large-scale timber leases explicitly demanded large-scale investment proposals. Improved utilization laws, for example, requiring wood chips be used to feed sawmills demanded further capital investments, and they further advantaged integrated firms.

Second, the provision of transportation and power infrastructure throughout the province broke down isolation, improving accessibility to resources and to markets, and permitting large-scale operations in all areas of the province, not just the coast. Indeed, the shift to fewer, larger operations was also associated with a concentration in larger communities (such as Prince George, Quesnel, Williams Lake, and Port Alberni) that provide basic services, labour pools, and road and rail transportation to a sufficiently large resource area, as well as to distant markets.

Third, economies of scale was an economic principle that translated as engineering common sense in building new mills in wood processing and pulp and paper. Radical technologies were eschewed in favour of adopting the latest, proven technologies that progressively expanded the idea of the best-sized mill. If the sizes of sawmills and pulp mills almost certainly exceeded the *minimum* optimal scale (Dobie 1971; Schwindt 1977), increased size and competitive advantage were clearly equated as a basis for corporate planning.

Fourth, the dominant firms in the forest industries perceived significant advantages by horizontally and vertically integrating pulp-and-paper operations with wood-processing operations. To some extent, vertical integration sought to realize technologically based economies of scale, such as in the continuous supply of pulp to paper-making operations to allow energy savings. Another important corporate motive for vertical integration was to provide greater security in accessing resources and markets while horizontal integration contributed to market power and firm-level economies of scale. These integration strategies directly replaced or preempted opportunities for SMEs.

Fifth, the high levels of foreign ownership in BC's forest industries in the Fordist period militated against entrepreneurship in a variety of ways. Typically, FDI was implemented by large MNCs implementing strategies of horizontal and vertical integration on an international scale. In many cases, entry into BC occurred by acquiring local firms. In addition, to the extent that foreign firms rely on parent companies for products and services, opportunities for generating similar business in BC are lost. Further, the mandate of many foreign firms was to secure bulk commodities for use in affiliated operations elsewhere. Such a mandate has few implications for spinoff benefits to SMEs in BC. It might also be noted that in the capital equipment supply industry, many of the larger firms came under foreign control during Fordism.

Sixth, the forest industries as a group became progressively more capital intensive, creating financial and technological barriers to entry.

Seventh, the forest sector, especially the dominant industries, soon became comprehensively unionized in the 1950s, and an adversarial form of collective bargaining was established. Workers received high wages and nonwage benefits, seniority, and job demarcation, and management got a stable, specialized, and productive workforce. This labour relations model, however, is not conducive to entrepreneurial activity. Workers themselves became locked in to a seniority-based system of cascading benefits that are forfeited when they leave the firm. Further, worker tasks were typically highly specialized and subject to close supervision, experiences not conducive to entrepreneurial attitudes. Middle managers, it might be noted, were similarly locked in to cascading benefits and were themselves highly specialized. Moreover, unionized operations are at least wary of, and sometimes antagonistic towards, business links with nonunion firms. Yet, SMEs are often nonunionized. In addition, during the Fordist boom, job shortages were generally not a problem, and the high wages offered by unionized forest industries were attractive to workers and difficult for SMEs to emulate.

Finally, forest communities had little incentive to engage in boosterism. During Fordism, the expansion of forest commodities created jobs with high wages. For lucky communities, jobs and a way of life were already provided.

Consequently, in a province and sector where entrepreneurialism had been important, indeed celebrated, a number of reasons combined to offset just such a culture. The relative importance of these reasons is difficult to assess. The evident lack of entrepreneurial activity is most forcibly expressed, however, in company towns where small market-size and isolation imposed structural constraints on the formation of SMEs.

But SMEs Didn't Go Away

The pronounced shift towards Fordism was not inevitable, at least to the degree it occurred. Different choices regarding forest tenure, foreign investment, and corporate concentration, for example, may have led to different structures and a greater role for SMEs. In any event, SMEs did not disappear. While many failed or were taken over, others survived and more were formed. Indeed, the logging industry continued to be dominated by SMEs. Thus, in early 1993, 2,164 firms were registered with the SBFEP, 16 percent of which operated small sawmills or remanufacturing plants, the rest being market loggers (Ministry of Forests 1993: 73). Statistics Canada estimates a similar number of logging companies for 1993 (1,974) and documented 3,297 establishments in the industry, some companies controlling several operations or establishments. A 1994 Forest Directory lists 138 logging companies, 771 logging contractors, and 525 companies in related logging activities, such as hauling, loading, and booming. In this latter regard, it might be noted that log trucking has emerged as a new, growing business mainly since the 1950s.

In fact, SMEs play a big role in the logging industry. For the province, contractor logging accounts for fully 83 percent of logging activity (Price Waterhouse 1995: 15). This role varies between the coast and the interior. Thus, contractors in 1994 accounted for 48 percent of logging on the coast and for 100 percent in the interior. While contract logging is protected by forest policy, the importance of logging contractors reflects their efficiency in relation to company logging. This efficiency is not simply a matter of lower labour costs, but it is rooted in their greater flexibility in making faster decisions, coping with highly varied conditions, and more fully utilizing machinery and equipment.

Even in manufacturing, SMEs did not disappear. Thus, bearing in mind that corporations control several operations, the size distribution of establishments

Figure 6.1

Employment size: Distribution of establishments in BC's forest industries

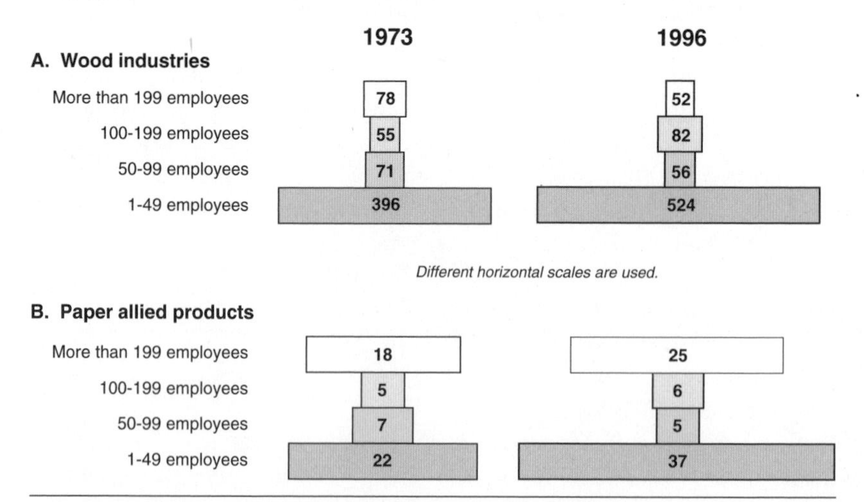

Source: Statistics Canada 31-203, Table 33 (1996), and Table 66 (1973).

in the logging, wood-processing, and paper-allied sectors broadly reflects population pyramids in which smaller operations are generally more common than large ones (Figure 6.1). Indeed, in a manner at least consistent with the flexible specialization thesis, various estimates suggest that the number of SMEs (and small establishments) in the wood industries have increased, especially from the late 1970s to the early 1990s in the so-called value-added sector (Rees 1993). Whether this growth has continued in the 1990s is debatable (Jordan 1998).

The Secondary Wood-Processing (Value-Added) Sector

For the forest industry to survive in a high cost-environment, competitive advantage must give increasing priority to value maximization. As Nakamura and Vertinsky (1994) note in the context of the Pacific Northwest,

Figure 6.2

Relationship between value-added shifts and consumer

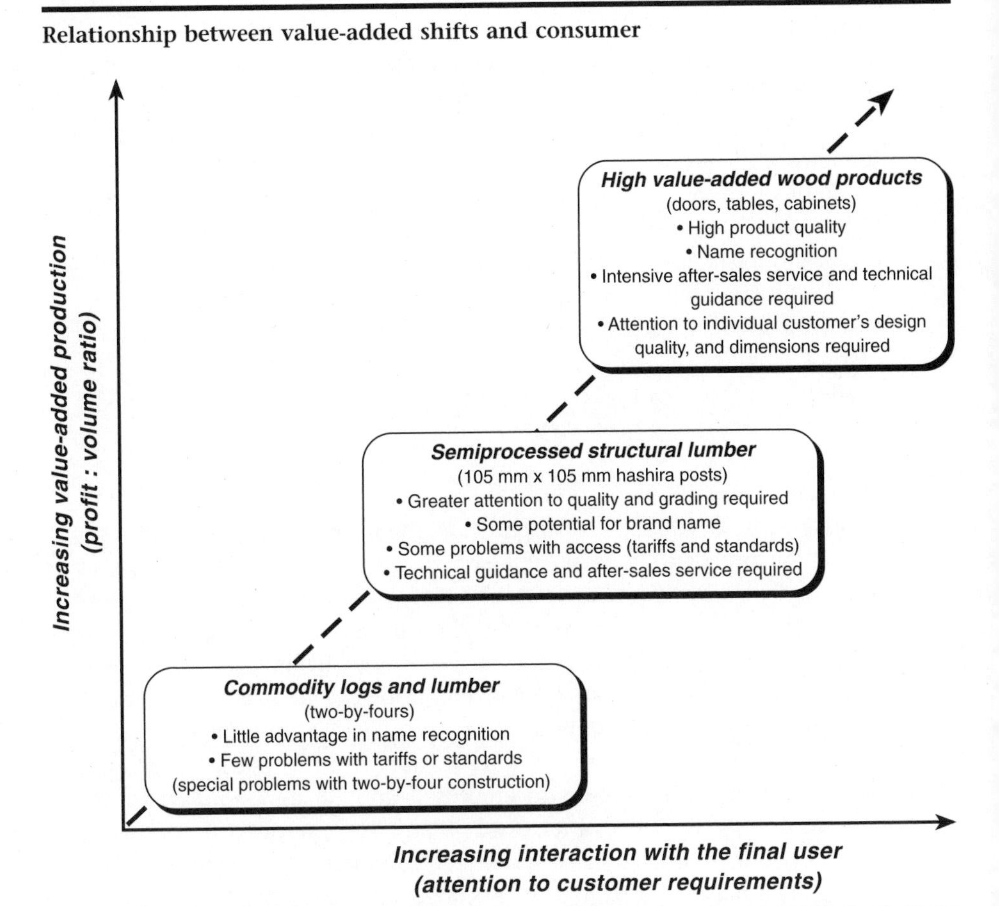

Source: Based on Nakamura and Vertinsky 1994.

competitive advantage demands a shift away from commodities to a more differentiated range of products, a shift that in turn requires increased market knowledge and more effective communications between producers and consumers (Figure 6.2). The Science Council of BC has likewise advocated a range of innovative, differentiated product initiatives across the forest industries (Ernst and Young 1998).

Although all manufacturing is value adding, these studies reflect the tendency in BC (and throughout the Pacific Northwest) to think of value-added activities as essentially noncommodity-based production. Activities include the manufacture of end-products such as doors, windows, and log homes that can be differentiated in a vast number of ways and the manufacture of components for these end-products that can be similarly differentiated. It can also mean technologically sophisticated structural products, such as engineered woods, as well as high-quality structural lumber cut to precise specifications, and the provision of specialized, closely controlled processes, such as kiln drying. Value added further implies a generalized shift towards increased commitments to quality control, precision manufacturing, uniform (and known) high-performance standards, value-based manufacturing, innovative design, custom-made products, and custom-cutting.

In large firms, the shift to value added is sought primarily within the framework of flexible mass production. Among SMEs, value-added shifts imply batch production and custom (one-off) production, as well as short runs of products with limited markets. In practice, no precise distinction exists between large-scale and small-scale production. Moreover, not all small-scale production reflects the intent of value-added shifts. For example, some SMEs provide relatively crude (imprecise) sawing services with little quality control. Even so, the shift to value added is strongly associated with SMEs, a reflection of the fact that market differentiation implies all kinds of niche opportunities for highly specialized producers.

SMEs and Flexibility

In the BC forest economy, consistent with the flexible specialization thesis, market dynamics and increased tendencies towards product differentiation, in part enabled by the microelectronics revolution and the associated creation of small market niches, provide opportunities for SMEs. Generally, in comparison to bigger corporations, SMEs are more flexible because of their low overheads, hands-on decision making, lower labour costs, and more informal, cooperative, usually nonunion labour relations, attributes that allow SMEs to profitably produce small quantities as markets demand. Moreover, in BC, fibre-supply uncertainty is another important factor underlying the flexibility advantages of SMEs. Thus increased variations in wood fibre mix, quality, and quantity are placing a premium on the entrepreneurialism and flexibility associated with SMEs (Woodbridge, Reed 1984; Rees 1993).

A combination of market and supply uncertainties has underlined the distinctive contributions of flexible SMEs, a flexibility that is further encouraged by geographic concentration (see Figure 3.6). Thus, the biggest centre of wood-processing SMEs is the Vancouver area, and many firms trade with one another and engage in various kinds of informal information exchange (Rees 1993). Indeed, through these relations, SMEs have developed a collective flexibility that is the hallmark of flexible specialization. Isolated SMEs have also revealed abilities to develop production flexibilities in small towns. Thus, stimulated by the crisis conditions of late 1998, several small sawmills in the Columbia-Kootenay region, such as Kitwanga Lumber in Revelstoke and J.H. Huscroft in Creston, have maintained production and employment when the much larger mills of multi-plant corporations have closed (Hamilton 1998i). Although these firms stressed flexibility and niche marketing as key attributes, both indicated that there are limits to their ability to survive.

In commodity production, competitiveness rests on economies of scale. For many SMEs, competitiveness depends on economies of scope, or the ability to differentiate and increase the value of outputs, with given human and capital resources. The shift in emphasis from economies of scale to economies of scope is a key theme in the shift from Fordism to flexibility.

Size of the Secondary Wood-Processing (Value-Added) Sector

Assessing the size of the so-called value-added sector is not straightforward. Recently, the provincial government defined the value-added wood-product sector in BC as comprising "secondary wood product manufacturers that add value to commodity wood or wood-based material by further processing it into specialty finished or semi-finished products" (McWilliams 1991: 1). As defined in this report, the value-added sector includes a diverse range of activities that are classified into four industry subdivisions, namely, the remanufacturing, engineered building components, millwork, and other wood-products industries. In 1992, these industries included 565 firms and 11,000 full-time jobs, constituting 12 percent of total employment in the BC forest-product industry. Financially, this sector produced $1.3 billion in revenue in 1992 or 12 percent of the provincial forest industry total (Price Waterhouse 1992a).

Within the value-added sector, remanufacturing is the largest industry, representing 27 percent of the firms, 33 percent of employment, and 45 percent of annual sales in 1992 (Table 6.1). It principally comprises "specialty products from commodity lumber and panel products ... and includes manufacturers of lumber specialty products, fencing, specialty panelboard, and custom processing activities such as lumber drying. It also includes finishing departments of primary sawmills that make specialty products" (Price Waterhouse 1992a: 2). In 1992, the engineered building-components industry comprised 19 percent of firms and employment and generated 20

Table 6.1

Value-added industries in BC's forest economy, 1992

		Employment	
Industry	Number of firms	Total	Average per firm
Remanufacturing	153	3,700	24.2
Engineered wood	107	2,100	19.6
Millwork	119	2,000	16.8
Other wood industries	186	3,200	37.2
Total	**565**	**11,000**	**19.5**

Source: Price Waterhouse 1992a.

percent of sales. This industry "contains a diverse group of companies ... manufacturers of laminated beams, trusses, pre-fabricated buildings, log homes, and wood treating plants" (Price Waterhouse 1992a: 8). The millwork industry had 21 percent of the firms, 18 percent of employment, and 14 percent of sales of manufactured wooden doors and windows, architectural woodwork, and turned wood for stair spindles. Other wood-product industry is an amalgam of cabinet and furniture manufacturers, pallet and container firms, and miscellaneous wood products such as chopsticks; it comprised 21 percent of sales, 29 percent of employees, and 33 percent of firms.

Recent reports differ on the industry's growth in the 1990s. Thus, one estimate suggested that, in 1997, 17,000 workers (or one-sixth of all forest-sector workers) were employed in value-added industry, not including remanufacturing, an increase of 20 percent since 1990 (Holm 1998: A14). This estimate, however, seems to imply a broad definition of value added. Another estimate, based on Price Waterhouse data, suggests that employment in the value-added sector has not changed in the 1990s and remains around 13,000 (Jordan 1998). The latter figure is more convincing.

Given that large firms are part of the search for value-adding activities in the context of flexible mass production (Chapter 4), most of the firms (and establishments) in the value-added sector of BC's forest industry are small; in 1992, average firm size amounted to 19.5 workers and $2.3 million of sales, and 16.6 percent of firms employed fewer than five workers, 56.6 percent between five and twenty-five workers, 15.8 percent between twenty-five and fifty workers, and just 10.9 percent more than fifty workers (Table 6.2). That is, the value-added sector has similar entrepreneurial characteristics to the logging industry and to the shingle-and-shake industry. It has also been a source of growth within the last two decades. The remanufacturing industry, for example, more than doubled between 1984 and 1992; that is, from 1,800 to 3,700 workers in this time period. In terms of Statistics Canada's category of sash, door, and other millwork industries,

Table 6.2

Size distribution of plants in secondary wood industries

Activity	Number of plants by employee size class 1-5	6-25	26-50	51-100	101+	Total
Remanufacturing	14	66	27	14	0	121
Engineered building components	9	52	10	7	3	81
Millwork	19	49	10	3	2	83
Cabinets	9	27	6	2	0	44
Furniture	6	15	4	6	3	34
Pallets and containers	6	5	3	0	0	14
Other value-added wood products	1	4	1	1	1	8
Total value-added	**64**	**218**	**61**	**33**	**9**	**385**

Source: Forintek Canada and McWilliams 1993: 8.

employment expanded from 3,483 in 1979 to 4,611 in 1996 and the number of establishments from 192 to 250 (see Table 3.1). This category accounts for about half of the value-added segment as defined here; other segments such as engineered wood expanded at a faster rate.

Examples of SMEs

SMEs are incredibly diverse in terms of technology, organization, geography, and functional role. In BC's forest economy, some SMEs do no more than chop timber into large blocks; others create custom-made houses for Japanese consumers. To provide a context for understanding the role of SMEs in value-added activities in BC's forest economy, Sarita Furniture, Paulcan, and Plenk's Wood Centre, illustrate key features.

Sarita Furniture was created in the 1980s by a husband-and-wife team. Both had worked (as research scientist and technician) for ten years at the Bamfield Research Station, a government marine-research laboratory in the small community of Bamfield, a few kilometres from Port Alberni. They had wanted to create their own business, and in 1987, they established a small home-based business (interior decorating and some furniture). The idea for Sarita Furniture evolved from there. In particular, the partners targeted quality cedar garden furniture as its market niche. Interestingly, they were aware of the revival of SMEs in Europe, especially the kind of cooperative behaviour among SMEs in Spain, and anticipated Sarita Furniture could be part of a similar trend in BC.

In 1989, Sarita relocated to Port Alberni, which offered a nearby wood supply (leftovers from local cedar mills) and was more accessible than Bamfield. The company began exporting to Germany after its attendance at a trade (wood home) show. Sarita contacted the Canadian Embassy to arrange introductions, and the company met a German-speaking wood specialist who became a good customer. Sarita regularly shipped six to eight containers of furniture (in knockdown form) a year to Germany and about 60 percent to 70 percent of Sarita's sales were to Europe. However, in 1997-8, the Asian economic crisis significantly reduced the price of Asian furniture, especially from Indonesia, and the German mark was devalued, so that Sarita had to find alternative markets. The company's sales and employment fluctuated. In early 1998, for example, Sarita employed seven workers, including Native carvers, in making furniture on site, as well as twelve contract workers at a local remanufacturing firm, a contract that was completed by early 1999.

To increase sales, Sarita diversified its range of cedar furniture products to around twenty-five, and in 1997, it began a new line of interior furniture products incorporating Native carvings. The hope was that these products had a "signature" that could not be copied and would be in demand. Sales are now targeted for specialty stores in the US that sell medium to high-priced distinctive furniture. Sarita has also considered species diversification, specifically to manufacture hemlock-based furniture and components. Thus, Sarita has developed and experimented with a variety of local connections, including local cedar mills, remanufacturers, an upholsterer, carvers, and salvage companies, in support of an unusually geographically diverse sales pattern.

Paulcan was established by an entrepreneur in the Chemainus Industrial Park in the early 1980s as a planing mill that was partly bought from MB's sawmill in Chemainus when it closed in 1982. Paulcan was the entrepreneur's second attempt at starting a business, the first having failed. The company, however, grew considerably in the late 1980s, at one time employing about 100 nonunion workers paid union wage rates. The firm added dry kilns and is primarily a subcontractor to MB. Much of its work is returned to MB, often for export to Japan.

The 1990s have been a volatile period for Paulcan, with employment fluctuating from twenty to fifty workers in its main planing mill. It has faced growing competition from other remanufacturers in the area and although it pays high union wages, a union local signed a contract with a rival for much less. In addition, its major corporate customers have invested in kiln-drying and planing facilities; for example, in MB's case, by the acquisition of Plenk's, located 100 metres from Paulcan. Partly in response to these problems and partly because of an interest in business creation, Paulcan's owner-entrepreneur built a hardwood sawmill in the same industrial park which by 1997 employed twenty people and used alder and maple.

An interesting feature of Paulcan is the extent to which it relies on used equipment (Table 6.3). Its principal equipment comprises several kiln dryers, a planer, and a trimmer. This equipment was all purchased second-hand, mostly at auctions, as was most of the firm's equipment, including small items such as welding equipment, and major items such as the firm's roof. Indeed, the main technology products bought new are the firm's

Table 6.3

Paulcan Enterprises, Chemainus: Source of plant and equipment, 1985-94

Item	Source
Main planer	MacMillan Bloedel, Chemainus (sawmill rebuilt)
Small planer	Fletcher Challenge, Youbou (sawmill rationalized)
Lumber sorter	Northcoast Lumber, Coquitlam (sawmill closed)
Strapper	MacMillan Bloedel, Nanaimo (sawmill rebuilt)
Beams/sawmill	City of Vancouver, Lion's Gate Bridge (maintenance)
Trimmer set	Champion Lumber, Libby, Montana (sawmill closed)
Beams/trusses	Canadian Pacific, Gold River, Tahsis cedar mill
Roof/main mill	Mayo Lumber, Nanaimo (sawmill rebuilt)
Boilers for kilns	Alberta Liquor Control Board, Edmonton
Kilns	Welland, Marysville, California
Tilboy system	Vancouver

Source: Hayter 1997a: 235.

computers. For virtually everything else, the firm is constantly scanning for auctions; it is prepared to travel as far west as Manitoba and as far south as California to fetch equipment even if this acquisition means hiring labour and trucks. The reason is cost. In August 1994, for example, Paulcan had just bought a trimmer at an auction in Libby, Montana, after a big lumber mill had closed. The cost of the second-hand trimmer, including delivery, was about $50,000, whereas a new one from a few kilometres away cost $450,000. Apparently, the firm usually buys second-hand items at around 10 percent of the purchase price. The only new equipment is the computer system. In the hardwood mill, the firm has again relied on second-hand machinery. In this case, the main equipment was Japanese and had been shipped to Merritt for a project that failed. The company bought the mill and hired a translator to interpret the machine's labels.

Plenk's Wood Centre, in the Chemainus Industrial Park, was founded by Walter Plenk in 1982, to manufacture top-of-the-line windows and doors. Its products are expensive and some are custom-made, including orders from a California-based architect who buys 10 percent of the firm's windows. In the 1980s, Plenk extended the firm's activities to include dry kilns, a planer, presser, laminator, finger jointer, and humidifier. By 1997, doors and windows accounted for just 10 percent of sales, with remanufacturing activities accounting for the remainder. Thus, Plenk's developed as a niche supplier of high-quality end-products and as a relatively diversified remanufacturer that supplies other firms. Most of its sales were local, although its exports to California have been supplemented by sales to Europe, including the first finger-jointed products to that market. Employment was typically in the range of sixty to seventy workers, but it dropped as low as three workers in 1998.

Within a short time following acquisition by MB in 1995, Plenk's experienced substantial change. MB required that jobs be defined, emphasized accountability, and developed teamwork, including investing in training. Thus, in this case, control by a large firm introduced more rules and procedures but also a stronger commitment to employment flexibility based on skill formation and job rotation. In addition, wages were increased dramatically, and workers given more autonomy and information about their jobs. On the other hand, managerial decision making, in terms of obtaining new equipment, is now slower. The benefits that accrued to employees have also proven to be short lived as most were laid off in the recession of early 1998. Plenk's no longer has the responsibility to buy wood, and it must now compete with other MB operations for remanufacturing work.

Summary Characteristics

Although each case is different, Sarita, Paulcan, and Plenk's (before 1995) offer important insights into what is typically implied by SMEs. These companies, which rarely employed more than 100 workers, are entrepreneurially run, independent, innovative, and willing to take risks. Plenk's is unusual because it was created by an immigrant entrepreneur. Most new firms are established in their founder's home environment, as was the case with Sarita and Paulcan. Sarita is unusual in that exports were important from its inception. On the other hand, as is typical of SMEs, Sarita is strongly connected to the local economy.

The three cases further demonstrate the competitive advantages of SMEs in relation to giant firms. Thus, the firms developed specific market niches, they profitably manufactured in small batches, they differentiated their product lines despite limited volumes, they coped with highly variable fibre supplies, and they revealed considerable versatility in the use of old equipment. They are also capable of making quick decisions. In summary, these SMEs draw considerable competitive strengths based on specialization and flexibility.

Paulcan and Plenk's reveal the interdependence of small and large firms. Paulcan served largely as a subcontractor to large firms and Plenk's was acquired by MB. Indeed, there are no rules that readily allocate production between small and large firms. Whether Plenk's is better served as part of MB or not, and whether society's interests are better served by this acquisition, is a matter for debate. Clearly, SMEs are vulnerable to takeover as well as to market forces. But in dynamic industrial contexts, where new knowledge is constantly created, the population of SMEs is constantly replenished by the formation of new firms that add value. Do wood products in BC offer such a situation? The recent expansion of SMEs building log houses, modular homes, and an array of products in the remanufacturing industry, including the remarkable example of Larrivée Guitars (Varty 1997), at least gives hope that the answer is yes.

Value-Added Potentials among SMEs

There are two main empirical reasons to suppose that value-added potentials

Table 6.4

International comparisons in value for jobs for timber cut, 1984

Region	Volume logged (million m³)	Value added ($/m³)	Jobs per 1,000 m³s
BC	74.6	56.2	1.05
Other Canada	86.3	110.6	2.20
US	410.0	173.8	3.55
New Zealand	5.3	170.9	5.00
Sweden	56.0	79.5	2.52

Source: Travers 1993: 196. Reprinted with permission of Harbour Publishing.

among SMEs exist in BC's forest industries. First, SME-controlled activities associated with the value-added label have grown steadily since the late 1970s. At least until the mid-1990s, in terms of employment, these activities offset the losses in commodity production. Second, in comparison to most other major forestry regions in advanced countries, the employment generated per cubic metre of logs harvested is relatively low in BC (Table 6.4). Thus, the jobs in 1984 created per 1,000 cubic metre of timber cut in BC is less than half the national average, and compared to other countries, the difference is greater (Travers 1993). The Pacific Northwest also generates more jobs per tree cut than BC. During Fordism, a declining ratio in jobs per tree cut readily translated as increasing productivity, consistent with cost minimization, mass production, and the adoption of the latest technology (Forgacs 1997). In the contemporary period of flexibility, however, to the extent that value maximization is the key motive, more labour-intensive SMEs have a potentially greater role in seeking out market niches as well as in generating process efficiencies.

Precisely assessing the potential for value-added investment is more difficult. At least three major impediments to the realization of value-added potentials should be identified.

First, historical legacy (that is, the institutional and infrastructural edifice of Fordism in BC) counselled strongly against entrepreneurialism in BC's forest economy. A high-stumpage regime alone is no magic wand by which to produce the "missing" entrepreneurs capable of constantly creating higher values to offset constantly increasing costs.

Second, other forestry regions in Canada, the US, and elsewhere similarly wish to add value. Although not a zero-sum game, there is competition for high-value as well as for commodity markets. Many SMEs in BC face stiff competition from SMEs in Washington and Oregon across a spectrum of forest-product markets in Japan, such as the construction frame, modular, and log-home markets (Hayter and Edgington 1999). Indeed, despite longer time lines and greater cost, Swedish and Finnish value-added producers are starting to tap into these same markets.

Third, the realization of value-added potentials depends on the entry of SMEs with innovating products and markets. There are innovative entrepreneurs in BC's forest industries, as demonstrated by Sarita, Paulcan, Plenk's, and other examples. But the size of the entrepreneurial pool is uncertain, as is the ability of SMEs to realize value-added potentials, and SMEs are vulnerable to market forces including takeovers by large firms (as Plenk's illustrates). Thus, Travers's (1993) argument – that a proper (and higher) valuation of the forest resource will lead to increases in the value of resource utilization as a result of enhanced supply of SMEs – is a leap of faith that is incorporated in present forest policy (Chapter 3). Indeed, anecdotal evidence suggests that recent provincial forest policy initiatives are counterproductive in stimulating SMEs (for example, by demanding high stumpage from innovative logging practices, thus undermining the very incentive to innovate) (Jordan 1998; Hamilton 1998i).

Policies Stimulating SMEs in the Forest-Product Industries

SMEs in logging have long been supported by policies that require TFL licensees to use independent contractors for at least 50 percent of logs harvested from the tenured area. Other forms of licence have typically stipulated use of contractors. With this important caveat to the logging industry, neither forest policy in particular nor economic policies in general paid much attention to stimulating SMEs in forest-product manufacturing until the late 1970s. Then, consistent with trends in other industrialized countries, the provincial and federal governments introduced programs that offered modest support to (very small) SMEs in virtually all sectors of the economy. A sectoral initiative targeting the forest industry also stemmed from the Pearse Commission (1976).

Following Pearse's recommendations, the Forest Act of 1978, the Small Business Enterprise Program (SBEP) and after 1987 the Small Business Forest Enterprise Program (SBFEP), provided 9 million cubic metres of timber to SMEs (loggers and manufacturers) that did not have long-term licences to log Crown land. Initially, this wood supply was taken from nonquota supplies available in the old PSYUs. SMEs were allowed to bid competitively on a portion of the AAC. The Forest Amendment Act of 1998 reduced the AAC of established (and replaceable) timber licensees by 5 percent so as to increase the cut available in the SBFEP. By 1993, SBFEP licensees held 14 percent of the allowable annual cut (see Table 3.2). This program provides timber to logging contractors and manufacturers, although the lion's share has been allocated to the former.

Recently, however, the AAC available to small firms under the SBFEP has not been fully used, contradicting one of the stated goals of the provincial government's Jobs Accord initiative. In general, harvest levels have declined as a result of a combination of increased wood costs and poor market conditions; these factors apply to SBFEP participants as well as to large firms.

Another goal of the accord is to divert wood within the SBFEP and from other sources to remanufacturers, because these firms, typically SMEs, are thought to have high potentials for creating value and jobs. For example, according to a government press release, the "wood secondary manufacturing industry" (labelled "wood remanufacturing" in the accord) has the potential to create an additional 6,500 direct jobs and a similar number of indirect jobs (Ministry of Forests 1997).

The basis for these employment projections is not made clear. In the Jobs Accord, the government states that the remanufacturing sector will generate an additional 5,000 direct jobs (accounting for most of the 6,500 jobs in secondary wood manufacture) and a similar number of indirect jobs (Ministry of Forests n.d.). But whatever their basis, these expectations reflect a widely held view that job potentials in BC's forest industries are concentrated among SMEs and in the remanufacturing or value-added (or secondary wood) sector. It is further assumed that these two sectors overlap considerably. At the same time, the Jobs Accord, and specifically the government's press release (Ministry of Forests 1997), recognizes wood-remanufacturing firms in BC face fibre shortages. Indeed, the government wishes to divert timber within the SBFEP program to wood remanufacturers, providing them with more timber (700 million board feet) in the Wood Fibre Transfer Program and ensuring that the AAC under the SBFEP is fully used.

The Jobs Accord further specifies that 18 percent of sawn timber on the coast and 16 percent in the interior must be made available by companies from "replaceable licences" to the "remanufacturing sector," which is identified as the most important job generator in the program. Thus, of the additional 20,000 direct jobs anticipated by the Jobs Accord in BC's forest economy, at least one-quarter of these jobs are expected to be in remanufacturing (and an additional 1,500 in other firms in the SBFEP) by the year 2001. While the value-added activities by large firms can count towards these targets, value added by SMEs and "independents" is clearly seen as important within the accord.

Apart from redirecting timber supply, provincial and federal governments have helped fund the British Columbia Wood Specialties Group (BCWSG) since its startup in 1989. The BCWSG, which draws 50 percent of its funding from member companies and 50 percent from government sources, was set up to serve the collective interests of the growing number of SMEs in value added. As of 1998, the BCWSG claimed its 120 member companies employed 10,000 workers. One particular theme of the BCWSG's activities is to facilitate market intelligence for high-value products, including information regarding entry into the Japanese market. A second important thrust is to train workers; following the opening of its training centre in Abbotsford, funded by Forest Renewal BC, 1,000 students had been trained between late 1996 and mid-1998 (Holm 1998: A14). The government also helped fund the Centre for Advanced Wood Processing at UBC, which opened in 1995.

If some policies have sought to ameliorate the wood-fibre problems of SMEs in the forest industries, especially those seeking higher-value markets, the high-stumpage regime has raised costs for all firms. SMEs may enjoy great flexibility in seeking out higher values, but even their flexibility has limits.

Paradoxes for Government Policy

One way to obtain a proper valuation of BC's resource is to privatize the resource and have proper markets for logs. Travers (1993) notes that after markets for logs are opened for small firms, the price has increased. But will the price increase now? And if the entire BC harvest, or large portions of it, enters open markets, who will be the likely buyers? Will such a policy increase or decrease corporate concentration? Can foreign buyers be prevented from buying BC logs in privatized and open markets? The dilemma facing the provincial government, at least in the context of this chapter, is that higher valuations of timber and/or market valuations of timber, may not be consistent with the growth of SMEs in the sector, a goal of the provincial government since the 1980s. The fact that the province's largest forest-product corporation, MB, is a leading proponent of privatization, suggests that the corporate sector will not feel disadvantaged by such a policy. It cannot be assumed that SMEs can generate values that will allow them to outbid large corporations for wood.

Another dilemma facing the provincial government stems from the fact that the small-firm sector in the forest industries has always been less unionized than the corporate sector. After all, this sector is associated with entrepreneurialism. SMEs almost inherently oppose bureaucratic interference and are often, but by no means completely, antagonistic towards unions. Thus, policy initiatives that criticize large corporations and seek to promote SMEs and entrepreneurialism, on the one hand, and strong, adversarially minded unions, on the other hand, are in conflict. Possibly, the government failed to anticipate that large corporations would move out of logging and release union workers in response to the high-stumpage regime. The government, in turn, has sought to use Forest Renewal BC to train laidoff workers in silviculture and to promote unionization among silvicultural firms. Yet, SMEs in silviculture depend on an extremely flexible workforce, and government policy may only serve to push up costs.

The provincial government also faces an emerging dilemma regarding the issue of "eco-certification" as it seeks to stimulate SMEs. Eco-certification is a system of environmental auditing of logging and manufacturing processes by internationally recognized ("third party") agencies that is designed to ensure sustainable forest management (Wallis et al. 1997). Those activities that meet the desired standards are certified. Eco-certification is in its early stages in BC, and the hope is that the initiative will establish transparent standards for business that will be supported by environmental groups, thus obviating the need for ongoing environmental opposition (Chapter

10). The certification process, however, imposes costs on firms, notably those related to improving ("greening") behaviour, the process of certification (payments to certifiers), identifying and monitoring inputs that are used, and labelling. In the US, these costs may range from $0.05 to several dollars per hectare per year and, importantly, they exhibit "substantial" economies of scale, that is, average costs of certification were much lower for large firms than for SMEs (Cabarle et al. 1995: 14). Consumers may not be willing to pay a premium for a certified product, simply because it is from a SME.

A final dilemma, as already noted, is that openness to FDI undermines SMEs in various ways. The potential of SMEs for adding value and creating jobs, however, suggests that addressing these dilemmas is worthwhile.

Wood Remanufacturing

By processing wood from commodity lumber into specialty items, the remanufacturing industry acts as an interface between primary manufacturing and manufacturers of finished wood products (Industry Science and Technology Canada 1989; McWilliams 1991; Price Waterhouse 1992b, 1992c). The main products include window and door components, interior and exterior panelling, decking, and lumber of various dimensions for manufacture by the customer. As shown by Rees's (1993) study, which is the basis for the rest of this chapter, remanufacturers are small and serve niche demands, producing small batches of products to meet the specific requirements of the customer. In addition to facing differentiated market conditions, wood-remanufacturing firms typically operate in conditions of considerable uncertainty over timber supply (see Rees and Hayter 1996).

Market and Supply Uncertainties

Remanufacturers serve increasingly diverse markets. Provincially, while domestic consumption of remanufactured products increased in importance between 1984 and 1992, accounting respectively for about 35 percent and 43 percent of sales (Figure 6.3), overseas markets have almost doubled from 17 percent to 31 percent of sales. Concurrently, exports of remanufactured goods to the US fell from 48 percent to 26 percent, a relative decline caused by recessionary conditions and the introduction of export tariffs on high-value Canadian wood products. This uncertainty regarding exports to the US encouraged BC remanufacturers to target other markets. In terms of exports, the most significant market development for remanufactured products has been the rise of Pacific Rim markets, particularly Japan, which in 1992 accounted for 14 percent of sales, overtaking Europe as most important overseas market. This export link is clearly encouraged by the revaluation of the Japanese yen and growing problems of timber and labour supply within Japan. These exports are largely controlled by Japanese traders and wholesalers (Reiffenstein 1999). Provided the remanufacturers make products of high value and quality to meet specific consumer tastes, Japanese

Figure 6.3

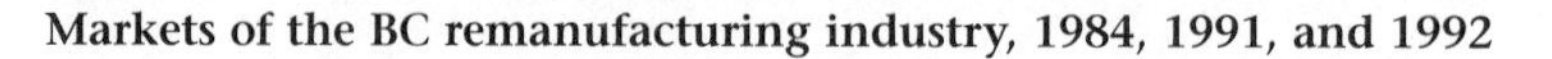

Markets of the BC remanufacturing industry, 1984, 1991, and 1992

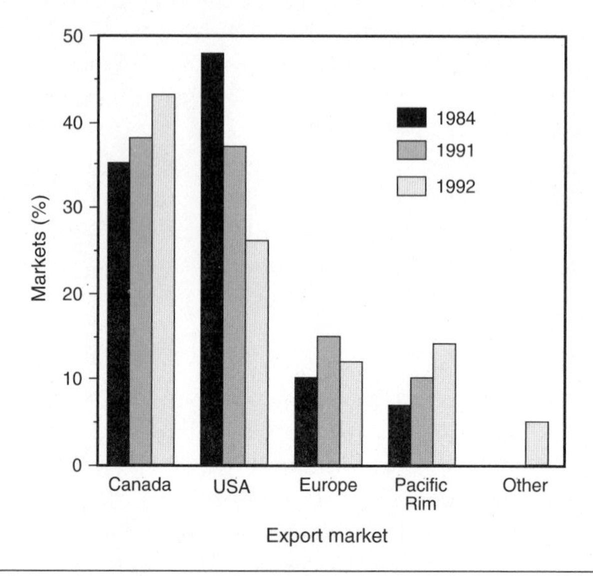

Source: Rees 1993: 74. Reprinted with permission of Canadian Association of Geographers.

demand is likely to grow, while Taiwan is a market for semifinished goods for remanufacturers and acts as a stepping-stone to Japanese markets.

Remanufacturing firms also face supply uncertainties. At the provincial scale, the remanufacturing industry uses a diverse range of species (Figure 6.4). On the coast, the dominant species are western red cedar and western hemlock, whereas interior remanufacturers use local supplies of white spruce, lodgepole pine, and alpine fir. However, in the 1990s, environmental pressures, Aboriginal land claims, and increased reliance on second-growth timber, mostly in the coastal region, have led to a decrease in the relative importance of hemlock and cedar. Furthermore, in the forest industry as a whole, competition has increased for wood supplies, especially for the "shop and better" grades of red cedar that remanufacturers prefer. Indeed, with primary manufacturers attempting to add value to their own lumber rather than release it to remanufacturers, the cost of red cedar has risen dramatically relative to other species, rising by 60 percent between 1988 and 1992 (Rees 1993: 68). The availability of hemlock, preferred by Pacific Rim consumers, has also become a problem in supply terms, while outside of the Pacific Rim, remanufacturers have differing opinions on hemlock's market value.

Enterprise Strategies in Vancouver Metro's Remanufacturing Industry

If collectively small, the firms that populate the value-added sector in general, and remanufacturing in particular, vary considerably in function and

Figure 6.4

Fibre used by the BC remanufacturing industry by species, 1984 and 1991

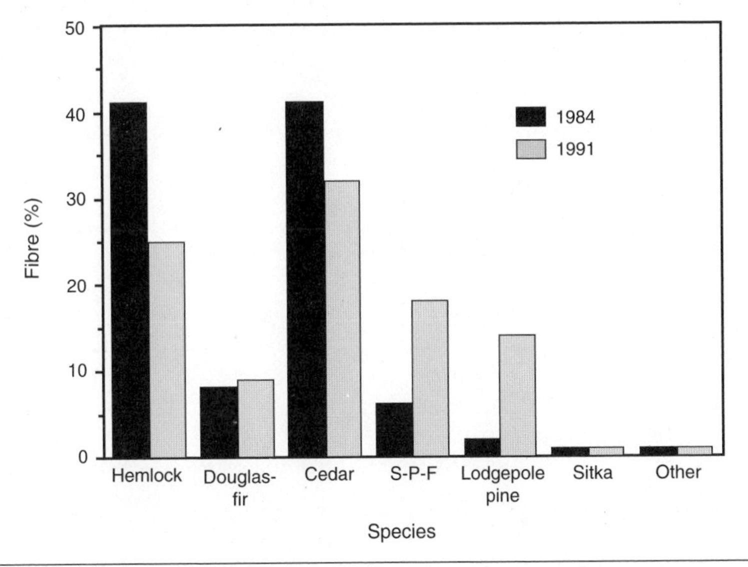

Source: Rees 1993: 66. Reprinted with permission of Canadian Association of Geographers.

organization. As Rees (1993) notes, the remanufacturing industry comprises three main types of SME: isolated remanufacturers, contractors, and custom remanufacturers. The latter type consists of capacity subcontractors and specialty subcontractors. A few branch plants are also controlled by large forest-product corporations.

Within the SMEs, contractors and custom remanufacturers dominate the industry and are linked to form a social division of labour. Contractors externalize production by paying local subcontractors to perform specific processes, whereas subcontractors provide other remanufacturing firms, sawmills, or distributors with the service of remanufacturing lumber and depend largely on work contracted out from other local firms. The capacity subcontractor is contracted out by remanufacturing firms to perform processes that involve both relatively large-volume production, such as the initial breakdown of lumber into grades and dimensions, or low-volume production of specialty tasks that the contracting firms can do in-house but, for various reasons largely related to cost, prefer to externalize to some extent. The specialty subcontractor performs specialized processes that the contracting firm does not have the technological capability or qualified labour to perform. In the remanufacturing industry, such processes tend to be at the high-value end of production, such as kiln drying, moulding, or laminating, requiring a relatively high level of capital investment and/or several years of labour training.

In contrast, the few isolated remanufacturers scarcely participate in the social division of labour, and they invest in a variety of processes to permit custom-made manufacturing or produce a single high-value product. Branch plants also typically operate in isolation from the rest of the industry while remaining integrated within parent-company operations.

As an illustration of this variability, recent case studies of six remanufacturing firms by Rees (1993) are instructive (see also Rees and Hayter

Table 6.5

Production characteristics of six value-added operations

Firm (type)	1992 output (value/unit)	Product or functions	Equipment	Interfirm relations
A (isolated)	20 ($1,000)	Window, door parts; panelling, siding, and skirting	Varied: includes small computerized dry kiln	Mostly in-house; 15% subcontracted
B (contractor)	12 ($900)	Window, door parts; panelling; 1 × 2 decking	Limited: initial log breakdown	50% to 60% subcontracted: sophisticated work
C (capacity subcontractor)	11 ($800)	Initial log breakdown; some products (decking and siding)	Limited: but includes resaw and planer	Relatively long runs (one to two days) common; no subcontracting itself
D (specialty subcontractor)	55 ($1,500)	High-quality drying; moulding, packaging, window, door parts, siding, fencing, shed parts	Varied: includes five large computerized dry kilns	60% subcontracted, 40% in-house products
E (branch plant)	2 ($1,800)	Laminated and finger stock for windows	Varied: but special purpose, one computer-assisted laminator	Specialized high-volume production, all in-house
F (branch organization)	13 ($1,000)	Window, door parts, sauna blanks	n/a	All work is subcontracted

Note: Output is in million board feet measure and value per unit is $ per 1,000 board feet.
Source: Rees and Hayter 1996: 210. Reprinted with permission of Canadian Association of Geographers.

1996). These cases studies are not randomly selected and their behaviour is not statistically representative of the population of firms as a whole: two of the firms are branch plants, and all six are bigger than the industry average in terms of employment and production. They are also more export-oriented. Nevertheless, the case studies reveal the organizational variability that exists within the industry (Table 6.5). Thus, Firm A exemplifies an isolated firm that manufactures a wide range of window and door parts as well as various sizes of panelling, siding, and skirting; Firm B is a contractor that manufactures door and window parts, some panelling, and one size of decking; Firm C is a capacity subcontractor that performs initial log breakdown (and also manufactures some decking and siding); and Firm D is a specialist subcontractor that performs various functions such as drying, moulding, and packaging as well as manufacturing window and door parts, siding, fencing, and shed parts. All of these firms are SMEs that buy fibre from, or process the fibre of, other firms. Firms E and F represent branch-plant operations of large corporations with extensive timber tenures to support sawmilling and pulp-and-paper operations.

Although four SMEs were in production before 1981, the two branch plants of tenured firms are relatively recent. Thus, Firm E was set up in 1991 to produce a narrow range of products (specifically, laminated and finger-jointed stock for windows and laminated beams for construction) as a specialty complement to the parent company's sawmill output. Similarly, Firm F was set up in 1989 as a specialty-product division. However, Firm E is based on an internal division of labour, whereas Firm F subcontracts its entire production of window and door parts and sauna blanks.

Fibre Supply and Marketing Patterns

The six firms represent in broad terms the timber-supply dynamics shaping the industry (Figure 6.5). For the four nontenured firms, red cedar was the dominant fibre in 1981, comprising approximately half or more of the total wood processed by each firm. After 1981, other species became increasingly important as red cedar declined. However, these firms responded to supply uncertainty in different ways. Thus, as a contractor, Firm A processes a wide selection of species, with hemlock and Douglas-fir increasingly important, and it tends to contract out processes involving red cedar. Similar to Firm A, Firm B processes a wide variety of species, increasingly spruce, pine, and fir (S-P-F), and it contracts out high-value-added processes. On the other hand, as subcontractors, Firms C and D, processed mainly or only red cedar in 1981. While Firm C has managed to retain complete reliance on this species, Firm D, a larger operation, has progressively shifted towards using more yellow cedar (for the Japanese market).

In contrast, tenured Firms E and F do not use cedar species and rely almost exclusively on hemlock, supplemented with Douglas-fir and S-P-F. Thus, through more homogeneous and predictable timber supplies, Firm E, as a

Figure 6.5

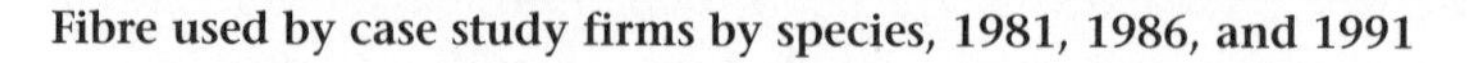

Fibre used by case study firms by species, 1981, 1986, and 1991

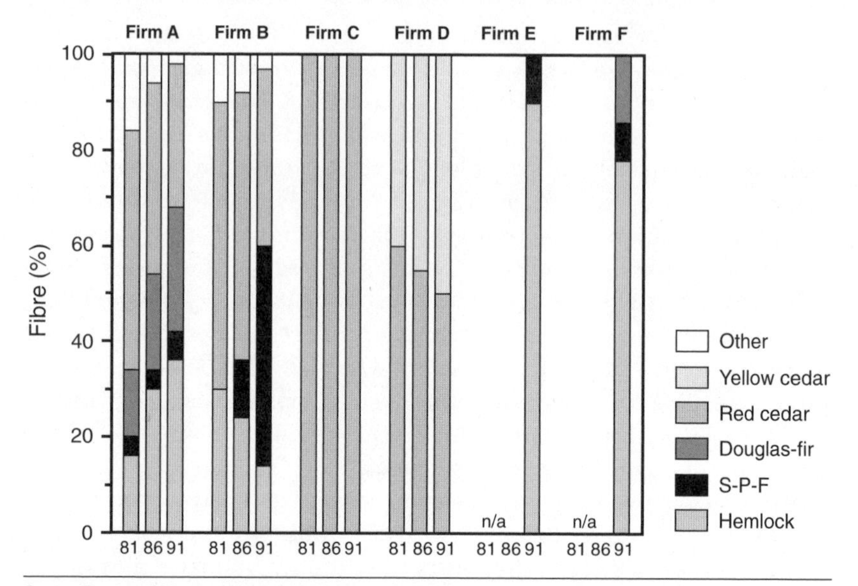

Source: Rees and Hayter 1996: 212. Reprinted with permission of Canadian Association of Geographers.

tenured firm, benefits by manufacturing products of consistently higher quality and through investment in specialized, sophisticated equipment.

Fibre-supply variability is mirrored by the diversity of markets served by the six firms (Figure 6.6). Given a general shift from traditional North American to Asian and European markets (see Figure 6.3), one obvious distinction is between the product manufacturers and subcontractors; the former (Firms A and B) serve more diverse markets than the latter (Firms C and D). The independent Firm A and the contracting Firm B are also closer to final market demands and readily react to market dynamism. They rely on the subcontractors (Firms C and D) for inputs to serve uncertain markets. The geography of sales of Firm A has changed dramatically. In particular, while sales within North America fell from 90 percent to 56 percent of its total between 1981 and 1991, the importance of Europe increased from 10 percent to 25 percent, and Japan and other markets (mainly Taiwan) from 5 percent to 20 percent of sales, a transition facilitated by Firm A's ability to target markets by means of the production of particular products. Whereas Firm A has shifted away from North American markets faster than the industry average, Firm B has increased its sales within North America but changed the species it uses to better serve that market. Firm D, in purchasing yellow cedar fibre for processing in-house, emphasizes Japanese markets somewhat more while reducing its dependency on custom processing for the American market.

Figure 6.6

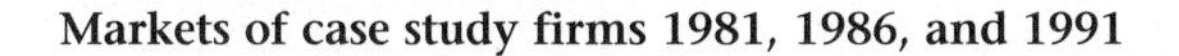

Markets of case study firms 1981, 1986, and 1991

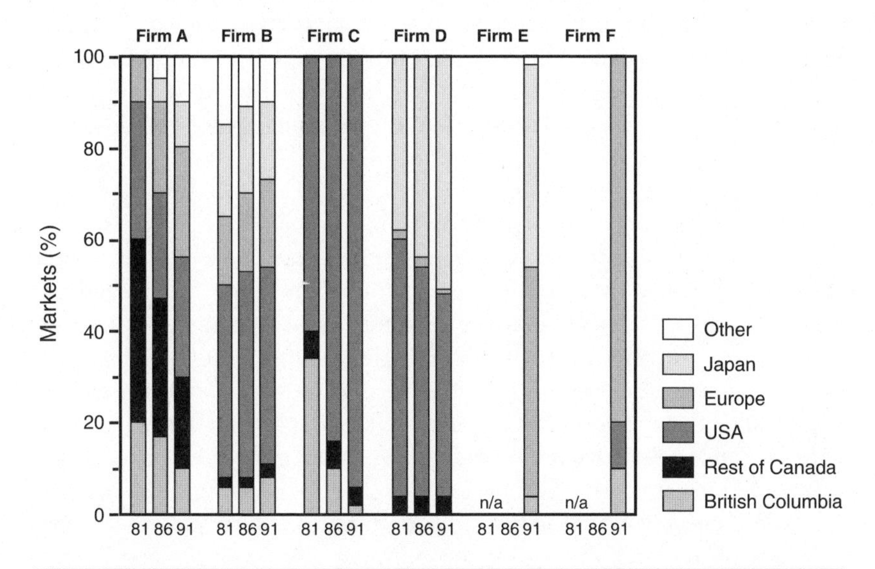

Source: Rees and Hayter 1996: 211. Reprinted with permission of Canadian Association of Geographers.

In the branch plants, although in early stages of production in 1991, Firms E and F are overwhelmingly export oriented and, to a greater degree than the others, supply the more difficult to penetrate but higher-value offshore markets in Europe and Japan (Figure 6.6).

Production Strategies of Nontenured Firms

The industrial organization of the wood-remanufacturing industry features vertical and horizontal disintegration in which the distinction between contractors and subcontractors is frequently blurred and the strategies of individual firms are dynamic. Firm F in 1992, for example, assessed the profitability of its role in the industry of relying completely on subcontractors; this assessment will determine whether it stays in the business, and if so, whether it will invest in production facilities to replace all or part of its present subcontracting linkages. In general, the production characteristics of wood remanufacturers vary considerably in the value and nature of products manufactured, degree of diversification, equipment, and interfirm relations. The six firms reveal this diversity (Table 6.5). An important distinction is between the SMEs (Firms A to D) that do not have forest tenures (nontenured SMEs) and the organizations (Firms E and F) that do (tenured organizations).

In the nontenured firms, the amount and logic of contracting out varies between contractors and subcontractors. Firm A, for example, prefers to retain high-value production in-house and to contract out initial log-breakdown processes. To support the in-house manufacture of a wide variety of high-value products, Firm A has purchased small, computer-controlled dry kilns with which to dry lumber in small batches for niche demands. It has also invested in basic, multipurpose machinery, notably a larger greenchain and resaw, to increase product mix and minimize the risk of market uncertainty by allowing small batch production of a range of products. Within the remanufacturing industry, Firm A enjoys a medium level of technological sophistication.

Firm B pursues a different production strategy from Firm A. In particular, Firm B relies far more on subcontracting and contracts out the more sophisticated production processes to specialty subcontractors while performing the initial (less sophisticated) stages of production in-house. By adopting such a strategy, Firm B does not require sophisticated equipment. Rather, the firm performs initial log breakdown on basic machinery purchased second-hand from sawmills, whereas it subcontracts kiln drying, laminating, edge gluing, and other value-adding processes. As the manager pointed out, "Probably the bulk of our remanufacturing is done on a custom basis by outside companies. What you see here is very basic. We have a sorting chain, a rip-saw and a chop-saw, that is it, we have no planer, moulders, resaws, dry kiln, any of the more sophisticated equipment ... It is not the best at anything but it can do pretty nearly anything, and when production becomes a major factor sometimes it is better to send it out elsewhere where we can get the gain of their productivity" (Rees and Hayter 1996: 212).

Through the externalization of specialized processes, Firm B attains flexibility in product mix, whereby customer requests are satisfied through the use of a social division of labour, each specialized process performed by the firm best suited to that task. The manager notes, "We think we are more flexible than our competitors because we can do a lot of different things. We are not necessarily doing them but we can arrange to get them done ... Certain plants are better at doing certain things than others, so we hone in on those ones" (ibid: 213). This use of subcontractors enables Firm B to react quickly to changes in markets and wood supply by changing or adding new subcontractors without the expense of investing in a wide range of machinery. According to one manager, "Before we [invest in new machinery] we would look at having somebody else do it for us ... [and this] gives us a great amount of flexibility. We can change our product mix by taking up and dropping subcontractors as we need" (ibid: 213).

Subcontractors tend to specialize in a limited range of processes central to the needs of contractor firms. Firm C, for example, focuses on initial log-breakdown work, and Firm D has invested heavily in kiln-drying facilities. At least 95 percent of all processes performed by Firm C are for contractors on a pay-for-service basis. Technology within Firm C is unsophisticated, limited to a greenchain (a conveyor belt used to sort lumber into various grades), a chop-saw, and a resaw. The firm has not invested in computers, and its managers have no computing expertise. In contrast to the flexible, short-batch production used by the other remanufacturing firms, production within Firm C occurs in relatively long runs (although compared to primary sawmills these runs would be considered short). Production is concentrated on a single product until all orders for that product are satisfied, at which time machinery is set up to run a different product. As the manager notes, "It could be a day's run, it could be two or three days' run depending upon how much volume there was ... If we have enough volume to run two shifts we would run right through with the same product until we have finished it, because every time you change you are losing half an hour in there. So once you get on a run you keep going until you complete it" (ibid: 213).

Firm D, a specialty subcontractor, is contracted by remanufacturers to perform highly specific processes generally at the high-value and consumer end of the production process. In contrast to the capacity subcontractor, Firm D has invested intensively in machinery to attract business through high-quality, specialized processes. Since the mid-1980s, Firm D has invested in five new computerized dry kilns, obtaining a drying capacity of 2.2 million board feet, and in moulding and packaging lines. Upon investing in computerized dry kilns, Firm D has increased the accuracy to which timber can be dried, allowing more specialized finished products to be processed, and it has increased its capacity to serve contractors. Several remanufacturers in Vancouver metro refer to Firm D as *the* firm with which to dry kiln lumber when narrow tolerances, in terms of variations in moisture content, are required.

Furthermore, Firm D supplements contracted work by purchasing lumber to achieve a degree of production autonomy and flexibility, using second-hand machinery such as planers, moulders, and resaws to produce a wide range of products. Its manager argues, "For the investment required to create a computer-generated line, you need big volumes of one thing to justify a $100,000 or half a million expenditure on making a line that is restricted to making maybe five products really well and after that you start making compromises. Our business is customer oriented and job and order specific" (ibid: 213). Overall, investment in Firm D has focused on medium- to high-technology equipment.

The technology employed by the nontenured SMEs reveals some important trends. First, basic multipurpose machinery to attain production flexibility continues to be used. This trend supports the claim by Schoenberger (1987) that sophisticated, computer-assisted manufacturing systems are of less use to small firms pursuing batch production and underlines the influence of uncertainty in promoting flexibility throughout the production system and within individual firms. Second, investment in technology is significantly influenced by the role of the individual remanufacturer within the production system; in comparison to the specialty subcontractor, contractors and capacity subcontractors have invested in relatively less sophisticated machinery, whereas the isolated remanufacturer has invested in a wide range of machinery to attain production flexibility internally. Third, within a flexible production system, the technological flexibility of an individual firm may be relatively low, as exemplified by Firm C. However, this firm plays an important role in promoting flexibility within the industry by enabling production smoothing by contractors who utilize capacity subcontractors (such as Firm C) when business is booming and drop them during recessions.

Production Strategies of Tenured Firms

The possession of tenure cutting rights allows more effective planning and influences production strategies. In Firm E, for example, relative supply security has permitted investment in expensive, sophisticated laminating machinery to create a product requiring fibre of high quality and consistency. While its final product is highly specialized, Firm E uses basic machinery to achieve flexibility in the cutting of wood in the initial stages of production. However, Firm E's strategy differs considerably from that of Firm F, notably with respect to subcontracting.

While Firm F contracted out its entire production as an experiment to test the nature of the business, Firm E does not contract out and only performs specific processes appropriate for the production of the specialized product. Investment in the remanufacturing plant reflects the highly specialized nature of its product and deficiencies in fibre supply. Because sawmills do not cut lumber to the exact specifications required by the specialty product division, the plant uses a bandsaw, rip-saw, and chop-saw to break down lumber and produce the specific dimensions required, and then a laminating machine, a finger jointer, and a moulder to manufacture its specialty product. The plant uses computer-assisted manufacturing on the laminating machine to obtain the precise pressure, temperature, and thickness of glue to satisfy certification standards, while attaining high volumes of a single product. Similar to some nontenured remanufacturers, Firm E uses basic multipurpose machinery to perform breakdown tasks that provide suitable grades and dimensions of lumber for laminating. Firm E is also distinctive in its use of sophisticated computer-assisted machinery that manufactures products of very high value.

In Firm F, flexibility is attained externally through the use of local subcontractors. In using a social division of labour, Firm F employs the technology of five core firms on a regular basis, the subcontractors performing specific tasks within the production system. Thus, the firm uses a specific remanufacturer to perform initial breakdown tasks, a specific firm to perform kiln drying, and a specific firm for planing and moulding.

Four of the subcontractors used by Firm F may be described as specialty subcontractors, used for specific processes in the manufacture of a product, with the other firm used as a capacity subcontractor to break down unsuitable lumber into sellable dimensions and grades. Firm F uses the flexibility of these small remanufacturing firms to produce the required product, varying subcontractors and therefore processes according to demand, wood quality, and price. The lack of investment by Firm F is a reflection of a specific strategy pursued by the parent firm. As the manager of the specialty product division explained, "We are learning about the markets, we are learning about how it is processed, we are learning what we need to do as a manufacturer to produce a piece of wood that is the right grade and size for that person" (Rees and Hayter 1996: 214). The result of this learning, of course, may be to radically change its production strategy.

The wood-remanufacturing industry of BC and Vancouver metro comprises many different types of firms pursuing specific strategies within the production system. Indeed, among these firms, the specialized and variable nature of production is matched by variability in employment strategies, a variability that has important implications for labour (Chapter 8). The remanufacturing industry also has claims as a flexibly specialized production system.

The Extent of Flexible Specialization

For Piore and Sabel (1984), the idea of flexible specialization is best represented by agglomerations of SMEs that use a common labour pool and interact with one another in a variety of ways, both cooperative and competitive (Hayter 1997a: 329-36). Ideally, flexible specialization also implies a commitment to innovation. According to Scott (1988: 53-4), flexibly specialized systems are disintegrated systems that are most likely to develop in highly uncertain economic environments "in which transaction costs are numerous, small in scale, nonstandardized, and unstable over space and time." In turn, a geographic concentration of linkages is both the cause and the effect of various forms of external economies of scale and scope, externalities expressed in various forms of interaction or networking. Following Rees's (1993: 136-63) analysis, the flexibly specialized nature of the remanufacturing industry in the value-added sector in BC is here summarized in terms of the characteristics of interfirm business transactions, the nature of cooperation and trust, and the role of industry associations.

Interfirm Business Transactions

The remanufacturing industry is populated by highly specialized SMEs that engage in a wide variety of complementary as well as competitive tasks, and subcontracting is an important feature (Table 6.5). Indeed, subcontracting patterns can be relatively refined, with contractors using different subcontractors for three or four different processes. In several other ways, the organization of business transactions conforms to a flexibly specialized industrial district. First, there are high levels of uncertainty, although the uncertainties experienced by SMEs in this industry relate to fibre supply as much as market dynamics. Moreover, the constant search for wood is becoming more difficult as highly automated sawmills increase their fibre demands. The remanufacturers even rely on sawmills making errors to provide them with a wood supply. Instability in wood supply is especially important for red cedar.

Second, for individual SMEs the cost of each transaction is high. Bearing in mind that the bulk of SMEs have annual sales of less than $2 million, one truckload of clear grade wood (at 1992 prices) can amount to $30,000. The value of product transactions is similarly high. Whereas sawmill-manufactured lumber averaged $350 per 1,000 board feet in 1992, remanufactured lumber averaged $1,000 per 1,000 board feet, and the most expensive – manufactured by Firm F – fetched $18,000.

Third, transactions are numerous. Timber products are continually exchanged among the remanufacturers, and as firms wish to reduce inventories, the need for transactions correspondingly increases. As Firm A's manager noted, "The whole method of distribution is changing so much, now everything is moving to more of a just-in-time type inventory; the pipelines have got a little thinner so now there is more of a constant flow of product" (Rees 1993: 139-40).

Fourth, transactions are small and nonstandardized. Remanufacturers are essentially specialized batch producers for specific consumers. Because the size of markets does not justify much investment in a wide range of specialized equipment, subcontracting is promoted. The scramble for small lots of timber further confines transactions.

Fifth, the effect of the high-cost, numerous, and nonstandardized transactions is to encourage geographic concentration. Indeed, in 1989, the Lower Mainland accounted for 62 percent of remanufacturing firms in BC. A third of these firms are located within 8 kilometres along the Fraser River in the New Westminster, South Burnaby, and North Surrey area. Two other clusters are found in the Cloverdale area (the intersection of Highways 1 and 15) and in the Fort Langley-Haney area along the Fraser River. These clusters reveal both the importance of transportation routes and the location of initial timber supplies; that is, sawmills along the Fraser and the need for access to one another. In practice, easy access reduces transportation costs for materials and facilitates person-to-person contact. As Firm F states: "The guys I do business with, I'm probably there two or three times a week. I am looking for quality control, making sure they have got the wood, and that they understand what they need to do with it" (Rees 1993: 142-4).

While the great majority of personal and material transactions occur within a relatively short distance, Rees (1993) notes two further observations regarding this pattern. First, there are occasional longer-distance transactions. One SME on the eastern edge of the Lower Mainland, for example, performs highly specialized processes on small-sized wood products.

Although use of this SME requires one-way truck hauls in excess of 40 kilometres, it has a reputation as a high-quality, reliable supply, and as of 1992, it was the only enterprise of its kind in BC. Second, there is considerable crossover subcontracting resulting from subcontractors subcontracting to other companies. Such behaviour also reflects a relatively dense production system.

In the Lower Mainland, where the economic viability of manufacturing is threatened by rising land costs, congestion, taxes, and planning regulations, a question arises about whether the remanufacturing industrial district can survive, let alone grow. Bramham (1993) specifically cited wood-product manufacturing as an industry facing pressure to move to smaller regional centres. Such pressures may well fragment the remanufacturing production system, and Rees (1993: 145), for example, notes that Firm C would like to relocate farther east into the Fraser Valley. On the other hand, Firm E still prefers its present location to another in Washington that is farther away from its fibre source.

Cooperation and Trust

Among remanufacturing firms, there is considerable competition for fibre and for markets in the remanufacturing industry and many firm-supplier (subcontracting) transactions. There would also seem to be plenty of potential for opportunism, especially because subcontracting is scarcely, if ever, underpinned by formal legal contracts. Purchasing and selling is by verbal agreement or fax, with purchase orders sent after the work is received, although such orders are not always sent. The absence of formal contracts is clearly conducive to highly flexible business relationships by reducing the level of commitment between contractor and subcontractor. If one supplier cannot provide a service, the contractor simply looks elsewhere. Moreover, any particular work order evolves with market demands (for example, for changes in specification or design requirements) without the inconvenience of a new agreement.

In this world of relatively informal, adjustable business agreements, trust is vital to maintaining overall stability and arresting hyper-competition and exploitative opportunism. Despite its inherently intangible nature, trust is neither a simple act of faith nor an expression of moral superiority. As the manager of Firm F notes, "There is an unwritten rule that if you don't trust the guy you don't do business with him, period." Such unwritten rules represent what Lorenz (1992) refers to as a norm of competition. In the remanufacturing industry, untrustworthy firms are penalized by loss of future business. Reputation is therefore an important consideration in choosing suppliers and word-of-mouth information may be reinforced by organizations such as the North American Wholesalers Lumber Association (NAWLA) that provide credit information to subscribers on remanufacturers, sawmills, and wholesalers. Moreover, remanufacturers enforce the norms of competition through collective action against noncompliant firms. As Firm B states, "Our industry is very truthful and honest, so if somebody is playing games and lying, we like to let everybody in the industry know about it, especially if he has hurt you in some way. So you warn them ... the goal is to get this guy out of here rather than let him go on and screw each of us one at a time. Let's stop him now. They don't last long because nobody gives them business" (Rees 1993: 149).

In turn, trust promotes a sense of community that is further enhanced through business networking (industry association meetings, seminar groups, Christmas dinners) and social networking (golf and so on). Rees (1993: 152) found that remanufacturers shared much "common information," particularly for machinery, equipment, labour relations, and wage levels, as well as about unreliable firms and customers. Such information sharing may be regarded as another norm of cooperation. On the other hand, cooperation does not extend to more strategic information on a specific and specialized niche that firms may be developing.

Industry Associations

Industry associations, maintained by the subscriptions of members, are an important institutional expression of cooperation. Until recently, two main BC-based associations specifically addressed the issues of SMEs in the remanufacturing industry, namely the British Columbia Wood Specialties Group (BCWSG) and the Independent Lumber Remanufacturers' Association (ILRA), whereas the Council of Forest Industries (COFI) serves the forest industry as a whole. Very few remanufacturers (mainly the branch plants of large corporations) belong to COFI, but the BCWSG and ILRA have substantial memberships at an annual (1993) cost of $2,000 and $600 per member, respectively. In addressing their mandates, both groups hold seminars and workshops, conduct market research, and lobby governments on behalf of their members (see Rees 1993: 154-63). In 1997, the BC Council of Value-Added Wood Processors (CVAWP) was formed to provide a uniform voice for six separate regional associations.

Clearly, such associations only exist to the extent that members wish to pay dues, although the BCWSG does receive government funding as well. Members that are critical of services rendered can withdraw, and for example, Firm B has threatened to withdraw from the BCSWG because of its lack of marketing effectiveness. In general, however, Rees finds the associations to be beneficial to SMEs in representing the industry on common problems (for example, lobbying over American protectionism and setting regulatory standards) and in promoting personal contact and exchanges of information on a variety of matters. Their impact on training was deemed less influential. Rees also stresses the importance of the NAWLA to remanufacturers, because it provides members with quick, accurate credit information and offers members prestige or "profile" throughout the industry.

Thus, industrial associations are an important expression of external economies of scale that serve to cement communication throughout an industry while providing services that would be uneconomic or at least not sensible for individual firms to provide themselves. At the same time, associations face limits in the role they can play, especially in such settings as the remanufacturing industry where individual firms strongly value independence.

Conclusion

Since the early 1980s, SMEs have become more important in the BC forest economy, especially in activities such as remanufacturing and log-home and prefabricated-home building. SMEs have demonstrated considerable flexibility in responding to difficult wood-fibre supply conditions and in seeking out market niches. There are even indications of the incipient emergence of flexibly specialized industrial districts. Moreover, the low job per cubic metre of wood harvested in BC suggests that the potential of SMEs remains considerable. In the 1990s, however, available estimates suggest that the shift towards higher-value production by SMEs has slowed, maybe stopped.

Provincial forest policy, specifically the high-stumpage regime, anticipates a stronger role for SMEs in adding value, and it assumes that SMEs have the flexibility to overcome increasing stumpage charges and more restrictive logging regulations. This assumption appears to have been stretched beyond its practical limits, however, and provincial forest policy now threatens this potentially more vibrant role for SMEs in the forest economy by imposing a too-high cost and regulatory burden. Provincial policy further hampers the emergence of SMEs by supporting the adversarial model of labour relations and the extension of unionization into a SME-dominated sector, specifically, silviculture. Unions serve a public purpose, but SMEs and unions do not mix well; provincial policy needs to be more nuanced in this regard. In the future, the anticipated shift towards eco-certification will also pose problems to SMEs, more so than to large firms, unless appropriate steps are taken.

Continuing support for FDI, as well as subsidies for large companies in small towns, further militate against a vibrant SME sector. SMEs continue to be acquired by foreign firms, a trend with mixed blessings for local development in BC. Small house-builders that have been acquired by Japanese firms, for example, gain access to Japan's market without learning much about Japanese demand (Reiffenstein 1999). Finally, as the next chapter intimates, trade friction with the US is now encompassing value-added activities and directly undermining SMEs throughout the province (Hamilton 1999b). This development contradicts another assumption of the high-stumpage regime, namely, that it would resolve these frictions. They have become worse.

More generally, economic development among western economies, in theory and policy, has been dominated by a dual model of the economy that distinguishes giant MNCs ("planning system firms") from SMEs ("market system firms"). Under Fordism, planning-system firms were assigned the central role (as hero or villain), and in the present era of flexibility, market-system firms have been reestablished as the key to economic vitality. Public policy has paralleled this shift. Thus, in the 1950s and 1960s, regional development policies emphasized attracting large firms and their branch plants; since the 1970s, more attention has been given to SMEs. The BC forest economy is an exemplar of this thinking, in theory and policy.

In practice, the size distribution of firms (and factories) is more varied than the dual model of business segmentation implies. Many firms do not fit into this duality. Slocan, for example, since its formation in 1978 has grown to a firm of 4,000 employees, all in BC; it is not a SME, but it is not a giant MNC either. In addition, interfirm relations among SMEs, large firms, and giants are extremely complex, partly competitive and exploitative, partly cooperative and reinforcing (Hayter, Patchell, and Rees, 1999). In the context of the BC forest economy, it is time to recognize the complexity of the size distribution of firms and to acknowledge that the potential to exploit economies of scale and scope varies considerably among firms. Public policy should support the realization of these benefits and give more recognition to complementary relationships among SMEs, large firms, and giants. More attention should be paid to linking the fortunes of firms of different sizes together. The problem is not lack of competition, as some SME advocates propose (Fulton 1999); it's lack of cooperation. Admittedly, the institutional mechanisms in an adversarial culture that promote balance and cooperation among the size distribution of firms, and between union and non-union workers, are unclear. Such institutional arrangements should be a policy priority.

7
Trade Patterns and Conflicts: Continentalism Challenged by the Pacific

For the BC forest industries, life on the geographic margin has meant a search for distant markets in the industrial powerhouses of the world, notably the US but also Europe and Japan. Accessing such markets has never been solely a matter of economics. Rather, politics has always been implicated, most obviously in the form of trade policies and relations. Indeed, in the present transformation of the forest economy from Fordism to flexibility, the politics of trade are of paramount concern (Percy and Yoder 1987; Hayter 1992).

Since the early 1980s, BC's forest-product exports have been embroiled in the politics of American protectionism, specifically as regards lumber, and the politics of environmentalism, an important theme of which is to encourage consumer boycotts in Europe as well as in the US. Until recently, the spectacular growth of lumber exports to Japan helped offset the problems of accessing American and European markets. The present Asian economic crisis, however, means that BC's forest producers are facing a triple whammy: explicit opposition in American and European markets by powerful industry and environmental lobbies, and reduced demand in Asia. The extreme export dependence of BC's forest economy (see Table 3.3) combined with an emphasis on commodities, has again proven to be an extremely vulnerable condition.

This chapter interprets the export dynamics of BC's forest industries from a political economy perspective. Canada-US trade relations provide the basis for this discussion because the US has been BC's dominant export market for most of the twentieth century. As with Canada's economy as a whole, BC's forest industries largely evolved within the implicit but powerful guidance of the principles of continentalism, which closely tied BC's fortunes with the those of the US. In the early 1980s, these ties came under the unexpected threat (and reality) of protectionism; paradoxically, the 1989 Free Trade Agreement (FTA) included Canadian acceptance of restrictions of softwood lumber exports to the US. In fact, if protectionism is the antithesis of free trade, it is not necessarily inconsistent with the principles of continentalism that assign hegemony to the US. However, protectionism did stimulate interest in the Japanese market, hitherto considered remote, difficult, and highly protected. This interest was not dissuaded by looming environmental boycotts in Europe.

In outline, this chapter first briefly reviews the growth of BC's forest-product trade under continentalism. The second and third parts examine the nature and implications for BC of the softwood lumber dispute with the US (Percy and Yoder 1987; Hayter 1992). Finally, the growth and organization of the Japanese connection is explored (Edgington and Hayter 1997; Hayter and Edgington 1997; Hayter and Edgington 1999). Given the focus on the nature and implications of BC's trade imbroglio with the US, European trade and environmental politics are only given superficial treatment (see Wilson 1998; Chapter 10). Because the federal government is ultimately responsible for trade, BC's trade relations are placed in the national context.

Continentalism: Trade Imperative for BC's Forest Industries

In Canada, since before Confederation, debates about development have invariably engaged the proponents of continentalism – whose priority is the creation of linkages with the US – and the critics of continentalism, the proponents of nationalism – whose priority is the creation of linkages within Canada. Indeed, the pros and cons of continentalism, or "reciprocity" as it was called, is a recurring theme in federal elections. Following Confederation, for example, Macdonald's conservatives failed to obtain a free trade agreement with the US, and subsequently won an election on the promise of the National Policy of 1879, comprising a tariff on manufactured goods, stimulation of agricultural settlement in the west, and the building of transcontinental railroads. The 1911 election was again fought largely on the basis of a proposed free trade treaty with the US. The conservative party, representing the forces of nationalism, won again.

Yet, in practice, the 1911 election was the watershed for the forces of nationalism (Hayter 1993). The extraordinarily rapid industrialization of the American economy soon engulfed the Canadian economy and resources (Marshall et al. 1936). Indeed, it is difficult to think of any other nation that is so tied to a single dominant trading partner (Hayter 1990). These ties imply a distinct political economy.

Principles of Continentalism: General Reflections

Continentalism implies for Canada, including BC, a distinct model of international relations, a distinct set of economic beliefs, and a specific spatial division of labour between the two countries. In particular, continentalism assumes the political hegemony of the US and Canada's integration with American defence planning; economically, continentalism means a commitment to free trade and the free flow of investment; and in terms of the spatial division of labour, continentalism casts Canada in the role of resource supplier to the American industrial machine. Defined in these terms, geopolitical connotations frame the support for continental free trade (Clark-Jones 1987; Haglund 1989). Thus, for the US, Canada offered vital resources that were considered politically and economically strategic, and American trade policy was designed to access these resources without compromising American industrial strengths. Thus, while Canadian resources were permitted to enter the US without tariffs, higher-value goods were restricted. This policy reached its zenith during the 1950s and 1960s, that is, during Fordism and the height of the Cold War.

For its part, Canada sought the security of American military defence, as well as access to American markets, by becoming a stable and secure resource supplier. In short, Canadian foreign policy was seen as a complement to American geopolitical responsibilities and Canadian governments *wanted* Canadian resources to be considered as domestic American resources (Clark-Jones 1987: 118).

Within Canada, including BC, the subsidization of infrastructure, often on a massive scale, the sale or lease of resources at a low cost or "rent," and facilitation of the entry of foreign capital were policies consistent with Canada's role as the continent's resource supplier (Aitken 1961; Clark-Jones 1987; Marchak, 1983). Only rarely have Canadian governments sought to limit resources to use in domestic manufacturing. Restrictions on log exports, introduced in the late nineteenth century, are exceptions. Moreover, nationalistic opposition to continentalism in Canada is increasingly ineffective. In the context of trade, the idea of the "third option," based on a trade diversification with the US, was mooted in the 1970s without results. The economic problems of the early 1980s encouraged Canada to publicly examine its global role. After much debate and a federal election in 1987, the Canadian government chose to confirm continentalism with the 1989 Free Trade Agreement (FTA) with the US. The US soon parlayed FTA into a North American Free Trade Agreement (NAFTA) to include Mexico. The consequences for BC have contradicted all expectations.

BC as a Stable and Secure Supplier of Forest Commodities

Newsprint, pulp, and softwood lumber were among the key staple industries in the forefront of north-south continental integration during the twentieth century. In the pulp-and-paper industry, continentalism's influence was established in the first two decades of the twentieth century. Thus, in twenty years, the Canadian industry expanded its output over ten times from 181,000 to 1,902,000 tonnes, and changed from a small, diverse domestic industry, with 79 percent of its output sold locally, to a specialized large-scale commodity industry with 69 percent of its output exported, largely to the US (Uhler et al. 1991: 108). The basic commodity mix of the contemporary industry was essentially established by 1920: in 1920 and 1987 about 75 percent of the industry comprised newsprint and market pulp, most of it for export to the US.

Table 7.1

American pulp-and-paper tariffs

Year	Newsprint	Bleached pulp	Printing and writing (cut)	Envelopes
1894	15%	10%	20%	20%
1909	17.5%	$5.00	$60.00+15%	20%
1911	free	$5.00	$60.00+15%	20%
1930	free	free	$60.00+22.5%	5%
1947	free	free	15%	2.5%
1972	free	free	7.5%	8.75%
1987	free	free	3%	3.5%

Notes: Dollar amounts indicate dollars per short ton. Also note that the tariff on bleached (and unbleached) pulp was removed in 1913.
Source: Uhler et al. 1991: 120-1.

Escalating American demands, at a time when American softwood forests were being rapidly depleted, stimulated the northward migration of the pulp-and-paper industry to Canada; first, by the introduction of restrictions on raw log exports by provincial governments between 1900 and 1910, and then by the American decision to remove tariffs on pulp and paper, a decision encouraged by newspaper interests within the US. While before 1939 most pulp-and-paper mills were built in Ontario and Quebec, several were established in BC, beginning with the 1912 opening of the Powell River newsprint mill.

In fact, since 1911 there has been virtually no American tariff on Canadian newsprint imports, and by 1913, the tariff on pulp (bleached or unbleached) had been removed (Table 7.1). In contrast, tariffs on other paper products, such as printing and writing papers and envelopes remained relatively high; in the case of envelopes, the tariff was reduced during the 1930s and 1940s but was substantially increased after 1947. Such tariffs, however, have clearly helped constrain the Canadian industry to a commodity orientation of newsprint and pulp (Uhler et al. 1991).

The situation for wood products is broadly similar. In 1909, the Dingley Tariff reduced the American tariff on imported dimension lumber from $2 to $1.25 per 1,000 board feet, and in 1913, this tariff was eliminated. In BC until that time, the markets for lumber were relatively diverse, as city and railway construction and prairie settlement generated substantial domestic demands to supplement a variety of offshore markets (Chapter 2). By the 1920s, however, the US had become the dominant market for BC lumber. Although the US reimposed tariffs during the 1930s Depression, and Canada entered a Commonwealth Preference Scheme, by 1935, a reciprocity treaty between Canada and the US led to the rapid elimination of the tariff on dimension lumber. As with pulp and paper, tariffs on more-processed wood products remained.

Table 7.2

Geographic distribution of British Columbia's exports of forest products, 1962 and 1987 ($ million)

	1962		1987	
United States	422.3	(73.6%)	4,788.0	(51.5%)
United Kingdom	69.8	(12.2%)	486.7	(5.2%)
Japan	18.2		1,769.4	(19.0%)
Australia	14.3		206.7	
Germany	6.1		in "other EC"	
China	in "others"		199.1	
South Korea	in "others"		115.8	
Other EC	in "others"		1,169.0	(12.6%)
Others	43.3		553.5	
Total	**574.0**		**9,288.2**	

Source: British Columbia, Government of, 1963 and 1988.

Fordism: The Handmaiden of Continentalism in BC

During the 1950s and 1960s, the explosive growth of BC's forest industries was narrowly based on a few, specialized commodities to access American markets (Marchak 1983; Table 7.2). Powell River's newsprint sales in 1953 (and 1954 plan), for example, illustrate a sole concern for western continental markets, with over one-half of its sales in California, accounting for 35 percent of the state's total newsprint consumption (Hayter and Holmes 1993). For Powell River, perhaps the world's largest newsprint mill in the late 1950s, this pattern of sales remained stable for almost three more decades. Lumber, the province's biggest forest industry, overwhelmingly relied on continental, especially American markets. Indeed, the rapid adoption of large sawmill technology in the provincial interior after 1960 featured dedicated machinery designed to cut a limited range of construction timber, especially two-by-fours, linked to continental housing needs. These so-called high-volume spaghetti factories illustrate the essence of Fordism. Kraft pulp, the fastest growing commodity, was not as tied to American markets, but these markets were, nevertheless, important.

Among BC's main commodity industries, softwood-plywood is an exception to dependence on American markets. Softwood-plywood manufacturers supplied construction sheathing mainly to domestic producers with some important offshore links. Similarly, the province's corrugated-box and paper-bag industry was relatively small and oriented to western Canadian markets.

This continentally based "core-periphery" pattern of trade, in which the BC periphery fed the American industrial metropole, reached its zenith in the Fordist period and was buttressed by government policy. While American tariffs, and nontariff barriers, on newsprint, lumber, and pulp remained low or negligible, more-processed goods faced tariffs. While these tariffs, under the influence of the General Agreement on Tariffs and Trade (GATT), were lower than before 1939, they remained significant and reinforced established production structures that emphasized commodities in BC and value-added products in the US. These production structures were further reinforced by the federal government's encouragement of a high level of foreign investment in BC's forest industries (Chapter 5). Facing no investment barriers, the motives of foreign firms were consistent with trade regulations. The softwood-plywood industry was not rationalized along continental lines for trade; and the American and Canadian governments provided the industry with tariff protection.

The Recessionary Crisis of the Early 1980s: Continentalism Revised

Perhaps the least appreciated impacts of the recessionary crisis of the early 1980s relate to trade. That recession severely hurt American and BC producers, but for the Americans an obvious remedy was at hand: limit BC (Canadian) exports. The consequences of this "remedy" – and the failure of Canadian governments to effectively respond – haunt the industry to this day.

The defining moment, when the industry's ingrained trade patterns and relations began to be fundamentally challenged, occurred during the 1982 recession when US-based lumber interests sought protection against Canadian (but specifically BC) exports. Following lengthy discussions, the American and Canadian governments signed a memorandum of understanding (MOU) in December 1986 that imposed an export tax on Canadian lumber exports to the US. Ostensibly, American actions were motivated by protectionism and rationalized by claims that BC stumpage constituted a subsidy. As Percy and Yoder (1987: xxvi) advised, with considerable prescience, the MOU "is historic with enormous implications for the Canadian forest-product industry and Canadian resource industries, in general." Percy and Yoder particularly note that the lumber debate and the MOU marked the first time that American governments had directly challenged the resource-management policy of another country and that the debate was resolved primarily on the basis of legal-political arguments rather than on economic grounds.

Moreover, the American trade challenge targeted BC. Whether or not the initial American motivation was solely that of protectionism, the ongoing inheritance of the MOU, which was grandfathered into the Canada-US Free Trade Agreement of 1989, is legitimization of American interference in the question of stumpage, and therefore, in the industrial use of BC's forests. Within BC, the added costs and uncertainties of accessing American markets created by the MOU and subsequent modifications of this deal, encouraged lumber producers to look for alternative markets, notably Japan, a serendipitous shift because Japan began to look for additional supplies of lumber at more or less the same time. However, the MOU and related deals have raised far-reaching questions, beyond how to replace exports to the US with exports elsewhere. For BC, the MOU has heralded a period of increasing continental influence on the province's forest-resource management, while trade ties remain vulnerable.

The Softwood Lumber Dispute, 1982-98

In outline, the lumber dispute centred on two countervail actions initiated by American lumber interests (specifically, the US Coalition for Fair Canadian Lumber Imports or CFCLI) against Canadian exporters in 1983 and 1986 (Table 7.3). This dispute was formally resolved by the two federal governments in December 1986 by a memorandum of understanding (Percy and Yoder 1987: Appendix B). The MOU imposed an export tax on Canada's lumber exports to the US while allowing for higher stumpage payments to provincial governments in lieu of the export tax, an option that BC subsequently accepted. The MOU was then grandfathered within the FTA. In fact, the lumber controversy was cited by the Canadian federal government as a reason for entering into free trade talks with the US, so that the MOU became an implicit condition for the FTA. In 1996, the MOU was revised and replaced by the Softwood Lumber Agreement (SLA) that featured export quotas and a system of penalties for exceeding the quota. The five-year SLA is already in dispute.

Bargaining over Continentalism: Round 1 – BC and Canada Win

The US Coalition for Fair Canadian Lumber Imports, an organization representing eight trade associations and 350 lumber companies from all regions of the US, began the softwood lumber dispute in July 1982 by petitioning the International Trade Commission (ITC) to countervail, that is, to establish tariffs on Canadian lumber imports (Table 7.3). The CFCLI (whose membership is kept secret for reasons unknown) was formed in 1981 following a report commissioned by the Northwest Independent Forest Manufacturers (NIFM) that blamed "subsidized" Canadian imports for low prices and loss of market share by American producers. Low stumpage was targeted as the subsidy in question. The years 1981-2 were a time of recession and collapsing lumber prices in the US and Canada, and in the Pacific Northwest many firms were locked into high-priced timber contracts on public lands they had bid for in the 1970s boom. Although the American government enacted legislation in 1984 that excused firms from these contracts and high stumpage, countervail action against Canadian producers, led by Senator Packwood of Oregon, was already under way. This action was the first time that another country's natural resource policy had been attacked by American countervailing duty law.

In the event, following determinations by the ITC, the International Trade Administration (ITA), and the Court of International Trade (CIT), the US Department of Commerce (DOC) in late May 1983 found in favour

Table 7.3

Chronology of American countervail action against Canadian softwood lumber imports, 1981-96

Dispute 1	
1981	NIFM report alleges Canadian imports are a major cause of job loss in the Pacific Northwest; CFCLI formed; Senate Finance Committee to direct ITC to investigate if Canadian stumpage subsidizes exports
1982	ITC's report released
October 1982	Countervail action launched
May 1983	US Commerce Department rejects countervail
Dispute 2	
September 1985	Canada requests (general) free trade treaty with US; CFCLI lobbies for second countervail as a condition for support in Congress for free trade
May 1986	CFCLI files second countervail; rationale is the same
June 1986	Canada protests to GATT
August 1986	Canadian position disintegrates; federal request to provinces to increase stumpage is refused
October 1986	US Commerce Department accepts countervail charge; suggests a tariff of 15 percent
November 1986	Canada proposes export tax
December 1986	The Softwood Lumber MOU imposes a 15 percent export tax, reducible by corresponding increases in stumpage; Canada's petition to GATT withdrawn
January 1989	FTA passed by Canada; trade dispute mechanisms included
Dispute 3	
October 1991	Canada (legally) terminates MOU; American administration self-initiates countervail; CFCLI claims Canada's log export restrictions are a subsidy; USTR threatens tariffs
November 1991	Canada takes dispute to GATT
March 1992	US Commerce Department's preliminary determination finds Canadian stumpage is a subsidy
July 1992	ITC determines lumber imports had materially injured American industry; Canada requests a FTA dispute panel to investigate dispute
February 1993	GATT's report partly supports both sides
May 1993	FTA panel decision supports Canada and rejects subsidy argument

►

◀ *Table 7.3*

July 1993	FTA panel rejects American argument of "injury"
September 1993	US Commerce Department rejects first panel decision and suggests higher tariff; the panel retains its position, not unanimously
October 1993	ITC rejects second panel decision
December 1993	CFCLI alleged Canadian members of panel were in conflict of interest and requested an "extraordinary challenge"
August 1994	Extraordinary Challenge Committee of three upheld panel's decision, but not unanimously
September 1994	Coalition challenges constitutionality of the softwood lumber decision and panel process in US Court of Appeal; successfully changes American countervail law to make implementation easier
December 1994	WTO legislation passed in US that reduces American discretion over definition of subsidy; Canada reimbursed for import duties; a consultative process established; CFCLI abandons its constitutional challenge to panel mechanism
1995	Consultations become negotiations; fourth countervail looms; a 25 percent tariff is proposed
April 1996	The Softwood Lumber Agreement commences for five years; this agreement imposes quotas beyond which tariffs are calculated

Note: Special thanks are due to Mike Apsey for help in compiling this table.

of Canadian interests, ostensibly represented by the Canadian Forest Industries Council (CFIC) but in practice led by the Council of Forest Industries (COFI), one of eighteen trade associations in the CFIC. (COFI's membership accounted for roughly 90 percent of BC's forest-product output at that time.) In Keohane and Nye's (1977) terms, Canada won this round of the conflict because the outcome coincided with Canada's objective – prevent the countervail.

Round 2: The Deal Reopens – The US Wins

The "once and for all" determination noted above proved short lived. In early 1985, the first of several bills requesting punitive tariffs on Canadian lumber imports for up to five years was introduced in Congress, and the CFCLI re-petitioned their case in May 1986 (Table 7.3). Preliminary decisions by the ITC in June and the ITA in October favoured the petitioners and a final decision was required by the end of 1986. In fact, since 1983 the political economy of the lumber dispute had evolved considerably. According to American trade law, tariffs can be imposed if imports are causing injury to American industry and/or if they are subsidized. Clearly, the stimulus to the lumber countervails lay in Canadian lumber's share of the American market, which increased from an already impressive 28.1 percent in 1980 to 30 percent in 1983 and 33.2 percent in 1985 (Table 7.4). In this period of depressed markets, these imports added to the difficulties of the American lumber industry. Whether the increase in market share resulted from market forces or underpriced timber, however, is contentious.

Whereas the CFCLI claimed Canadian stumpage was too low, COFI argued that the principal determinants of Canada's growing market share were a depreciating Canadian dollar that benefited Canadian exporters receiving revenues in American dollars, and superior Canadian productivity resulting from aggressive modernization programs in mills traditionally oriented to international markets. COFI's position was supported by Percy and Yoder (1987), who argued that the relatively lower Canadian stumpage, compared with the US, is explained by differences in species composition, accessibility, the size, quality, and density of timber terrain, and timber-pricing mechanisms, and by increased restrictions on timber supply in the US in certain national forests.

In the hearings, economic explanations of increased market share and of stumpage variations were largely ignored, especially for the second countervail. Rather, the discussion focused on legal reasoning. Moreover, in 1986, unlike 1983, the CFCLI was effective in claiming low Canadian stumpage as a form of subsidy because the interpretations of the key concepts in American countervail procedures evolved, and in effect, became looser and more arbitrary under the influence of new trade legislation and court rulings. Details of this complex discussion are available elsewhere (Percy and Yoder 1987). In brief, in 1983 the ITA concluded that stumpage practices were not countervailable, in large part because stumpage programs did not automatically restrict timber specifically to the lumber industry (rather, Canadian timber in principle was "generally available"), and because stumpage prices did not discriminate unfairly by providing timber to exporters at preferential rates. Indeed, BC's method of assessing stumpage was judged reasonable. In contrast, in 1986, the ITC accepted the view that lumber and other forest industries were essentially the *same* and that stumpage was provided specifically to the lumber (forest) industry.

Table 7.4

Canada's share of American lumber markets and US-Canada exchange rates in the 1980s

	1980	1983	1985	1989
Market share (%)	28.1	30.0	33.2	28.1
$US per $Can	0.85	0.81	0.73	0.85

Source: Hayter 1992: 159.

As Percy and Yoder note, "It is difficult to imagine how any government program designed to allocate natural resources to the private sector could not be in violation of this interpretation of the specificity test" (1987: 100). Moreover, the ITC accepted a new and inflated definition of the value of Canadian timber that included, in part, both the revenues received by governments from the forest industry and forest management costs incurred by governments. Some double counting of timber values led the ITC to conclude that Canadian stumpage, whose precise calculation is typically related to assessed value, was too low, and therefore awarded at "preferential," that is, subsidized, rates.

The Memorandum of Understanding (MOU)

Under American trade law, following the ITC's preliminary decision in June in favour of the petitioners, Canadian interests had exactly six months to appeal; otherwise, tariffs would be imposed. After intense negotiations, during which the Canadian federal government replaced COFI in representing Canadian interests, on 30 December 1986, hours before the deadline, the two federal governments signed the memorandum of understanding and the countervail action stopped.

The MOU required the Canadian government to impose a 15 percent tax on lumber exports to the US. It also stated that this export tax could be reduced or eliminated only if stumpage, or related charges, was increased by provincial governments and agreed on by the two federal governments. The MOU also required Canada to provide the American government with all the information necessary to monitor the effectiveness of the agreement. In December 1987, the MOU was amended, most notably with measures to replace the export tax. In part, the amendment clarified the intent of the MOU towards provincial replacement measures by incorporating the following paragraph: "The Government of Canada agrees to monitor the operation of replacement measures adopted by the provinces. The Government of Canada agrees to reimpose the export tax to the extent necessary to offset fully any reduction that has occurred in the value of the replacement measure. Calculation of the value of the reduction will be subject to agreement between the two governments" (quoted in Hayter 1992: 161).

This same amendment permitted BC to replace the export tax with the higher stumpage rates it had introduced in October 1987 and by the full costs of forest renewal, the responsibility for which it had largely transferred to long-term industry timber licensees. In another amendment, Quebec obtained a partial reduction of the export tax as a result of higher stumpage, whereas other provinces, notably Alberta and Ontario, have preferred to pay the export tax. (Atlantic Canada was exempted from the MOU.) If Canadian negotiators claimed the export tax at least kept revenues – that would have been lost with a tariff in place – in Canada, in Keohane and

Nye's (1977) terms, the outcome of the second countervail was much closer to the American objective of protectionism.

Round 3: Revising the MOU – The US Wins Again

COFI, primarily concerned about the impact of the MOU on stumpage, wanted the MOU reopened. This lobbying was initially ineffective, in part because the Canadian government did not wish to derail the negotiations leading to the FTA. On the American side, there was no support for change. In late 1990, for example, American federal officials indicated that, even if they received a formal request from Canada, they did not intend to reopen MOU, not least because of the political difficulties it would create (Canadian Press 1990). Indeed, CFCLI's chair stated that, although the MOU is "acceptable," it should have been "stronger"; that exchange rates are irrelevant to the principle of the lumber tax; and that the MOU "helped the provincial governments in Canada, which were virtually giving away standing timber, and it helped the US industry as well" (Durrant 1990b: 33; see also 1990a).

Yet, with the FTA almost two years old, and possibly stimulated by criticisms over loss of autonomy in forest management and the growing crisis in the BC lumber industry, Canada withdrew in September 1991 from the MOU. The basis of the Canadian action was that the intent of the MOU had been met: in BC, stumpage levels imposed higher costs than the export tax and further increases in stumpage were planned. The FTA and the trade-dispute mechanism were also in place. In addition, the MOU allowed either party to withdraw if thirty days' notice was provided. In fact, in extensive consultations, Canada provided considerable documentation, using American accounting systems, to show that stumpage was not a subsidy (while pointing out the large number of below-cost timber sales in the US). Senator Packwood, and other senators supported by the CFCLI, chose to retaliate, and in October 1991, the American government accepted their concerns. In a hitherto unprecedented action, the American government "self-initiated" a third countervail against Canadian lumber (see Table 7.3). In late 1991, the American government imposed a bonding requirement (tariff) on imported Canadian lumber.

In self-initiating the third countervail, the DOC allowed for a third change in the definition of a subsidy based on comparisons between the two countries. Essentially, by reducing the costs of raw logs, the CFCLI now identified Canadian log-export restrictions as a subsidy (even though the US had similar restrictions). This argument still has significant implications for BC.

Over the next five years, American and Canadian arguments and counterarguments occurred in various forums and progressively took on new dimensions (see Table 7.3). In outline, Canada took the case to GATT in late 1991, and in February 1993, GATT's report was equivocal. In the meantime, the DOC, hearing its own case, now concluded, contrary to its conclusion of 1983, that Canadian lumber was subsidized (because of export log restrictions). After GATT, Canada proceeded to the FTA's Chapter 19 Binational Panel Review, which had been conceived with the 1986 softwood lumber case in mind. In 1993, Canada won major victories in two separate decisions; the FTA panels rejected the subsidy argument and the American claim of injury. Later in the year, the DOC and the ITC rejected the panel's decisions but could not get them reversed (although the American panellists changed their minds). Subsequently, at the end of 1993, the CFCLI initiated an "extraordinary challenge" on the basis that the Canadian members on the dispute-mechanism panels were in conflict of interest. The CFCLI lost this challenge in two separate courts, although in one court, the dissenting American judge forcefully opined Canadian panel members were biased.

Undeterred, in 1994 the CFCLI embarked on a constitutional challenge that argued Chapter 19 had denied them due process and was in violation of the American Constitution. Because this challenge threatened the integrity of FTA and NAFTA, the American government balked at providing support, and the challenge was dropped. Instead, the CFCLI sought to amend American law, specifically for the World Trade Organization (WTO) agreements that stemmed from the Uruguay Round of multilateral trade talks. The results were blurred. On the one hand, the US Congress criticized the FTA panels and passed amendments to American countervailing duty law to facilitate the identification of a subsidy. For example, the DOC no longer has to determine economic effects (injury) within the US. On the other hand, the WTO Subsidies Agreement amends the idea of a subsidy to mean a financial contribution by government that confers a benefit. Such a definition should mean log-export controls are not subsidies. While the American government apparently disagrees with this interpretation, the WTO legislation was passed by Congress in late 1994, and the softwood lumber industry entered into the final phase of Round 3.

The Softwood Lumber Agreement of 1996

In early 1995, the American government fully reimbursed the Canadian industry for duties collected since 1991 of approximately $2 billion. The two sides also agreed to engage in a consultative process to avoid further costly litigation. At the same time, the CFCLI abandoned the constitutional challenge. Yet, with the support of the American government, the CFCLI continued to press the charge that Canadian lumber was subsidized, and another tariff was threatened. In late 1995, one bill proposed a 25 percent tariff. In Canada, attempts to obtain an agreement with province-specific annexes proved impossible. To preempt another American tariff and further costly, exhausting negotiations, the two countries in 1996 capped lumber exports from Canada to the US in a five-year agreement (Table 7.5).

Table 7.5

Main features of the US-Canada Softwood Lumber Agreement, 1996

(1) The SLA is a five-year agreement commencing 1 April 1996.

(2) The SLA imposes an export tariff-rate quota. In particular, Canada is allowed to export 14.7 billion board feet of lumber to the US free of tax. An additional 650 1,000 board feet is subject to a tax of $50 per 1,000 board feet. Shipments above that level are subject to a tax of $100 per 1,000 board feet.

(3) If the price of lumber exceeds a trigger price for an entire quarter, Canada is entitled to ship an additional 92 million board feet tax-free in the four quarters following that quarter.

(4) Lumber will be placed on Canada's Export Control List, and export permits are required so that the province of origin and volumes can be tracked.

(5) Disputes about volume are referred to the dispute settlement mechanism under NAFTA Chapter 20.

(6) Other aspects of the SLA include that the American government agrees not to launch further trade action against lumber during the life of the agreement, and the Canadian government agrees not to reduce costs of timber without American approval.

Note: The advice of Mr. Mike Apsey on the SLA (and the entire trade dispute) is gratefully acknowledged.

The Softwood Lumber Agreement (SLA) is an export quota agreement. The agreed-upon level is 14.7 billion board feet of lumber permitted to be exported duty free from Canada to the US. Within Canada, BC received 59 percent of the quota, and firms in the coastal region received a relatively smaller quota than interior firms because of the former's exports to Japan. BC hoped the SLA, although fundamentally contrary to the idea of free trade, would bring trade stability to the lumber industry. This expectation has not been met. Within two years of the signing of the SLA, the Japanese market collapsed, and BC lumber firms found themselves out of quota in the US, lacking compensating Japanese markets and with higher wood costs. Several coastal-based BC firms have expressed strong opposition to the SLA and indeed against NAFTA itself (Hamilton 1998g). Meanwhile, American firms have continued to harass Canadian producers by broadening the definition of lumber; according to one forestry consultant in June 1999, US Customs is targeting at least sixteen products previously exempt from the SLA, threatening 1,000 jobs in BC and $2 billion worth of Canadian exports (Gibbon 1999a). In response, forest-product firms are requesting the federal government take legal action against the American reclassification.

Reflections on the Canadian Approach to the Softwood Lumber Dispute

After almost two decades of trade conflict, the governments of British Columbia and Canada have failed to eliminate American protectionism aimed at the Canadian softwood lumber industry, and they have been unable to prevent American influence over domestic forest-resource policy. In an era of so-called free trade, politics has dominated the new trade environment of the softwood lumber industry in Canada, especially in BC where it is concentrated. In 1980, BC softwood lumber firms had tariff-free access to the American market, market share depended on competition among firms, and no one questioned the provincial government's autonomy in setting the price of its own timber. In 1996, export volumes of softwood lumber to the US were limited by political agreement. Within Canada and BC, the quota is politically adjudicated, and the provincial government feels obligated to submit its timber-pricing (stumpage) policies to the Americans. The unified stance of the Canadian forest industry before 1983 now features bickering and personal attack; meanwhile, American interests relentlessly use the SLA to further harass Canadian exporters. The situation is a mess.

For the Canadian and especially the BC softwood lumber industry, which are committed to the ideals of free trade, limited access to the American market and more political interference are trends to be deeply regretted. What went wrong?

One perspective, comforting to the Canadian side, emphasizes the David and Goliath aspect of the conflict. In this view, the US is simply too powerful and – for whatever reasons – was determined to limit lumber imports from Canada. Such a David and Goliath perspective is comforting because it blames Canadian problems on American politics and law. As such, it is only half the story. Continentalism has never been solely about free trade; it has always implied American hegemony. American markets, up to a 25 percent to 30 percent share, have long been available to Canadian lumber producers and still are. Should Canada realistically expect more? In the context of the Canada-US lumber trade dispute, the US – that is, Goliath – clearly has much more bargaining power. Moreover, the US is not particularly "sensitive or vulnerable," in Keohane and Nye's (1977) words, to Canadian pressure. The US has a large domestic industry, and Canada was not in a position to hold back supply. In such a bargaining situation, following Keohane and Nye's advice, Canada – that is, David – had to develop a highly coherent position to have any chance of winning. This was not to be.

The Collapse of Canada's Bargaining Position

In Round 1 of the dispute, Canada's success rested on the coherence of its bargaining position. Industry and the federal and provincial governments shared the same definition of national interest, the objective was clear, and COFI was able to orchestrate Canada's negotiations. In doing so, COFI elicited considerable support from three important American interest groups: US-based multinationals with lumber interests in Canada, the US National Association of House Builders, and the North American Wholesale Lumber

Association. The last two associations, representing American lumber-consuming interests, lobbied particularly hard on COFI's behalf. The two federal government bureaucracies also favoured maintaining free trade.

To some extent, Canada's bargaining position deteriorated for reasons beyond its control. Thus, before the ITC's preliminary decision, members of the US Senate and Congress, representing the CFCLI, successfully politicized the lumber conflict, making it highly "visible," most notably by linking the conflict with the embryonic FTA talks. Specifically, they argued for a lumber tariff before agreeing to the FTA, an argument they reinforced with the threat of an omnibus trade bill that included a natural-resource subsidy provision (Percy and Yoder 1987: 145-7). The limited time (six months) to appeal the ITC's preliminary decision also placed pressure on Canada. Even so, the Canadian federal government chose to accept the linkage between the lumber dispute and the FTA, naively arguing that the FTA (and the associated dispute mechanisms) would ensure similar disputes did not occur in the future.

By summer 1986, Canadian views fragmented and were overwhelmed by the urgent need to negotiate an agreement by American-imposed deadlines and by Canadian willingness to accept a lumber agreement as a precondition to the FTA. Subsequently, with COFI now replaced by the federal government as chief negotiator, disagreements among the Canadian representatives were aired publicly. The most obvious split occurred when the federal government rejected COFI's wish to continue fighting the countervail and to prepare a request for conciliation negotiations at GATT in case the final decision went against Canadian interests. Moreover, COFI did not anticipate that federal and provincial governments would agree to raise stumpage rates. Indeed, industry vehemently opposed incorporating stumpage into a trade agreement because this compromise tacitly accepts the subsidy accusation and because higher stumpage raises costs throughout the industry, not only on lumber exports to the US.

The original suggestion (then secret) for an export tax came from COFI as an attempt to evade imposition of higher stumpage. Subsequently, a new provincial (Socred) government in BC unexpectedly announced in fall 1986 a review of forest management practices and a request that this review be completed before any final agreement between the two federal governments on the tariff issue. In turn, the Canadian government rejected this request and made what was called a final offer to the Americans to increase stumpage in four provinces – at the same time stating that an export tax would be unfair. The American industry rejected this offer, and in October, the DOC issued its preliminary determination of a 15 percent tariff on lumber.

Canadian provincial and federal ministers met to consider a response, and BC considered making its own deal. In the event, the federal government proposed to the US that Canada impose a 15 percent export tax on lumber on the condition that the countervail be withdrawn. If the federal government eventually accepted COFI's idea of an export tax, COFI was still effectively excluded from the final negotiations. COFI was unaware of the MOU's contents until the agreement was signed, and then it was dismayed to discover the linking of an export tax with stumpage. COFI's concern was that although an export tax only applies to US-bound exports, stumpage affects costs industrywide.

In effect, the motives underlying Canada's approach to the softwood bargaining issue became confused after Round 1 when federal and provincial governments became the leading players. Thus, the federal government was primarily motivated by the desire to ensure success of the overall FTA. (The motives of the FTA may have been changed by the softwood lumber dispute. Canadian arguments initially emphasized trade expansion with the US, but they then shifted to emphasize the need to defend existing trade patterns. Certainly, the dispute mechanism was a much-touted aspect of the FTA that directly stemmed from the softwood lumber experience. In the meantime, American interests emphasized the need for Canadian negotiators to compromise on softwood lumber to ensure that the FTA was not derailed.) If the extent to which this linkage actually shaped the MOU is not known, the Canadian approach to the softwood lumber dispute was clearly informed by divided agendas and mixed motives. Provincial governments were attracted by the opportunity for higher revenues from an export tax; in BC, the American countervail was not seen as a threat but as support for the high-stumpage regime. Meanwhile, industry was motivated by its long-standing desire to access the American market as easily as possible. Astonishingly, none of the parties on the Canadian side seemed to grasp the implications for Canadian sovereignty arising from the inclusion of stumpage in the MOU.

The Sovereignty Problem

CFCLI claimed Canadian stumpage to be a subsidy as a way of seeking a tariff. After COFI's initial successful defence of this argument, the willingness of the Canadian federal government, fully supported by provincial governments, especially in BC, to incorporate higher stumpage in the MOU at least appeared to give tacit support to the American view. In the context of the MOU, Canadian governments only perceived higher stumpage as a substitute for the export tax (and COFI only saw higher stumpage as an increased cost). Yet, an enduring, apparently unforeseen implication of incorporating stumpage in the MOU has been to encourage American interest in Canadian forest policy, hitherto considered an area of provincial autonomy. Moreover, American concerns about forest policy, especially in BC, have continued after Canada withdrew from the MOU. These concerns underlay the SLA of 1996, and since that agreement, BC's forest policy initiatives are still haunted by how the US will respond.

Yet, it is worth emphasizing that the principle and contents of the MOU, and the SLA, were negotiated and involved Canadian choices. In the MOU discussions, for example, COFI's original position could have been retained and the case appealed to GATT if necessary. Alternatively, Canadians could have simply accepted a protectionist bill – a leading related suggestion was for the imposition of a three-year (12 percent) tariff starting in 1986. Canadian interests were adamantly opposed to a tariff and saw the revenues associated with either export taxes or higher stumpage as the preferred alternative. But this alternative has ensnared the BC industry in a cycle of costly, frustrating negotiations, market access is still a problem, bickering is rife, and the autonomy of forest management is compromised. A tariff would have imposed clear up-front cost penalties within a time horizon, costs would have been borne by American consumers, the industry would have been more free from politics, and governments would have retained autonomy over stumpage.

The Canadian side still had the option of forcing the US to use its tariff threat in the negotiations leading to the SLA. This option does not appear to have been given any thought. Moreover, according to Tom Stephens, now MB's CEO but once part of the American lobby, the Canadian side too readily acquiesced to the idea of a quota and then failed to effectively bargain for a bigger quota (Hunter 1999). Indeed, Stephens linked the strategies of Canadian softwood lumber negotiators with the appeasement tactics of Neville Chamberlain towards Hitler's Germany. The underlying motive of this comment appears to be to reopen the SLA so that MB can obtain a larger share for the quota, or to encourage the BC government to pursue a policy of privatization to create markets for timber (and even forests) that would remove the basis for the concerns of American protectionists while offering considerable benefits to MB. Other industry leaders, especially in the interior, sharply rebutted Stephens's comments and defended the SLA. (After all, they have a bigger quota.) If this debate is unusually personal, it reveals that the fragmentation of the Canadian side created by the trade dispute now publicly includes industry leaders who hitherto had always presented a united front. Indeed, quota arrangements have inherent tendencies to politicize business relationships, precisely the opposite of what free trade promises. The debate also reflects on how the trade dispute has undermined the once-sacrosanct view of provincial autonomy over the forest resource. There is every reason to agree with the view that from 1986 to now, Canadian negotiators have been overly passive in the face of American aggressiveness.

Low Stumpage Is Not Necessarily a Subsidy

As the ITC initially ruled in 1983, low stumpage does not inevitably imply a subsidy. According to Percy and Yoder (1987), low stumpage is explained by such considerations as species mix and distance to markets. More generally, it is important to bear in mind the historical context of low-stumpage rates – as an enticement to invest in export-oriented large-scale development in a peripheral and sparsely settled territory. In the evolution of BC's forest industries as specialized, low-cost producers of bulk commodities serving world markets, low stumpage is mirrored by a low-value commodity-oriented industrial structure.

As a commodity-based industry selling in global markets, the forest industry has competitive pricing as a norm. However, a low-stumpage regime in support of a low-value (and volatile) production structure does not equate to a subsidy providing unfair market advantage, regardless of whether it is good social policy. BC's stumpage formulas traditionally have been sensitive to cost and price fluctuations and, historically, have not allowed higher profits for the BC forest industry (Schwindt 1987). In addition, Copithorne's (1979) admittedly debatable view – that the provincial government loses stumpage (rent) through high wages to workers – does not provide grounds to interpret stumpage as an export subsidy. This argument simply says that the industry is paying relatively more to labour than to the provincial government, an outcome that may be socially beneficial. Moreover, if the decline in the value of the Canadian dollar has been extremely advantageous to exporters, it is scarcely controlled by the lumber industry.

Ironically, forest-product firms elsewhere in Canada have received considerable subsidies. One federal program in the early 1970s and 1980s, for example, targeted the modernization of the pulp-and-paper industry in Ontario and Quebec, while explicitly excluding BC and rejecting BC's plea for reforestation funds. Other provincial governments have often provided subsidies. Moreover, the massive growth of forest industries in the American South has been supported by a wide range of subsidies, including capital grants and tax waivers. It is likely that the BC forest industries are the *least* subsidized within North America.

BC's provincial government may have good reasons for establishing a high-stumpage regime. But recognition of those reasons does not require acceptance of the argument that low stumpage is therefore a subsidy.

Implications for BC

Provincial responses to the options allowed by the MOU varied. Alberta and Ontario accepted the tax on lumber exports to the US; Quebec obtained an amendment that permitted progressive reductions in the export tax to 6.2 percent in November 1990 (and to 3.0 percent in 1992) in association with higher stumpage; Atlantic Canada won a complete exemption from the MOU; and BC obtained an amendment that entirely replaced the export tax with the higher stumpage introduced in October 1987. These differences are noteworthy because stumpage is part of the total industry cost structure, and the export tax is restricted to American exports. Thus,

Table 7.6

BC stumpage charges per cubic metre, 1987-97

	October 1987	December 1993	May 1994	December 1997	June 1998
Coast	$10.59	$17.20	$28.03	$33.80	$25.00
Interior	$8.59	$15.17	$27.47	$28.37	$21.40

Source: Hamilton 1998e.

Alberta lumber producers gained an advantage over those based in BC in serving offshore markets.

In BC, the changes stimulated by the trade dispute were especially profound. Successive right- and left-wing provincial governments were motivated by the opportunity to increase stumpage (and their own revenues), and quick agreement was obtained on replacing the export tax by higher stumpage (Table 7.6; see also Table 3.10). Indeed, a widely held view in the late 1980s and early 1990s was that the US had done BC (and Canada) a favour (Richards 1987) by forcing more realistic valuations of the Canadian forest resource, a plea made in several studies (Copithorne 1979; Gunton and Richards 1987; Percy 1986). As noted, the provincial government used the American softwood lumber trade imbroglio to help justify the high-stumpage regime of the 1990s, for both the change in formula and the stumpage escalator (see Figure 3.8). This same rationale also firmly placed its own autonomy on the bargaining table.

Indeed, the BC (and Canadian) government expressed concern that proposed reductions in stumpage in 1998 may not be acceptable to the US, which still has a role in BC forest management. This role results from a clause in the SLA preventing provinces from making changes that cause a net reduction in timber cost (Hamilton 1998d; Table 7.5). Clearly, the government is in a considerable bind, albeit one it has encouraged. Moreover, this bind has prompted MB to suggest the privatization of forests as a way out of the dilemma. Privatization would provide MB (and other firms) with incentives to invest in forest management and flexibility in forest utilization. Privatization could offset American countervail threats because the buying and selling of timber would be driven by market forces. Privatization is also consistent with the government's high-stumpage regime and its concern to realize full (market) values for timber, hitherto suppressed by the low-stumpage regime. Privatization, however, was not part of the government's agenda and may help resolve one dilemma while creating others. Thus, the open timber markets, in association with privatization, will place further pressure on log export restrictions, encouraging American firms to argue, as they have, for a right to bid on BC timber and export it in raw form. The government, however, anticipated that the high-stumpage regime would encourage value-added exports rather than log exports. In any

Table 7.7

COFI's get smart strategy

Year	Activity
1972	COFI implements new export assistance program
1973	Japanese Ministry of Construction recognizes platform frame construction (PFC) for residential buildings; COFI distributes technical information on PFC in Japan and hosts three missions from Japan to study PFC techniques
1974	COFI establishes permanent office in Tokyo and organizes the Canada-Japan Housing Committee to exchange technical information and develop exchange programs
1975	Continuing Canadian missions to and from Japan
1980-5	COFI increases promotional efforts for two-by-four housing, and sits on committees and working groups in Japan dealing with building and fire codes, standards
1987	COFI certified as a foreign testing organization for plywood to Japan Agricultural Standard
1988	Japanese building codes amended to accept three-storey wood-frame structures with vertical separation
1989	COFI sponsored demonstration two-by-four housing projects in Kobe and Yamaguchi
1990	COFI accredited as a foreign testing organization for traditional and new products
1991	COFI participates as technical experts in the Japan/Canada/United States technical committees on the use of wood in construction; approval of new standards for machine stress rating and structural finger-jointed lumber; changes in the Japanese Building Code to permit construction of three-storey wood structures for apartments and mixed commercial residential use
1992	COFI facilitates Japanese delegation of prefabricated-housing executives to Canadian promotional programs for BC coastal sawmill posts; assists with the BC Trade Development Corporation showcase of Canadian building products
1994	Coordinated three-storey apartment demonstration project in Japan with the Osaka Housing Supply Corporation

Sources: Interviews with COFI in Japan and British Columbia, 1994; Cohen and Smith (1992).

case, any provincial government support for privatization, and there is some, will be opposed by environmentalists and Aboriginal peoples, as they have already made clear (Hunter and McInnes 1999).

The softwood lumber trade dispute has been massively disruptive to the BC forest economy, and because of the way it has been handled by Canadian negotiators, it has contributed significantly to compromising provincial autonomy over the forest resource, a right conferred by the Canadian federation. One positive effect of the trade dispute was to encourage BC firms to look at other markets, notably Japan. Indeed, the high prices received in this market provided the basis for the high-stumpage regime. The Asian economic crisis of the late 1990s has further exposed the vulnerability of this regime, at least from an industrial perspective.

The Japanese Connection

The costs and uncertainties created by the MOU and subsequent negotiations sent strong signals to forest-product firms in BC, especially lumber firms, to diversify markets by geography and product. Even before American countervail action, the recession of the early 1980s imposed a significant cost-price squeeze: costs in BC were high and prices were softening in the US. In fact, the BC-based Council of Forest Industries anticipated the potential of the Japanese market in the early 1970s (Table 7.7).

Thus, in the early 1970s, COFI established a Tokyo office and began concerted lobbying of the Japanese government to reduce the nontariff barriers (NTBs), notably complex building and fire codes, facing BC sawmillers in traditional and nontraditional housing markets. COFI engaged in extensive promotional work in Japan (for example, demonstrating Canadian housing and building techniques), and in 1991, it became the first organization to be accredited by the Japanese government as a foreign testing organization (FTO) for a number of traditional and value-added lumber products. Under this licence, COFI is authorized to certify member mills in BC with Japan Agricultural Standard (JAS) quality assurance stamps (Cohen 1992; Cohen and Smith 1992), and the need to regrade in Japan was eliminated. In recent years, COFI's efforts have been supplemented by those of the BCWSG whose activities focus more on SMEs.

COFI's sustained and patient pressure can be contrasted with the more get-tough strategy adopted by the American government in its negotiations to reduce Japanese tariffs on sawn lumber through the provisions of the

Table 7.8

British Columbia's lumber exports by volume and value, 1985 and 1992

	1985		1992	
	Volume (%)	Value (%)	Volume (%)	Value (%)
United States	81.1	73.6	72.8	58.8
Japan	9.9	13.6	17.2	27.2
Europe	4.6	6.1	6.4	10.0
Elsewhere	4.4	6.7	3.6	4.0
Total	**100.0**	**100.0**	**100.0**	**100.0**

Source: Hayter and Edgington 1997: 196.

Log sorters and graders are recalled for work in 1999, partly to supply the log export market. Does this make sense?

1990 Super 301 trade bill (Hayter and Edgington 1999). At the same time that the US is placing political pressure on Japan to import more American forest products, the Pacific Northwest states have become more proactive in accessing the Japanese markets. BC can expect increasingly severe competition from its Pacific Northwest neighbours.

The recession of the early 1980s was the stimulus to the surge in BC's exports to Japan. While prices were falling and protectionist sentiments increasing in the US, Japanese lumber prices were high and increasing. Moreover, at a time when BC firms were looking for alternative lumber markets, Japanese organizations were looking for more lumber imports. Indeed, apart from MB's initiatives, the rapid growth of BC's lumber trade with Japan has been largely organized by Japanese interests (Edgington and Hayter 1997).

Preeminence of the Coastal Region

By 1996, Japan was a significant market for the BC forest industries (Table 7.8; see Table 3.4). By value, over 27 percent of wood products shipped from BC went to Japan; volumes of newsprint and of pulp, paper, and paperboard to Japan represented 11 percent and 15.7 percent of total production; and shipments of pulp and paper to other Asian markets were substantially greater.

The shift to Japanese markets has been led by the lumber industry in the coastal region. By the early 1990s, many coastal sawmills were partially or

Table 7.9

British Columbia lumber shipments from the coast and interior, 1971 and 1989

	1971	1989
Coastal shipments		
Total (MBF)	4,171.5	4,203.0
Percentage to Canada	17.1	21.7
Percentage to United States	53.7	32.4
Percentage elsewhere	29.2	45.9
Interior shipments		
Total (MBF)	4,785.5	11,039.8
Percentage to Canada	19.9	30.1
Percentage to United States	71.0	59.6
Percentage elsewhere	9.1	10.3

Source: Hayter 1992: 165.

totally renovated, and many were exporting to Japan, which, for the coastal region, had become as important a market as the US (Table 7.9). Although the quantity of shipments remained more or less the same, offshore exports, especially to Japan and the European Community, substantially exceeded those to the US by 1992 and accounted for almost half of total shipments. For the interior industry, American markets remained paramount. In this regard, the more profound restructuring of the coastal region, compared with the interior, reflects its earlier development and older plants, higher-quality but higher-cost timber, and tidewater access to timber supplies and markets. Even in the interior, however, non-American markets have grown relatively fast, most notably domestic markets but also Japan.

These highly generalized trade dynamics have two main value-added implications for production. First, sawmills have become more concerned with quality as Japanese customers in particular have distinctive preferences that are reflected in rigid technical standards and aesthetic characteristics, notably for wood grain and colour. Second, offshore markets, again especially Japan but also the European Community, require a different and a wider range of lumber dimensions compared with the American construction market.

Japan's Global Supply System in Forest Products: An Overview

BC forest-product exports to Japan are part of a global supply system. Since the late 1960s, Japan's domestic forest industry has provided less than half of domestic consumption, and forest-product imports have grown rapidly (Yamakawa 1993; Edgington and Hayter 1997: 153). Since 1980, for example, imports have accounted for at least 65 percent of total forest-product consumption, and in 1993, almost 76 percent of Japan's consumption of more than 111 million cubic metres of timber were imported. Since at least 1960, Japan has been the world's largest importer of logs and second only to the US as an importer of lumber (Kato 1992: 93-5).

Geographically, Japan's timber imports are highly diversified, with different sources specializing by species and commodity type (Edgington and Hayter 1997). Japan is the world's biggest importer of logs, drawing hardwoods from Southeast Asia and softwoods from the US and the former USSR, although such imports are declining. Canadian exports principally provide lumber, and Canada is the largest lumber exporter to Japan. In 1993, for example, Canadian exports of lumber to Japan amounted to 5.2 million cubic metres, over three times US levels. In the same year, US log exports to Japan were 7.5 million cubic metres (Edgington and Hayter 1997: 153). Within North America, virtually all of Japan's log and lumber imports come from the Pacific Northwest; in Canada's case, approximately 98 percent of timber bound for Japan is from BC (COFI 1992). In the 1960s, the availability of softwood logs from the US encouraged Japanese firms, particularly the *sogo shosha,* to favour Washington and Oregon as their principal North American source of supply. Since then, lumber exports from BC to Japan have grown steadily, and they spurted rapidly after the mid-1980s. At a time when BC firms were looking for offshore markets, Japanese buyers were looking for lumber rather than log supplies.

Japan's Demand for Lumber

The basis for BC's timber exports to Japan is Japan's huge demand. Japan, with approximately half the population of the US, is the world's largest wood-based housing market. Since 1970, in most years the number of housing units built in Japan has been larger than in the US. In 1994, 1.5 million units were completed in Japan, about 100,000 more than in the US. In comparison to North America, Japan's traditional houses are more wood intensive, and construction methods are more labour intensive, skilled, and varied. Thus, traditional construction (*zairai kuko*) refers to Japanese-style houses built on site, using post-and-beam construction techniques in which intricate notching is required to connect the large wood structural members (the posts and beams) to a wide variety of other wood components, measured on a metric standard, under the direction of skilled carpenters. Moreover, traditional construction exhibits considerable regional variation within Japan (Toyama 1993).

In contrast, in North America, platform frame construction (PFC), based on a two-by-four imperial standard, is the most common house-building method in which each floor acts as the platform and structural support for the next floor, reducing the need for large timbers. In addition, North American houses use an average of twenty-eight types of lumber, while traditional Japanese homes, by contrast, use around 150 different types of lumber. As a result, erecting homes takes twice as long in Japan, is more costly, and involves carpenters and framers whose training is much longer in comparison to their North American counterparts.

In Japan, shortages of skilled labour have become increasingly acute in sawmilling and construction. Compared with 1985, far fewer skilled carpenters are able to recognize and use BC timber in Japanese traditional housing according to the old "sight and fingers and thumb" techniques of selecting timber for strength and appearance according to different uses in house construction. Work inside sawmills, as well as on building sites, is increasingly shunned by young Japanese. One projection is for the number of construction workers in the year 2000 to decline to around 75 percent of the 1990 level (Toyama 1993). Moreover, the rising value of the yen – which became especially sharp after the 1985 *endaka* (the rise in the value of the Japanese currency) – dramatically increased domestic prices of lumber and land until 1996, thus increasing the attractiveness of imported timber and related products.

As housing costs have escalated, Japanese consumers, especially younger consumers, have begun to shift preferences from traditional to nontraditional types of housing, especially since the mid-1980s. Nontraditional housing accounted for 22 percent of the wood-housing market in 1992, whereas traditional wooden construction has fallen from around 46 percent of annual starts in 1985 to roughly 40 percent (Edgington and Hayter 1997). In this regard, the appeal of nontraditional wood housing – the two main forms of which are prefabricated construction (including standardized two-by-four houses) and precut factories – is that it reduces costs by using less wood per house and more imported wood, and requires fewer workers and less building time because most of the components are manufactured in automated factories (Yamakawa 1993).

In summary, the spurt in BC timber exports to Japan beginning in the mid-1980s reflected the coincidence of *endaka,* a Japanese housing boom, and high prices and a severe recession in BC. At the same time, log supply limitations became increasingly apparent in Washington and Oregon as a result of environmental constraints, while marketable reserves from Russia and Southeast Asia declined. Consequently, Japanese importers switched to BC mills as an area of supply growth and offered a higher price. In support of this shift, the Japanese government also reduced its tariffs on imported lumber and gave JETRO, an agency mandated to promote Japanese exports, a new department to encourage the rationalization of traditional distribution channels, widely recognized as a major nontariff barrier for exports to Japan, as well as to support more standardized building systems to help reduce the final price of housing.

The Organizational Dynamics of the BC-Japan Lumber Supply Chain

Although driven by powerful economic forces, the growth in the level and value of BC's lumber exports to Japan has been a difficult, contentious process. Japan is an extremely complex timber market with an internal distribution of building products that stretches through a pyramid-shaped hierarchy of several tiers of wholesalers, each ranked by size and financial power, and many end-users at the market base comprising individual carpenters and house builders (Toyama 1993). Moreover, demands for housing materials are extremely varied by region, and many Japanese are still sceptical of using foreign products for fear that they are not suited to "unique" Japanese conditions. In addition, the BC industry had to overcome its own tradition as a commodity supplier to the US and develop new manufacturing and marketing capabilities. The development of BC's timber exports to Japan has also required new forms of organization that in turn led to increasingly complicated production chains (Edgington and Hayter 1997).

In the late 1960s, BC's relatively small-scale timber exports to Japan were logs and cants (large sections of timber, one step along the value-added chain from logs) in a variety of long lengths and a relatively crude form of "Japanese" squares or "tinmancles" (but still cut in BC according to imperial measurement). The trading companies sold these to wholesalers and sawmills in Japan where they were remanufactured into a multitude of finished products (for example, posts and beams, mouldings, doors, and windows). Subsequently, the BC sawmill industry has increasingly added value to its sales to Japanese customers. Thus, during the 1970s, the industry began to export four-by-four, rough-sawn, green "baby squares" used as posts (called *hashira*) in traditional Japanese post-and-beam construction. After 1974, kiln-dried two-by-four timbers were also exported in small quantities, initially from the interior region of BC (and Alberta) to Japan. By the late 1970s and early 1980s, some firms in BC had begun to supply posts and beams in accurate metric sizes according to Japanese specifications (for example, 90 mm and 105 mm baby squares) and to focus on grain orientation, qualities of far more importance and value in Japan than in North America. In recent years, this trend has continued, and exports of more finished-wood products have begun, including interior dimension lumber for shoji screens, engineered building products such as finger-jointed lumber, and even log homes and prefabricated housing (Cohen 1993; Toyama 1993; Reiffenstein 1999). However, BC still exports large volumes of timber in large sizes that are processed further in Japan.

The *Sogo Shosha*-Dominated Model of the Early 1970s

Until the early 1970s, the Japanese *sogo shosha* organized almost all of the timber trade from BC to Japan (Figure 7.1). Essentially, the *sogo shosha* took the initiative and tapped into the BC timber lumber industry, which was then overwhelmingly oriented towards serving the commodity needs of American markets. At this time, the dominant role played by the *sogo shosha* reflected several considerations. First, the Japanese timber supply market was traditionally characterized by fragmented markets and a lack of

Figure 7.1

The BC-Japan timber-production chain in the early 1970s

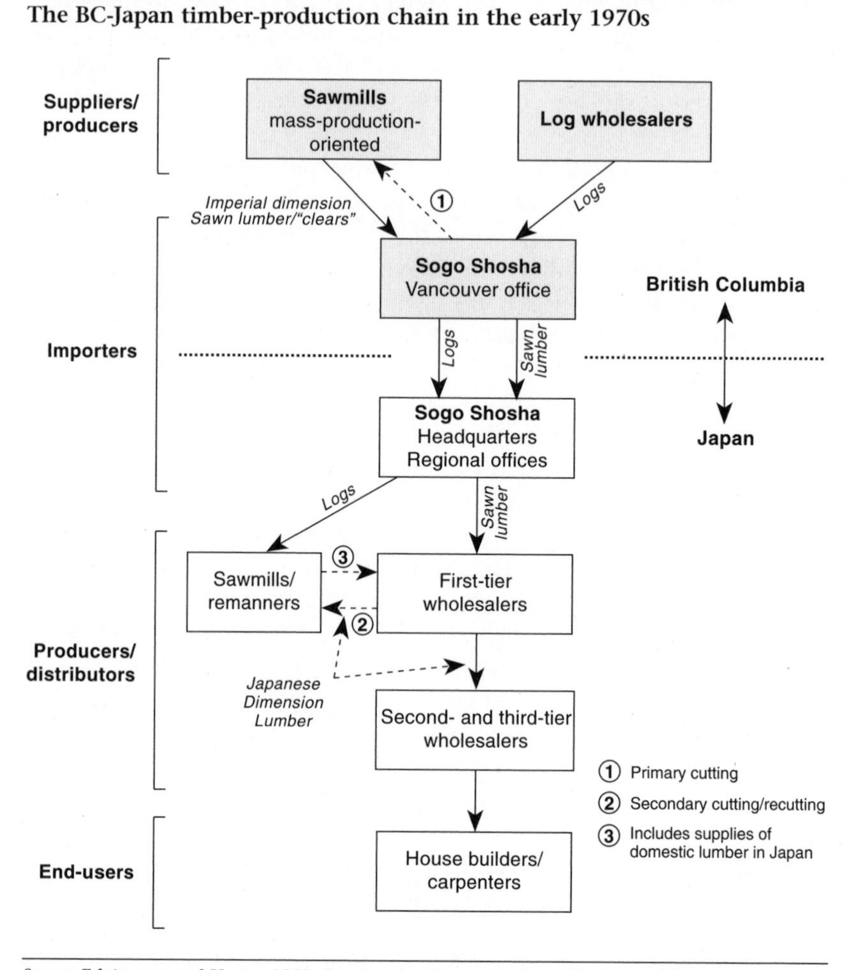

Source: Edgington and Hayter 1997. Reprinted with permission of Taylor and Francis.

standardization that did not allow for economies of scale for individual builders. The *shosha* played an important role in aggregating large numbers of small customers and a wide variety of suppliers, all with their own niches and facilities. The *shosha* assisted in reducing transaction costs for these firms through their various operations, such as breaking down and repacking imported timber into different sizes, cutting Japanese house-builder sizes where required in BC, and passing the timber on to Japanese wholesalers to arrange further repackaging and recutting if necessary in Japan.

Second, the Japanese payment system in the timber industry is customarily carried out through promissory notes giving three to five months' credit to a customer. Traditionally, the *shosha* played the critical function of immediately paying the Canadian supplier in dollars but then providing credit for the Japanese purchaser. In this way, most first-tier timber wholesalers in Japan were tied to the *shosha* through credit, giving the *shosha* a privileged position in the timber trade because they acted in a way comparable to a merchant bank. Similarly, being so much in debt, their Japanese clients usually could not disengage from the relationship.

A third factor relates to the superiority of the *sogo shosha* in overseas dealings. Thus, up to the end of the 1970s, the *shosha* were the only Japanese companies that could afford to travel and invest time in understanding currency exchange problems and local cultures (Young 1979). Moreover, BC suppliers until recently did not have the interest or sophistication to deal with the complex Japanese market, and they were willing to concentrate on manufacturing, leaving the marketing in Japan to the trading companies. In sum, the early trading patterns were characterized by business continuity, including the fact that each *shosha* had a fixed number of clients in Japan (mainly, the major wholesalers in each region) as well as an established route for obtaining Canadian supplies. In the 1970s, it was virtually impossible for small wholesalers or house builders to buy lumber other than from the hierarchical distribution chains of wholesalers and trading companies.

The original *sogo shosha*-dominated timber production chain, however, became subject to forces of change occurring in Japan and BC. In Japan, wholesalers, house builders, and even large sawmills began to seek ways to reduce the costs of timber supply and to develop new forms of house construction. Once exclusively dependent on the *shosha* for imported timber, these organizations sought ways to rationalize supply lines, including circumventing the *shosha*. For their part, the *shosha* have had to depart from their traditional role as intermediary and paymaster in the timber distribution system. Meanwhile in BC, forest-product firms, desperate for new markets in the 1980s recession, sought to add value and, in MB's case, make more direct contact with consumers down the production chain. As a result, significant challenges to the original *sogo shosha*-dominated model have been mounted.

Towards a More Complex Production Chain

The evolution of a larger and more diversified timber-supply production chain between BC and Japan is characterized by a growing internationalization of production as Japanese and BC organizations have become increasingly interdependent through strategies of horizontal and vertical integration. These strategies have sought to widen market penetration and increase control of linked stages in the production chain in search of profitability and security. By the early 1980s, the *sogo shosha*-dominated production chain had been considerably modified (Edgington and Hayter 1997: 157). By the early 1990s, this production chain had become even more

Figure 7.2

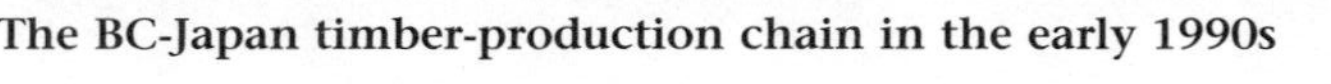

The BC-Japan timber-production chain in the early 1990s

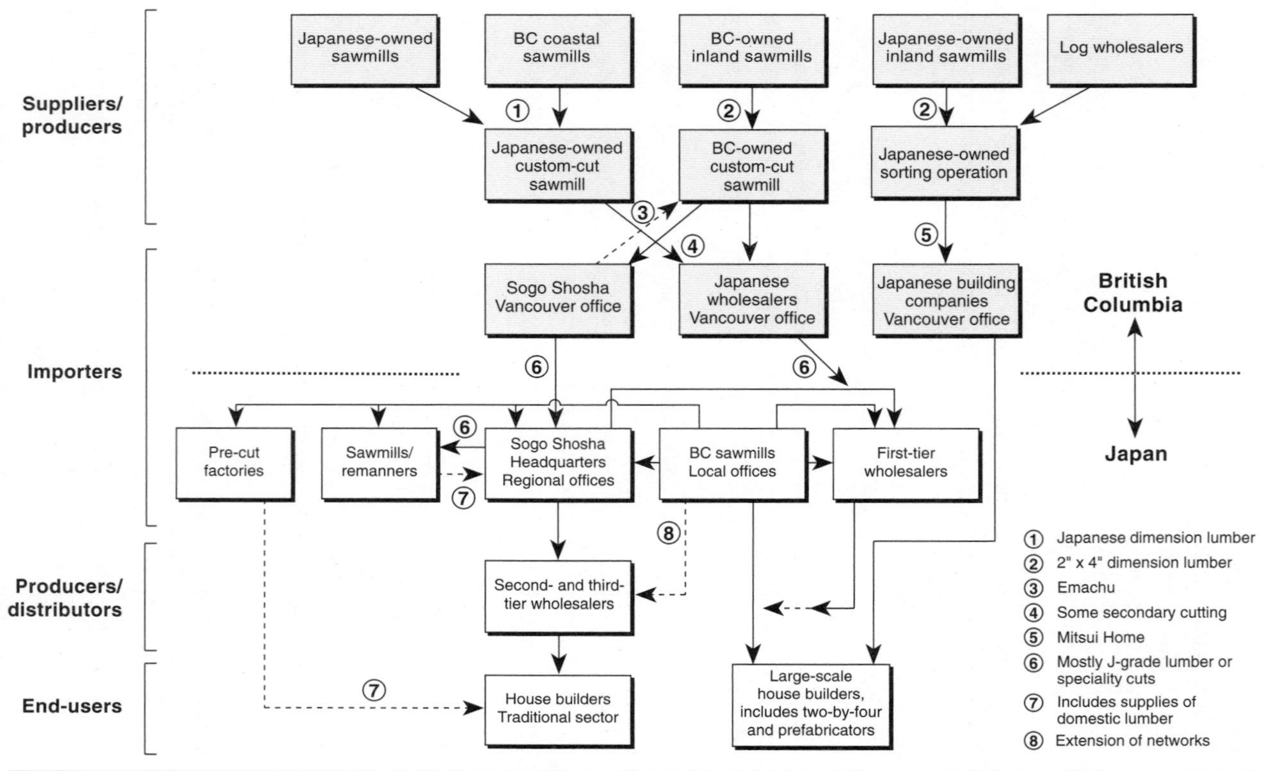

Source: Edgington and Hayter 1997. Reprinted with permission of Taylor and Francis.

complex and the *sogo shosha* even less important (Figure 7.2). In the evolution of this more complex model, several broad trends should be noted.

Japanese Importers Shifted Closer to Their Suppliers
A few wholesale companies have established offices in BC to save on the trading companies' commissions and to more effectively serve sawmills in Japan that now import dimension lumber, rather than logs, for further recutting in Japan. For these wholesalers, a Vancouver office served to systematize direct trade with BC, helped enforce the annual contracts made with BC suppliers, allowed for checks on the quality of purchases, and facilitated transport and customs arrangements. These local offices also conduct market intelligence. Thus, for the routine trade in either Japanese or North American dimensions, first-tier wholesalers have increasingly replaced the *sogo shosha*. In addition to trade in structural timbers, Japanese wholesalers operating in Vancouver have tried to move into the more lucrative custom-cutting of large timber sizes (known as "clears") for specific customers such as temples or smaller wholesalers, or for "remanners" (remanufacturing or secondary sawmills) to resurface for particular house-building customers (for example, jambs for Japanese shoji sliding screens). In a few cases, high-quality logs are custom-cut in BC to irregular sizes for Japanese furniture, panels, windows, and so on.

A number of prefabricated house-building companies have also set up procurement offices in BC. Once again, a basic motivation is to reduce the role of the *shosha*. Mitsui Home, for example, pioneered the use of two-by-four lumber in frame construction in Japan, beginning in 1975, shortly after such construction was authorized by the Japanese government. Initially, Mitsui imported two-by-four dimension lumber from inland BC as

Japanese wood auction at Port Alberni.

well as Alberta sawmills through the *shosha,* who also financed the trade. In 1980, the company set up a Vancouver office to buy from the BC and Albertan sawmills directly. It originally focused on buying logs from Washington and Oregon, but since the mid-1980s, the company has targeted Canadian sawn lumber, given environmental controls in the United States. In 1993, Mitsui Home further vertically integrated operations in Vancouver by building a sorting plant in Langley to allow the purchase of cheap Canadian lumber classified at slightly below JAS grade. This lumber is then re-sorted into grades that meet the official JAS standards, thus saving on tariffs, before its dispatch back to Japan. Other prefabricated builders, such as Sanwa Home, have not invested in manufacturing or presorting, preferring to continue to import J-grade timber through the *shosha.*

BC Suppliers Shifted Closer to the Market

For BC suppliers, the arrival of Japanese wholesalers and builders increased competition for wood, encouraging higher prices. For several large coastal mills, Japan became their principal lumber-export market. In BC, these mills have adapted in many ways to Japanese demands, notably by introducing extensive quality control systems in forest harvesting, especially through better initial sorting of logs to be cut for the Japanese market, by better sorting of logs by diameter, length, and quality at mill sites, and by more precise timber cutting. Coastal timbers for Japan are also cut in designated mills, and much lumber exported to Japan has to be kiln dried for new precut factories there.

Apart from increasing the quality generally, two BC forest-product corporations (MB and Interfor) have developed their own distribution channels in Japan to circumvent the *shosha* (Chapter 4). While MB opened MBKK, its Japanese subsidiary, in 1985, Interfor opened its local marketing office in 1992 after being part of the Seaboard group for many years (Sorensen 1990). A primary purpose of MBKK and Interfor's Tokyo office is to move downstream towards their end-users to obtain information and to influence wholesaling and distribution functions. By the mid-1990s, MBKK had begun to stock product in Japan to meet customer just-in-time delivery requirements. It might also be noted that Fletcher Challenge's acquisition of Crown Zellerbach Canada and British Columbia Forest Products led to a reorientation of their markets from the US to around the Pacific Rim, for lumber as well as pulp and paper.

Sogo Shosha *Adjustments*

If the *sogo shosha* are less important than they used to be in the BC-Japan lumber trade, the ten largest *sogo shosha* reportedly still control about 60 percent of timber imports to Japan. For the two-by-four business, about 50 percent of these timbers are sold through the *shosha,* whereas the other 50 percent are sold directly to major end-users such as Mitsui Home. Stimulated by these competitive threats, the *shosha* have adjusted their role within the production chain through strategies of horizontal and vertical integration.

In the pursuit of horizontal integration, the *shosha* have increased their presence in Vancouver to cope with the increased demand for BC lumber, while reducing dependency on trade in logs (now handled almost exclusively by the Seattle offices of the *shosha*) and traditional post-and-beam dimension lumber (four-by-four timbers). The *shosha* have targeted a greater variety of timber products, including new value-added directions in the trade such as two-by-fours for the increasingly popular platform housing, or the initial cutting of specialty timber products from clears that are then sent to Japanese wholesalers for sale to specialized remanufacturers in Japan. As this initial cutting used to occur in Japan, this trend represents a shift in value-added production towards BC.

Another horizontal move by the *shosha* strategy is to negotiate dimension timber for large-scale prefabricated builders in Japan, including special Canadian building materials – such as finger-jointed and glue-laminated lumber for posts and beams in nondomestic construction, oriented strandboard (OSB) for packaging, flooring, and roofing to replace the more expensive plywoods, or even for windows and door frames. These higher-value items have traditionally been produced in Japan and acquired from Japanese wholesalers, but since *endaka,* builders are more willing to source them from overseas suppliers, and they often need trading company assistance. A further strategy has been to mediate kiln-dried timber for precut factories in Japan. Marubeni now claims that around 25 percent of its exports from BC are higher-value-added timbers or building products.

In the pursuit of vertical integration, the *sogo shosha* have broadened their role within the production chain, most notably by becoming suppliers themselves rather than merely intermediaries (Figure 7.2). Itochu was the first *shosha* to pursue such a strategy by investing in the dimension sawmill CIPA Lumber Co. in 1968 as a joint venture with BC-owned Pacific Logging (Chapter 5).

The *shosha* have therefore integrated activities to regain a competitive edge over wholesalers and sawmills and to match wholesaler and builder interest in obtaining custom-cut timber and building products. In most cases, wholesalers have not had the capacity to move into the supply business to the same degree as Marubeni and the other *shosha,* and they have also encountered problems with direct sales channels between BC and Japan. Specifically, direct trade has required them to resolve quality problems among BC suppliers, including rejecting poor-quality timber. Traditionally, the *sogo shosha* took care of such responsibilities by having access to alternative customers. BC suppliers have been attracted to trade directly with Japanese wholesalers and builders, a trend that requires they make greater efforts to ensure better quality and faster order turnaround than the previous trade with the *sogo shosha*.

Japan and the Pacific: A Path to a Value-Added Nirvana?

Resnick (1985) wondered about the long-term benefits of BC's growing lumber trade across the Pacific and whether it would prove to be any more stable than continental trade with the US. The recent Asian economic crisis gives further cause for concern. For BC, the development of the lumber-supply production chain to Japan provided profits and diversification to replace traditional reliance on American commodity markets, especially in the coastal region. The Japanese connection has also required BC producers to become more flexible and quality conscious in meeting a wider range of more precise demands. This shift towards value-added production has primarily comprised trends towards flexible mass production (Barnes and Hayter 1992), although SME-based batch exports have also increased, including exports of log homes and prefabricated housing.

Direct selling and distribution in Japan by small suppliers is extremely difficult (Reiffenstein 1999). SMEs that do export to Japan still largely rely on Japanese organizations, and in some cases, joint ventures have been formed. In addition to language and cultural barriers, a specific problem relates to the medium-term credit expected by purchasing clients, especially "mom and pop" wholesalers, sawmills, and builders. Moreover, unless BC suppliers carry stock in Japan, they are unlikely to service small builders by providing the specified mix of Japanese dimension lumber sufficient for just one house, on site, and on a daily turnaround. BC suppliers, such as MBKK, still rely for direct trade on the larger, financially secure housing companies and the larger wholesalers; direct servicing of the smaller wholesalers and builders has proven almost impossible.

SMEs have achieved considerable success in tapping Japanese markets. Primex is often cited in this regard: it has actively promoted links with Japanese customers and has upgraded production and made specialty products, but so far it has been content to do so from a Canadian base with occasional visits to Japan by its marketing managers (Lush 1992). Primex believed that if it was dealing merely with a few housing prefabricators, then a Japanese distribution office might make more sense. Although Primex has some direct links with Japanese wholesalers, it still relies extensively on the services provided by the *shosha*. Few SMEs have generated autonomous links in Japanese markets.

Just how far BC can increase forest-product exports to Japan, especially for value-added activities, is difficult to say. The level of BC's secondary-wood exports lags behind those from the Pacific Northwest, and both regions are in direct competition. Moreover, the control of the production chain is still dominated by Japanese organizations, even if the *sogo shosha* are comparatively less important than they once were. Few BC firms have developed permanent bases in Japan, whereas the Japanese organizations that have developed stronger ties to BC have also initiated links with other supply sources, including some that are clearly more costly (such as

Scandinavia) and others that are much cheaper (such as Siberia and Chile). Thus, maintenance of Japanese markets is not assured, and it may be that the Japanese connection will turn out to be a reed rather than a lifeline. Until 1997, however, the Japanese market offered substantial diversification opportunities for the BC forest-product industries. Despite the market declines of 1997 and 1998, BC can ill-afford to neglect its Pacific opportunities when the forces of continentalism are sending such mixed messages.

Conclusion

Situated on the geographic margin and heavily dependent on exports, the BC forest-product industries have long espoused the virtues of free trade. Indeed, it is conventional wisdom to interpret the evolution of the BC forest economy as a prima facie case of a free trade economy, especially when considered within continental terms of reference. Thus, the Shearer et al. (1973) classic study of the BC economy concluded that if Canada negotiated free trade with the US, such an agreement would make little difference to BC because its forest industries had already developed in a continental free trade environment. Conventional wisdom in this regard is only partially correct. Politics have always conditioned BC's forest-product trade with the US. During the heyday of Fordism, and before, BC's forest commodities were exported to the US on a free trade basis, but free trade even then did not apply to value-added products. Moreover, trade was always subject to the discretion of BC's powerful trading partners. The 1990s have witnessed powerful use of this discretion, and the signing of the FTA and NAFTA has not offset the increasingly politicized nature of BC's forest-product trade environment.

American protectionism is often presented as an external force over which BC (and Canada) has little control. This conclusion is the wrong lesson to learn from the softwood lumber dispute. Rather, the dispute provides a textbook case of how a smaller country (and region) should not negotiate with a superpower. Canada (and BC) lacked a coherent, long-term view of objectives that reflected long-term national and regional interests. If the provincial government was naive, the federal government was inept. The results are there for all to see. In an age of free trade, BC's softwood lumber exports to the US are formally restricted for the first time since the 1930s. The SLA of 1996, designed to provide at least five years of stability, has opened up public and even personal bickering in BC's forest industry and encouraged American protectionists to expand their definition of what needs protecting while endorsing their right to interfere in BC's forest management. The provincial government, which somehow managed to think that the Americans were doing "us" a favour, can now scarcely change forest policy without looking first over its shoulder southward. In terms of the experience of softwood lumber, the FTA and NAFTA are disasters. The view – now expressed by the CEO of Canfor, a large BC-based forest firm and doubtless once a major supporter of free trade – that Canada should withdraw from NAFTA, has merit. It also reveals the extent of the disruptive nature of the dispute.

If the Japanese connection helped ameliorate the trade dispute for ten years, the Asian economic crisis has again emphasized the importance of trade relations with the US. BC's bargaining position would clearly be greatly improved by better Japanese markets. But, surely, BC needs to make better efforts to develop stronger, long-term provincial and national markets for its products?

If the present federal government cannot tear up NAFTA, which it once said it would, Canada should drop out of the SLA. A long-term priority of the Canadian government should be to eliminate American interference in provincial forest policy. If the US threatens to impose tariffs, so be it. A tariff is at least honest and visible. It would also allow BC to regain autonomy over its forest management; to reestablish the argument that it does not favour subsidies (which, relatively speaking, it doesn't); and to lobby American consumers hurt by higher timber prices. Tariffs can also be removed easily, should the US so decide. Indeed, it might be speculated that if the US had imposed tariffs in the 1980s, they might have been imposed for a limited period. Tariffs would now present hardships for BC firms. Nevertheless, tariffs are preferable alternatives to a quota system, especially a system that does not work. In the meantime, US pressure on BC to remove log export restrictions will doubtless increase. Such pressure should be resisted. As this chapter has revealed, BC's forest-product markets are not "perfect." Rather, BC's forest-product trade has been manipulated by the tariff (and other) policies of the major powers, especially the US but also Japan and the EC, and controlled by MNCs in these core countries. BC's log export restrictions are a minimalist policy that seeks to ensure minimal local development. Log exports also export jobs, skills, diversity, and hopes for innovation. BC's interests are to shift increasingly away from log and indeed commodity exports.

8
Employment and the Contested Shift to Flexibility

The recession of the early 1980s marked a turning point for employment and labour relations in BC's forest industries. For the three preceding decades of Fordism, sustained growth of employment and labour productivity was underpinned by collective (and adversarial) bargaining between management and unions that realized improved wages and nonwage benefits and organized work based on the Taylorist principles of seniority and job demarcation. In this period, several recessions took place and blue-collar workers were laid off. But these layoffs were temporary, and workers were rehired according to seniority. The impacts of the recession of the early 1980s were different, and they stimulated ongoing changes. Since then, layoffs have been permanent and have involved management as well as labour. Taylorism has also been challenged, specifically by demands for flexible operating cultures.

In the BC forest economy, as elsewhere, the search for more flexible workforces is a contested and ambiguous process. The search is contested because the demands for worker flexibility are usually part of employment downsizing and require changes in established, legally sanctioned agreements. The search is ambiguous because the concept of employment flexibility is multifaceted and no consensus has been reached on an appropriate definition. Thus, flexibility may be realized through a low-wage, low-skilled, unstable workforce, or by a high-wage, multiskilled, stable workforce, or by other combinations of characteristics (Anderson and Holmes 1995). Moreover, definitions of skill and stability, not to mention high or low wages, involve judgment, and neat, theoretical models of employment flexibility inevitably become blurred in the real world of shop-floor relationships.

Among unionized industries, flexibility has to be negotiated (Hayter, Grass, and Barnes 1994; Hayter 1997b). Inevitably, it is easier for management to achieve the employment flexibility characteristics they desire by building plants in greenfield locations (Clark 1981). Such a geographical solution to the labour-relation problems is not always available, however. In BC's forest industries, for example, firms are tied to established locations by sometimes massive capital investments, human resources, and access to resources, all of which are relatively immobile. The in situ transformation of labour relations, however, is a problematical process, especially in BC's forest industries where Fordist collective bargains are entrenched.

This chapter analyzes the search for labour market flexibility in BC's forest industries. The discussion is in four main parts. First, the evolution of collective bargaining in BC's forest industries is presented as a classic example of Fordist labour relations. Second, the shift from Fordist to more flexible labour bargains following the recession of the early 1980s is described. Third, case studies of labour market flexibility in the corporate sector are outlined. Finally, the varied and inherently flexible labour relations within SMEs are noted. The chapter draws on recent studies of employment change in BC's forest economy (Grass and Hayter 1989; Barnes, Hayter, and Grass 1990; Hayter and Barnes 1992; Hayter, Grass, and Barnes 1994; Hayter 1997) that in turn draw on models of labour market segmentation outlined elsewhere (Doeringer and Piore 1971; Atkinson 1985, 1987; Streeck 1989; Marshall and Tucker 1992).

Collective Bargaining and Fordism in BC's Forest Economy

Given that labour history in BC needs a proper accounting (Seager 1988: 118), Marshall and Tucker's (1992) characterization of employment and labour relations in the US during the twentieth century, which eventually centred on the Fordist labour bargain and Taylorism, reflects in broad terms the experience of BC's forest industries. Before the 1940s, employment was characterized by hard physical work, low wages, and poor working conditions. Management had considerable bargaining power in relation to labour; unions were weak and usually broken; and workers had little protection if management chose to be capricious. Until the late 1930s, union organization was a clandestine activity, meetings were often held at night in local forests, and suspected union leaders often fired. Powell River's experience is perhaps not unusual. When its newsprint mill came on stream in 1912, two unions (the International Brotherhood of Papermakers and the International Brotherhood of Pulp, Sulphite and Papermill Workers) tried to organize production-line and trades workers, respectively (Lundie 1960). Neither local survived long, and it was not until 1937 that both unions were reestablished at Powell River, following a vote supervised by management and union representatives.

In line with continental trends, during the first half of the twentieth century, labour relations in BC's forest industries increasingly incorporated the principles of scientific management or Taylorism, which offered a method of realizing high productivity from workers with little or no formal qualifications. Thus, Taylorism defined jobs as narrowly as possible into the simplest tasks to be repeated rapidly and efficiently. Under Taylorism, worker responsibility is confined to simple, repetitive tasks, with managers responsible for decision-making tasks and for controlling workers. Under this system, bureaucratic control is justified by the productivity gains realized from specialized workers. In practice, the problems of boredom, turnover, low wages, and capricious management combined to undermine productivity and encourage unionization as a way of enhancing and protecting the well-being of workers.

The idea that workers should have the right to be represented by unions – and that unions would serve the interests of productivity – finally gained political approval in the US in the late 1930s and in Canada shortly after. The grafting together of unionized collective bargaining and Taylorism provided the basis for so-called Fordist labour relations in the major mass production industries of North America, including the BC forest industries. If the idea of "one big union in wood" (Seager 1988: 123) was not realized, a significant concentration of union power did occur. Thus, in pulp and paper, the Canadian Paperworkers Union (CPU) by 1957 had gained leadership of the workers in the pulp-and-paper industry, its position until the early 1990s when the CPU merged with the Communication Workers of Canada (CWC) and the Energy and Chemical Workers (ECW) to form the Communications, Energy and Papermakers' Union (CEPU). Among the wood industries, the International Woodworkers of America (IWA) quickly established control after 1945, its position until the present time. Indeed, labour relations by the early 1950s in the main forest industries of BC were dominated by bargaining between large corporations and large unions. Both unions, represented by the IWA and CPU, and management agreed to industrywide collective bargaining.

The Dual Segmentation Model

The collective bargains developed in the BC forest industry during Fordism were explicitly based on the principles of seniority and job demarcation. These principles were seen as advantageous to both management and unions. For unions, seniority and job demarcation create order, stability, and dignity in the workplace because each worker has clearly defined job roles and expectations. These principles address basic union philosophy, namely, the need to reduce competition among workers and the implications of such competition for degraded working conditions and declining wages. For management, seniority and demarcation have two main benefits. First, they encourage a stable workforce because seniority, and the associated accumulating wages and nonwage benefits, progressively lock workers in to the firm, and job roles are precisely demarcated. Second, job demarcation implies Taylorism, already a firmly established managerial belief. Moreover, under Taylorism, the promotion of workers on the basis of seniority rather than merit is not a major problem because obedience and discipline are considered more important than intellect and initiative.

Collective bargaining occurred every two or three years. Over time, collective bargains became more complex to create "cascading benefits" for wages, nonwage benefits, and working conditions, including grievance procedures (Clark 1986). Indeed, this system worked well during Fordism. Forest-product workers received high wages, improved working conditions, and relative stability, whereas management gained productivity and relative stability. In fact, labour markets in BC's forest industries are well portrayed by Doeringer and Piore's (1971) dual labour market segmentation model (Figure 8.1).

According to Doeringer and Piore (1971), the labour market has a primary and secondary segment, which in turn are organized by a dominant Fordist sector comprising a few large, capital-intensive, oligopolistic firms that engage in mass production and serve relatively stable markets, and a competitive sector made up of many small firms primarily supplying small, fluctuating markets. The primary segment features "high wages, good working conditions, employment stability, chances of advancement, equity, and due process in the administration of work rules." Doeringer and Piore further subdivide the primary segment into a primary independent segment (management, research, and development workers) and a primary subordinate or dependent segment (production workers, tradespeople, and office workers). The former generally enjoy higher levels of remuneration, greater

Figure 8.1

Labour segmentation under Fordism

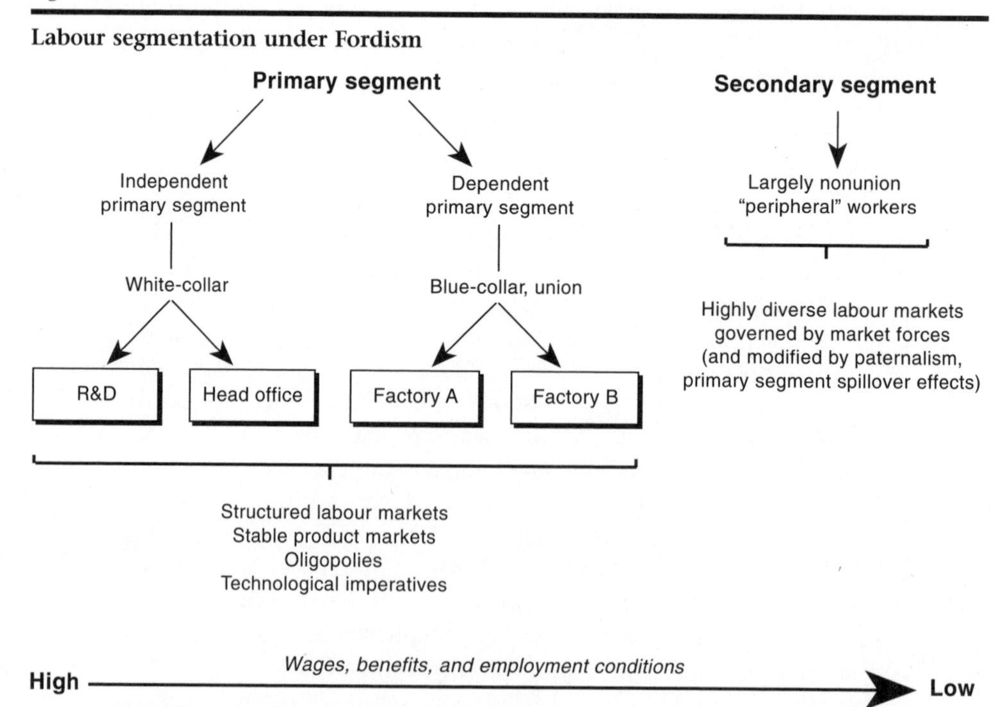

Source: Hayter 1997a: 297. Reprinted with permission of John Wiley and Sons.

employment stability, and better nonwage benefits than the latter. In contrast, jobs in the secondary segment tend to have "low wages and fringe benefits, poor conditions, high labour turnover. Little chance of advancement, and often arbitrary and capricious supervision" (Doeringer and Piore 1971: 165). In comparison to the primary sector, workers in the secondary sector are more likely to be nonunion and female, and to belong to a visible minority. Workers in the secondary sector are hired and fired according to competitive market conditions.

This model effectively represents labour market conditions in BC's forest economy during Fordism. Thus, the internal labour markets of large firms of the dominant forest industries illustrate the primary segment. White-collar managers occupied the primary independent segment, and blue-collar workers defined the primary dependent segment. The former generally enjoyed higher wages and greater employment stability than the latter. For blue-collar workers, seniority and job demarcation resulted in closely structured jobs, working conditions, and mobility. Thus, hiring was limited to the lowest grades and positions, and higher-level vacancies were filled by internal (vertical) mobility according to seniority that accumulates once workers fulfil probationary periods and are considered full-time employees. Wage rates, promotion, and layoffs were regulated by seniority. Seniority also structured how workers were laid off and rehired during recession and recovery. On the other hand, the labour markets of SMEs in BC's forest economy evolved according to Doeringer and Piore's (1971) secondary segment in which workers are often nonunionized, and wage levels and employment stability are typically less structured than in the primary segment.

In summary, specifically for BC's dominant forest-product industries, collective bargains during Fordism awarded union workers accumulating wage and nonwage benefits (statutory holidays, vacation rights, pension formulas, and various allowances). In return, management gained a specialized, efficient workforce. There were strikes, lockouts, and layoffs during recessions. But the recessions were temporary, and employment expanded and was generally stable. Indeed, in the early 1970s, the high noon of Fordism, labour supply not demand was the problem.

The High Noon of Fordism: Labour-Supply Problems in Remote Communities

Led by investments in kraft pulp mills, forest-product growth rapidly spread throughout BC in the 1950s and 1960s (Chapter 3). Employment levels expanded, and for those firms building mills in isolated communities, labour often had to be drawn from distant sources, including foreign countries. The nature of the labour-supply problem facing new mills located in remote communities is evident in three pulp mills built near the end of the Fordist period at Kitimat, Mackenzie, and Quesnel (Hayter 1979; Ofori-Amoah and Hayter 1989). The Kitimat mill was completed in 1970 by the

Finnish firm Eurocan, and the Mackenzie and Quesnel mills were both started in late 1972 by British Columbia Forest Products (BCFP) and by Cariboo Pulp and Paper (CP&P), respectively. BCFP was taken over by Fletcher Challenge of New Zealand in 1987, and CP&P is a joint venture between Weldwood (a BC-based and US-owned subsidiary) and Daishowa Paper and Marubeni of Japan (Chapter 5).

These firms preferred an industrially experienced and stable workforce. Before startup, however, the recruitment efforts of BCFP and CP&P were better organized than those of Eurocan. While the latter relied considerably on Canada Manpower and inquiries at the plant site, BCFP and CP&P advertised extensively in trade journals along with national and provincial newspapers, and they organized recruiting teams to interview candidates, in BCFP's case at locations across Canada. Candidates were interviewed in person and by questionnaires designed to assess overall aptitude, attitudes towards living in remote communities, and job-related skills. BCFP was particularly concerned about potential labour turnover problems and sought workers perceived to be stable, notably those who were married and from small western Canadian and Ontario communities whose climatic environments and living conditions were considered comparable to the town of Mackenzie. While CP&P and BCFP were able to pick and choose employees before startup from an applicant pool that was six times larger than needed, Eurocan was unable to exercise similar discretion.

The Startup Workforces at Three Pulp Mills

Clearly, the geographically extensive pattern of recruitment for the Kitimat, Quesnel, and Mackenzie mills was motivated by a search for workers with pulp-and-paper experience (Table 8.1). Thus, of the workers hired at startup from nonlocal sources (that is, beyond local commuting areas), these mills respectively hired 51 percent, 67 percent, and 73 percent of their startup workforces from other pulp-and-paper mills. Recruitment for the Mackenzie mill illustrates the tendency of the firms to recruit a few workers from many pulp mills located throughout Canada (Figure 8.2), reflecting the acceptance of an industrywide quasi-antipirating code that disallowed direct solicitation and newspaper advertising in local mill towns and discouraged raiding of individual mills beyond acceptable limits. Affiliated mills did supply large numbers of skilled and semiskilled workers; specifically, BCFP's Crofton plant and Eurocan's parent mills in Finland provided approximately 10 percent of the startup labour force at Mackenzie and Kitimat, respectively. Virtually all production-line employees were obtained from other pulp-and-paper mills.

The firms hired workers from within the industry, in part to realize full efficiencies as quickly as possible and to reduce the uncertainties of operating new equipment in new environments. Because of the workers' familiarity with pulp and resource town environments, they were expected to be

Table 8.1

Selected characteristics of workers hired at startup

	Kitimat		Quesnel		Mackenzie	
	No.	(%)	*No.*	(%)	*No.*	(%)
(1) Locational origins						
Local commuting area	84	(30.3)	63	(28.6)	20	(12.9)
Rest of BC	124	(44.8)	127	(57.7)	82	(52.9)
Elsewhere	69	(24.9)	30	(13.6)	53	(34.2)
(2) Industrial origins						
Pulp-and-paper mills	99	(37.4)	105	(46.9)	99	(63.9)
Other forest products	23	(8.7)	34	(15.2)	21	(13.5)
Other manufacturing	51	(19.2)	5	(2.2)	6	(3.9)
Construction	28	(10.6)	34	(15.2)	9	(5.8)
Other sectors	64	(24.2)	46	(20.5)	20	(12.9)
(3) Age						
<20 years	17	(5.8)	21	(9.2)	7	(4.3)
20-24 years	67	(22.8)	49	(21.4)	38	(23.5)
25-34 years	147	(50.0)	98	(42.8)	85	(52.4)
35-44 years	42	(14.3)	41	(17.9)	28	(17.3)
>44 years	21	(7.1)	20	(8.7)	4	(2.5)
(4) Marital status						
Single	80	(27.1)	51	(22.3)	16	(9.8)
Married	215	(72.9)	178	(77.7)	147	(90.2)
(5) Children						
0	127	(43.1)	79	(34.5)	51	(30.7)
1 or 2	115	(39.0)	102	(44.5)	78	(47.0)
>2	53	(18.0)	48	(21.0)	37	(22.3)
(6) Urban development						
Large town (>50,000)	36	(13.9)	17	(7.8)	19	(12.3)
Small town	223	(86.1)	202	(92.2)	136	(87.7)
(7) Job status						
Tradesperson	117	(39.7)	79	(34.5)	58	(36.3)
Production line	79	(26.8)	47	(20.5)	60	(37.5)
Other	99	(33.5)	103	(45.0)	42	(26.3)
(8) Skill						
Skilled	129	(43.9)	87	(38.0)	71	(43.8)
Semiskilled	107	(36.4)	67	(29.3)	54	(33.3)
Unskilled	58	(19.7)	75	(32.8)	37	(22.8)
(9) Formal education						
Grade 11 or more	173	(63.4)	135	(63.7)	75	(73.5)
Less than Grade 11	100	(36.6)	77	(36.3)	27	(26.5)
(10) Job changes in previous five years						
0	77	(29.1)	72	(32.4)	42	(28.4)
1	72	(27.2)	137	(61.7)	52	(35.1)
>1	116	(43.8)	13	(5.9)	54	(36.5)

Source: Hayter 1979. Reprinted with permission of Elsevier Science.

Figure 8.2

Mackenzie, British Columbia: Labour shed of pulp mill at startup, 1973

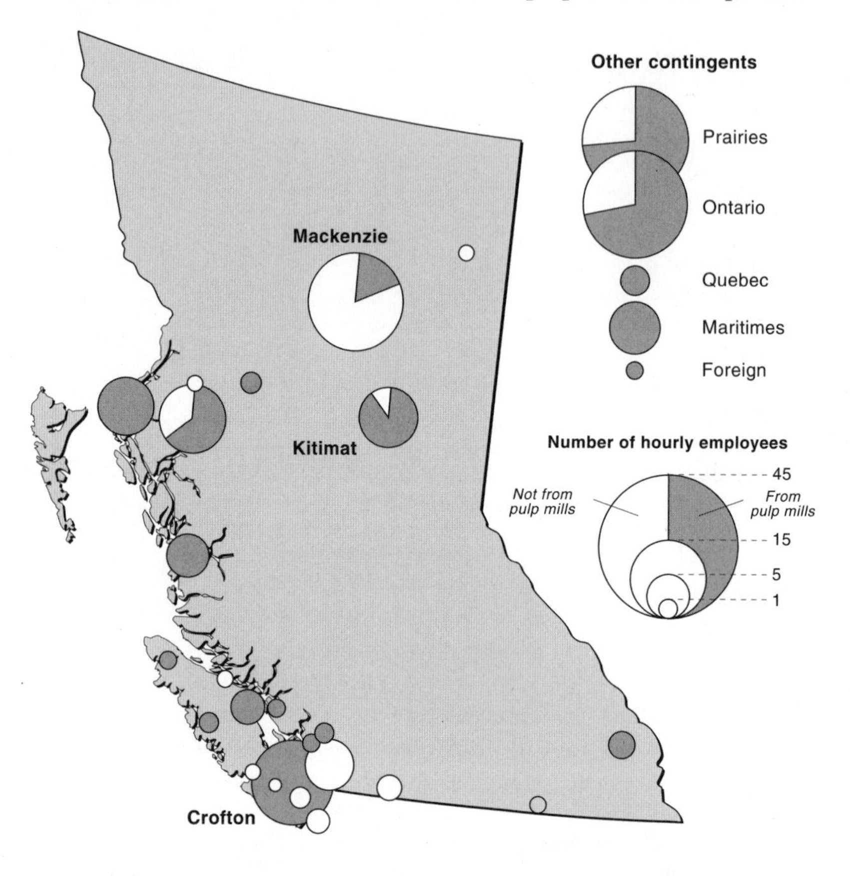

Source: Hayter 1979: 169. Reprinted with permission of Elsevier Science.

stable and unlikely to quit their jobs. Married men who were at least twenty-five years old were similarly perceived to be stable. Because workers with these characteristics were largely unavailable within local labour markets, recruitment that relied on an informal, passive, and local approach (including through Canada Manpower) was low cost but risky.

The workers hired were all unionized (mainly in the CPU) and internal labour markets were structured according to prevailing collective agreements. Consequently, while the firms enjoyed flexibility in the selection of employees at startup, employees would later be locked in by the seniority principle. As a result, these firms had strong incentives to invest in recruiting a workforce with the desired characteristics at startup. Similarly, from the perspective of employees who are locked in a firm's vertical line of progression, new plant construction provides specific opportunities for faster

Table 8.2

Number of quits among hourly workers at Kitimat, Quesnel, and Mackenzie[1]

	Kitimat	Quesnel	Mackenzie
Employed at startup	299	229	163
Losses of startup[2] employees after eighteen months	163 (50)	60 (14)	50 (12)
Turnover of startup labour force (eighteen-month period) (%)	55.0	26.54	30.67
Total employee quits[3]	558	94	99
Total skilled quits	186	24	12

1 The numbers recorded in this table are underestimates because of a small proportion of unusable employee records and because summer casuals are omitted.
2 The figures in parentheses indicate the number of skilled worker quits.
3 The Kitimat figures represent four-and-one-half years of operation compared to eighteen months of operation at Mackenzie and Quesnel.
Source: Hayter 1979. Reprinted with permission of Elsevier Science.

promotion, along with accumulated pension, vacation, and bonus rights. In fact, some employees recruited by Eurocan in 1970 moved to Mackenzie and especially to Quesnel in 1972, a community that may be considered more attractive than the others (Hayter 1979).

The Problem of Turnover

In the early 1970s, a major concern of forest-product firms was the high level of voluntary quit rates by workers. The quit rates were so high in logging and sawmilling, especially in northern regions, that the firms expressed concern about recruitment costs and the impact on productivity (Ross 1973; Pinfield et al. 1974; Hayter 1978c). Various social problems were also associated with turnover (Sinclair 1974). At that time, especially for young workers at the bottom of the seniority scale, quitting did not impose long-term penalties because other jobs were available.

While pulp-and-paper jobs were widely seen as more desirable than sawmill or logging jobs, Eurocan, BCFP, and CP&P all experienced turnover in their startup workforces (Table 8.2). For example, in the first eighteen months of operations, a total of over 100 hourly workers voluntarily left the Quesnel and Mackenzie plants, while at Kitimat, out of over 1,000 workers hired by Eurocan in the first four-and-one-half years of operation, almost 600 quit their jobs. Given differences in the length of operation of the three mills, the problem of turnover was greatest for the Kitimat plant and least for the Quesnel plant. The Kitimat mill had a significantly greater proportional loss of startup employees within the first eighteen months of operations and a higher percentage of skilled employees who quit.

If the greater quit rates at Kitimat compared to Quesnel can be partially explained by the relative appeal of the two communities, the problems

Table 8.3

Employment change in the forest-product industries of British Columbia

	1979	1982	Job change
Logging	24,474	18,000	-6,474
Wood industries	51,369	39,500	-11,869
Paper and allied	20,998	18,000	-2,998

Source: Grass and Hayter 1989: 244.

experienced at Kitimat also reflect Eurocan's lack of recruitment planning. Compared to BCFP, for example, Eurocan recruited proportionately fewer workers who were from pulp-and-paper mills, married, and older. These workers did in fact prove to be more stable; quit rates at Eurocan were greater among young, single workers hired outside the industry (Hayter 1979). Quit rates also tended to decline significantly after one month from being hired, reflecting the so-called induction-crisis effect (Hayter 1978b). This effect results from the shock many new workers experience when they enter an unfamiliar work environment that is not to their liking.

Whatever the problems of recruitment, turnover, and induction in the high noon of Fordism, they were turned upside down by the recession of the early 1980s. The problem of the voluntary turnover of young, single workers was quickly replaced by the problem of forced layoffs of older, married workers and managers with families.

Recessionary Crisis: Stimulus to Creating Flexibility in the Workforce

In BC's dominant forest industries, the immediate response to the early 1980s recession was to lay off labour and reduce the number of shifts (Grass and Hayter 1989). One estimate suggested the layoff of over 20,000 workers in the three years from 1979 to 1982 (Table 8.3). Layoffs frequently, but not inevitably, became permanent as firms rationalized plants (reduced capacity), closed mills, intensified work (maintained functions with a smaller workforce), and automated or modernized (invested in technology). In practice, these processes are interrelated and to an important degree involve technology.

Characteristics of Layoffs, 1981-5: A Provincewide Perspective

An analysis of a random sample of sixty-three surviving plants and eleven failed plants in 1981-5, stratified by industry (sawmilling, softwood-plywood, and pulp and paper), region (coastal and interior), and size of plant, provides insights into the long-term employment characteristics of layoffs (Grass and Hayter 1989). In 1981, this sample accounted for around 30 percent of sawmill employment, 60 percent of plywood employment, and 30 percent of pulp-and-paper employment in the province.

Table 8.4

Aggregate employment change among surveyed plants by gender and occupation, 1981 and 1985

	Administrative	Clerical	Trades	Production	Other	Total
Males						
1981	2,202	269	2,913	12,391	723	18,498
1985	1,769	234	2,614	10,362	661	15,640
% change	-19.6	-13.0	-10.2	-16.3	-8.5	-15.5
Females						
1981	65	329	1	298	11	704
1985	64	254	1	202	28	549
% change	-1.5	-22.7	0	-32.2	154.5	-22.0
Total						
1981	2,267	598	2,914	12,689	734	19,202
1985	1,833	488	2,615	10,564	689	16,189
% change	-19.1	-18.3	-10.3	-16.7	-6.1	-15.6

Source: Grass and Hayter 1989: 245. Reprinted with permission of Canadian Association of Geographers.

Between 1981 and 1985, employment declined by 3,013 (16.2 percent) among the sixty-three surviving plants (Table 8.4) and by 5,413 (25.1 percent) if the failed plants are included. The most significant layoffs among the sixty-three surviving plants took place between December 1981 and December 1982, after which employment declined more gradually. As of December 1985, the aggregate level of employment had not yet started to reverse its slide from the 1981 peak. Layoffs were concentrated in the coastal region. Throughout the interior, job loss was modest, and in one subregion in northcentral BC (around Prince George), employment actually increased in 1982 and in 1981-5 (Grass 1987: 44).

For the most part, among the sixty-three surviving plants, employment change by industry, but especially the largest employer (sawmilling), reflected these aggregate trends (Table 8.5). The main caveats are, first, employment declines in the softwood-plywood industry occurred mainly after 1982, and they were particularly strong in 1983 and 1985 (on the coast). Second, the interior's pulp-and-paper industry experienced employment increases. The eleven failed plants reinforce this interpretation: they were sawmills or plywood mills, and of the 2,400 jobs lost, 1,700 were located in the coastal region.

These aggregate employment declines were distributed, albeit unequally, among different labour market segments. In the province, for example, female employment dropped faster than male employment (Table 8.4). In the coastal forestry region, female job declines were especially fast (45.9 percent compared to a male decline of 26.3 percent); in the interior, female job loss amounted to 7.9 percent (compared to 2.5 percent for males). The region centred on Prince George that experienced employment growth involved

Table 8.5

Aggregate employment change among a sample of plants by industry, 1981 and 1985

Region	1981	1982	1985	% change 1981-2	% change 1981-5
All industries					
Coast	10,298	8,141	7,525	-20.9	-26.9
Interior	8,904	8,858	8,664	-0.5	-2.6
Total	**19,202**	**16,999**	**16,189**	**-11.5**	**-15.7**
Sawmills					
Coast	6,042	4,235	4,239	-29.9	-29.8
Interior	4,525	4,357	4,260	-3.7	-5.8
Total	**10,567**	**8,592**	**8,499**	**-19.5**	**-18.6**
Plywood					
Coast	1,629	1,614	1,127	-0.9	-30.8
Interior	1,696	1,625	1,557	-4.1	-8.1
Total	**3,325**	**3,239**	**2,684**	**-2.5**	**-19.2**
Pulp and paper					
Coast	2,627	2,292	2,159	-12.7	-17.8
Interior	2,683	2,876	2,847	7.2	6.1
Total	**5,310**	**5,168**	**5,006**	**-2.6**	**-5.7**

Source: Grass and Hayter 1989: 245. Reprinted with permission of Canadian Association of Geographers.

solely male employment growth (Grass 1987: 56). Given that female employment is a relatively small proportion of total employment in this industry, this trend should not be overstated. Nevertheless, the recession weakened the already marginal position of females in forest-product manufacturing.

Job losses varied by broad occupational grouping (Table 8.4). Thus, administrative, clerical, and production-line workers among the surviving plants all experienced above-average proportional job losses, with administrative workers declining the most (19.1 percent between 1981 and 1985). The reduced number of administrative workers among the sampled plants is all the more unusual given the predominance of males in such roles. Admittedly, these results do not adequately represent important groups of primary-sector workers, notably professional employees in specialized head offices and R&D laboratories. But other evidence reveals that provincial forest-product head offices were drastically cut back during the recession (Hayter 1987: 223). The main caveat to this trend is provided by a relatively small group of female administrators who, in sharp contrast to their male administrative counterparts, as well as to female clerical and production-line workers, experienced virtually no job loss between 1981 and 1985.

That the coastal region experienced greater job losses than the interior reflected greater coastal reliance on older, more labour-intensive technology designed to convert very large fir, hemlock, and cedar logs. During the

1970s, this large-log technology became increasingly obsolete because supplies of large logs declined and sawmilling became more capital intensive. In the interior, sawmills were already more capital intensive and oriented towards smaller-sized logs.

Grass also found a positive relationship between job decline and size of plant (1987: 69) and the size of firm. In particular, major job losses were more likely to occur in plants that were part of multi-plant firms. Coastal multi-plant operations were more prone to job loss than their interior counterparts or single-plant firms.

The Influence of Technology

Investment and technical change is a pervasive influence on employment change. Apart from the eleven plants that failed, only twelve of the surviving sixty-three plants that were surveyed did not incorporate new technology in 1981-5. Indeed, fifty-one plants invested about $672 million. For nineteen of these plants, mechanization was the most significant context for job loss. For eight of the surviving plants, job loss was primarily associated with rationalization; that is, with the partial closure of facilities. Yet, firms rationalize because technology creates larger, more efficient machinery. Technology also provides firms in other regions with competitive advantages in the marketplace, forcing less competitive firms to downsize. In addition, in the ten plants where job loss resulted from intensification, technological change was an underlying factor. At the same time, all eighteen plants that expanded employment in 1981-5 invested in new capacity and technology.

Technological change should not be narrowly equated with job loss. Modernization is vital to the long-term health of firms. If new technology replaces existing jobs, it also serves to maintain the viability of surviving jobs. In other cases, new technology creates jobs (Forgacs 1997). For example, among the sixty-three plants surveyed (with one exception), plans to diversify or specialize the product mix to more effectively serve market needs were all associated with technological change (Grass 1987). On the whole, plants that belong to multi-plant firms chose to specialize, and single-plant firms chose to diversify. Active product-market plans typically require investment in technological change.

Technological change also underpins shifts towards flexible operating cultures.

The Flexibility Imperative

In 1980, on the eve of recession, employment flexibility was not an issue for management and union. Indeed, even if the need for change in labour relations had been recognized, any challenge to the central principles of collective bargaining (namely, seniority and job demarcation) would have been counterproductive. But once the severity and permanent implications

Figure 8.3

Labour segmentation based on flexibility

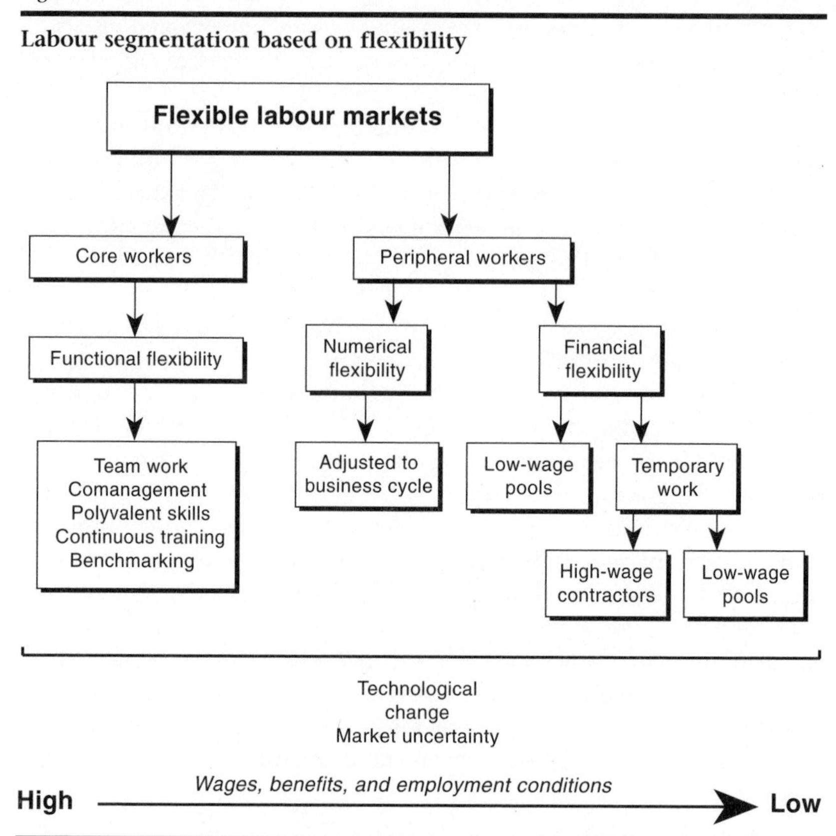

Source: Hayter 1997a: 295. Reprinted with permission of John Wiley and Sons.

of the early 1980s recession began to be appreciated, corporate restructuring – with gathering albeit ad hoc momentum – included the search for more flexible operating cultures.

The severity of the early 1980s recession created awareness that the conditions underpinning Fordist labour bargains in BC's forest industries had changed. The highly specialized, demarcated forms of work organization in support of mass production were being made obsolete by technical change, market differentiation, foreign competition, and increasing uncertainty. In response, corporations sought flexible operating cultures as a way of increased productivity and coping with more dynamic product markets. At existing mills, unions vehemently resisted this change that strikes at the heart of Fordist labour bargains: job demarcation and seniority. Today, the trend towards flexible operating cultures may have been slowed, but it has not been stopped.

Atkinson (1985, 1987) argues that firms seek to develop two types of worker flexibility: flexible core workforces and flexible peripheral workforces. The characteristics of flexibility differ in the two segments (Figure 8.3). Thus, core workers are functionally flexible, or "polyvalent"; that is, highly skilled and able to perform different functions as part of teamwork, responsible for decision making, and committed to developing further skills as needed and to benchmarking, which implies continual assessment of the best practices developed by rivals. In contrast to Fordism, flexible core workforces are not narrowly structured by job demarcation or by seniority. Rather, job categories are far fewer and broadly described. The advancement of workers depends on performance and the ability to perform tasks rather than solely seniority. Core workers initiate productivity improvements and self-supervise, and are rewarded with high wages, nonwage benefits, and employment stability.

In contrast, peripheral workers are numerically and/or financially flexible. The former are hired as needed, and they include part-time and temporary workers and "permanent" workers whose hours can easily be varied by adjustments in shifts and overtime, and by layoff if necessary. Such workers may be employed directly by firms or indirectly through subcontracting. Financial flexibility is achieved by hiring workers traditionally associated with lower wages (such as female employees) and used part-time or on contract as needs arise within the firm. Firms can also achieve financial flexibility by subcontracting high-value work to suppliers comprising highly skilled workers and professionals with the expertise that firms occasionally require (Hayter and Barnes 1992).

High Performance or Neo-Taylorism?

According to Marshall and Tucker (1992), functional flexibility and peripheral flexibility are key attributes of competing visions of labour relations that they label a high-performance model and neo-Taylorist model, respectively (see also Streeck 1989). The latter model extends Taylorist principles of scientific management, and emphasizes rationalization or cost cutting by simplifying worker tasks through automation and restricting worker control over production processes. In this model, low wages are a competitive advantage and employment flexibility is achieved by managerial discretion over hiring and firing, work processes, and training, and possibly, by multi-tasking of simple tasks but not multi-skilling.

In contrast, high-performance organizations emphasize "diversified quality production" (Streeck 1989) and the cause-and-effect relationships among high wages, job security, productivity, and innovation. In this model, high wages and job security stimulate, and are made possible by, highly skilled workforces committed to learning, high quality, and efficient production. In high-performance organizations, employment flexibility is achieved by the blurring of the boundaries between management and workers as workers take on decision-making responsibilities, and by the ongoing investment in skills development to create a polyvalent workforce.

For the BC forest industries, Marshall and Tucker's (1992) and Streeck's (1989) arguments imply that the high-performance organization model is the preferred flexibility option in the transition from Fordism. In their view, innovation is essential for long-run competitiveness in the high-cost North American economy and for stable and coherent community development. For firms, the advantages of high performance over neo-Taylorism are potentially threefold. First, the high-performing organization creates efficiencies by reliance on self-supervision and fully developing human capabilities. Second, because workers have greater skills and problem-solving abilities, the sources of innovation throughout the firm are broadened. Third, the model emphasizes quality work, thus reducing the cost associated with defects and increasing the price of outputs. For workers, high-performance flexibility promises interesting, varied, well-paid, and stable work.

A shift towards flexibility, whether high performance or neo-Taylorist, is not inevitable in BC's forest industries. As shown by the nine-month strike/lockout at Fletcher Challenge Canada mills – a dispute that began in July 1997 primarily over flexibility issues – the replacement of entrenched Fordist labour bargains with flexible operating cultures is highly contested.

The Shift to Flexibility in Union Mills

The shift to employment flexibility in BC's forest-commodity industries has occurred firm by firm, indeed mill by mill. In a few new greenfield mills, labour flexibility has been implemented relatively easily. Compared to the 1970s, union bargaining power has declined. For example, Trus Joist's Parallam plant in Delta is based on principles of flexibility *and* is nonunion, a development MB, who originally built the plant, would scarcely have contemplated in the 1970s.

At existing mills, the in situ transformation of Fordist to flexible labour relations has faced deep resistance. There are several related reasons. First, job flexibility attacks the basic principles of existing collective bargains, namely, job demarcation and seniority. Second, collective bargains are legal documents that defend worker rights and benefits gained over many decades through hard bargaining. Third, core workers under Fordism may not have the attributes to be core workers in flexible operating cultures. Fourth, a long history of adversarial bargaining, especially in the coastal region, had created tough, even poisonous bargaining environments by the late 1970s. Fifth, if flexibility is hardly likely to happen without the "stimulus" of job loss, relentless employment downsizing creates morale problems for negotiating flexibility.

Given these difficulties, it is extremely hard to say which model of flexibility – the high-performance model or neo-Taylorism – underlies changing labour relations in BC's forest industries. The theoretical distinction between high performance and neo-Taylorism is often blurred in practice. Neither term is part of actual negotiations. Innovation itself is an uncertain process, and neither management nor unions may be convinced that high performance is appropriate or attainable, especially if it implies the rationalization of old facilities. Given shareholder priorities for short-term returns, firms may also prefer the neo-Taylorist model emphasizing job (that is, cost) cutting.

The shift from Taylorism towards high performance requires managers and workers to learn new skills and new attitudes. For older workers, such demands may be seen as threats rather than as opportunities. In the past, firms and workers in BC's forest industries developed specialized, often machine-based skills. However, Taylorization deliberately limited both the *width* of worker experience (by narrowly demarcating job tasks and restricting job movement by the seniority principle) and the *depth* of worker experience (by assigning problem solving as a management prerogative). In contrast, in the high-performance model, worker skill is an asset, and skill formation across the width and depth of worker experience is emphasized. Moreover, in this model, technical skills alone are not the basis of polyvalence. Rather, a crucial work qualification for the functionally flexible worker is the capacity to acquire more work qualifications and to develop "even more unspecific, 'extra-functional' skills, that are essentially of an attitudinal and behavioural kind and which include individual characteristics like diligence, attention to detail, thoroughness and a willingness to carry responsibility" (Streeck 1989: 97).

There is no once-and-for-all debate over the kind of flexibility to be developed in BC's forest industries. In practice, flexibility is debated and introduced at the mill level, rather than comprehensively introduced through a new collective bargain. A few case studies reveal the anatomy of the process in long-established union mills.

Chemainus: The Pioneering (Sawmill) Model of Employment Flexibility

In 1980, the Chemainus sawmill employed 650 workers and produced 167 million board feet of timber (Table 8.6). The mill, however, was antiquated; its equipment was designed to cut very large timbers into low-value products while generating considerable waste. The 1982 recession provided the death blow, and as part of corporatewide restructuring efforts, the mill was closed and its entire workforce laid off (Barnes, Hayter, and Grass, 1990). The community and union reacted strongly to the closure, and representations were made to MB to reopen the mill. MB agreed, a decision doubtless aided by the mill's tidewater location and accessibility to high-quality resources and markets, and the availability of skilled labour and infrastructure. As well, MB owned the land on which the mill stood. The entirely new mill cost $22 million, and with state-of-the-art computerized equipment added the year after its 1985 startup, the mill produced smaller volumes of a wider range of higher-value products, most notably for the Japanese, other Pacific Rim, and European markets than had the old mill.

Table 8.6

**The Chemainus and Youbou sawmills:
Employment and production, 1980-9 and 1995**

	Chemainus		Youbou	
Year	Employment	Production (MFBM)	Employment	Production (MFBM)
1980	650	167	655	128
1981	550	135	615	113
1982	450	36	466	133
1983	0	0	413	150
1984	0	0	413	155
1985	125	69	360	154
1986	125	69	350	96
1987	130	102	350	159
1988	135	105	350	144
1989	140	101	176	140
1995	150	101	224	52

Notes: Employment figures are year-end totals, and MFBM refers to million board feet.
Source: Hayter, Grass, and Barnes 1994: 31, and personal research files.

Given MB's bargaining advantage over the union, stemming from the closure of the mill for two years and the over 2,500 applications for the 125 positions available at start-up, MB agreed to keep the union but required the union local, IWA-80, to accept the principle of teamwork as the basis of work organization. The union requested that seniority of laid-off workers be respected in its hiring policy. Of the 125 workers hired, 66 percent were former employees, although not all with seniority (Barnes, Hayter, and Grass 1990). In practice, the firm gave hiring priority to workers with a lumber-grading or first-aid ticket as evidence of "scholastic ability" relevant to a sawmill environment. Formal levels of education (or seniority) were otherwise not given much consideration. Prospective employees, however, had to undertake written tests, show willingness to work in a shared work environment, and agree in writing to participate in team- or module-based competitions with other workers in learning new skills. The training manuals for each module were created specifically for this plant by the Trade Advisory Committee, part of the provincial government, in association with the firm. Teams of four to seven workers select and rotate team leaders, and have weekly meetings to organize work patterns, including the details of job rotations, and to discuss problems. Supervisors can be invited to these meetings or not. The plant operates on a five-day, three-shift working week.

All the occupational groups in the old mill (Table 8.7) were drastically reduced, and the new mill is much more efficient, converts a higher proportion of timber into lumber, is more attuned to market demands, and has a more functionally flexible workforce. Although managers are fewer than in

Table 8.7

The Chemainus and Youbou sawmills: Occupational and gender profiles, 1980-1 and 1989

	Chemainus		Youbou	
	1980	1989	1981	1989
Administration	50	14 (3)	15	12
Clerical	14 (12)	4 (4)	10 (9)	4 (3)
Trades	80	19	135	45
Production	506 (8)	103 (1)	455	115 (2)
Total	**650**	**140**	**615**	**176**

Note: Figures in parenthesis refer to female employees.
Source: Hayter, Grass, and Barnes 1994: 32.

1980, the management group now includes three females and is more functionally and financially flexible, with greater marketing and production autonomy. The workforce is committed to teamwork and ongoing training, with its hourly wage-rate depending on the number of skill levels passed – regardless of the task the worker is performing at a particular time. Another incentive is relief from boredom. The plant has also reintroduced an apprenticeship program for tradespeople. The core group of permanent workers is supplemented by a small group of on-call (that is, numerically flexible) workers comprising the so-called spare board. These workers, thirteen in number in 1989, also have to pass entry-level tests, unlike in 1980 when 100 workers comprised the spare board, virtually all of them unskilled. This plant also subcontracts some functions, notably planing and kiln-drying activities formerly done in-house, to four small local operations.

The teamwork has its problems, though, as workplace democracy can result in heated arguments, and one team has stopped rotating. Union and management also failed to agree on how to build a new planer mill, so the project was scrapped. Nevertheless, the plant continues to innovate its labour relations. For example, a gain-sharing system was introduced in 1992 whereby employees were awarded a year-end bonus based on a formula-based share of production value. This bonus represented 24 percent of the 1992 payroll; the system is now being tried at other mills. In 1996, the trades agreed to spend two weeks each year learning each other's trade. The mill has been continuously profitable since its startup, wages are high, no temporary layoffs have taken place, employment has increased, and the management structure is flat and participatory. Within the large corporate sector, Chemainus is the pioneering example of flexibility that closely resembles the high-performance model; other sawmills within MB and beyond have sought to emulate developments at Chemainus; and the IWA has watched the experiment closely. Other sawmills that have sought to create employment flexibilities in situ have experienced considerable difficulties. Youbou is a case in point.

Youbou's Double-Decking System

Like Chemainus, Youbou developed as a large-log export-oriented sawmill in 1913. In 1980, its employment and output were similar to those of Chemainus (Table 8.6). Youbou workers also belonged to the same union local, IWA-80, as did Chemainus. Unlike Chemainus, Youbou is a nontidewater location, the only such survivor on Vancouver Island. For this reason, it has long been recognized as a marginal mill (Hardwick 1963). In the late 1950s, its then corporate owner, BCFP, contemplated a new sawmill alongside its new pulp-and-paper mill on tidewater at Crofton, 40 kilometres from Youbou. This plan was again set aside in the 1980s, and the firm sought to negotiate flexibility in situ. At Youbou, however, IWA-80 disagreed with the idea of teamwork; in fact, the local had no legal obligation to do so.

For whatever reason or reasons, BCFP and its corporate owners decided to adapt the Youbou sawmill in situ as part of ad hoc attempts at modernization. Thus, from 1978 to 1986, various parts of the Youbou sawmill were replaced, other parts were rationalized, and employment was substantially reduced with virtually all losses in the blue-collar segment (from 655 in 1980 to 176 in 1989). During this period, some flexibility was achieved by the creation of a double-decking system, the underlying principle of which is to use the *same* workforce to operate the two sawlines within the mill. Specifically, double decking enhances flexibility because each worker is required to perform tasks in two mills that had historically been performed by separate workers who, by union-management agreement, were differentially demarcated. In practice, double decking was negotiated and extended gradually from 1981 until 1986-7 when the system applied to virtually the entire workforce. The introduction of this system required agreement over a redefined and broader set of job categories (and associated implications for wage rates) and the development of different skills associated with technological change. As a result, the number of job categories dropped from 120 in 1980 to just 12 in 1988, and several new skills were introduced.

Within the context of job loss, and the real possibility of plant closure, IWA-80 was ultimately willing to accept double decking. But the system had to be negotiated. It is consistent with established notions of seniority, and senior workers are not required to perform "lesser" tasks or have their authority modified as, for example, might have occurred as part of teamwork. In instances where senior workers have been unable to master new tasks, special arrangements have been made, including early retirement for those aged fifty-five years or older. Even so, seniority remains the guiding principle of industrial relations at Youbou. As the workforce was reduced, it became older and the influence of long-established workers grew even stronger.

Certainly, double decking is a different type of employment flexibility compared to that achieved at Chemainus where teamwork is the cornerstone of new working conditions. At Chemainus, workers move "vertically" among tasks requiring different skills within groups, such as saw operators, and they can request to move between groups. At Youbou, however, worker movement is restricted to a "lateral" type whereby, for example, the headrig operator in Mill A performs the similar function in Mill B, but the operator does not exchange responsibilities with other types of sawyers. Moreover, the Chemainus firm achieved financial flexibility by hiring female managers and by contracting out work formerly done in-house to local planing and dry kiln mills.

It may be argued that as of 1989, the search for flexibility at Youbou resulted in maintaining more jobs (176) than at Chemainus, where 640 jobs in 1980 were reduced to 125 jobs in 1985 and only 150 in 1996. Yet, Youbou's partial modernization and limited flexibility has not eliminated questions about its survival. Between 1981 and 1988, the mill achieved profitability in only three of eight years, and in total, losses exceeded profits. In 1991, the mill again recorded a substantial loss and was closed in February 1992 for almost two months when it ran out of wood. In 1992 and 1993, the mill was closed twice more and then employed 135 workers who alternate shifts between Mills A and B.

Unlike Chemainus, the Youbou sawmill by the early 1990s could not be classified as high performance: employment flexibility was limited and innovation and job stability were not clear commitments. Possibly because its various corporate owners had been unsure about the mill's future, the incentive to push for flexibility was not strong for the firm or the union. The difficulties in bargaining for high performance in situ are well illustrated by the Powell River paper mill.

The Powell River Paper Mill in Transition

Since 1980, the Powell River paper mill, part of MB until 1998, has undergone tumultuous change. Employment levels expanded during the Fordist long-boom, and peak levels of employment at Powell River were attained in December 1973 when the mill employed almost 2,600 people, including 233 relief workers all of whom belonged to one of two unions (Table 8.8). The recession of the early 1980s, however, saw employment fall by over 500 employees by 1985, hourly and salaried jobs being equally adversely affected. Some employees were rehired, but in the peak recovery year of 1988, employment levels were less than they were in 1965. In the early 1990s, more substantial layoffs took place, and a further 700 jobs were lost by December 1994. At that time, total mill employment stood at 1,275 jobs, although an internal document discussed by both management and union recognized the mill would likely be cut back to 945 jobs (Hayter 1997b: 34). The union thought 800 realistic. MB's restructuring plans of January 1998 called for further reductions, but the mill was soon spun off to private investors.

Table 8.8

Powell River mill: Hourly occupations, 1974-94

Department	1974	1980	1986	1992	1994
Operating					
Lubrication	39	29	30	25	15
Power generating	14	14	14	11	11
Wood mill	373	327	215	220	177
Groundwood mills	248	135	71	37	28
Paper machines	228	212	136	91	87
Stock preparation	52	41	35	28	19
Kraft mill	53	51	47	48	46
Technical	32	19	19	16	12
Finishing	130	86	63	45	34
Warehouse/railway	112	50	46	33	26
Shipping	see Whs	45	50	45	46
CTMP	0	0	0	11	12
Steam plant	65	67	58	52	49
Clothing crew	0	0	21	15	8
Total	**1,346**	**1,076**	**805**	**677**	**570**
Maintenance					
Total	**509**	**533**	**443**	**345**	**254**
Services					
Technical	21	16	14	14	12
Mill stores	21	18	15	11	10
Security/yard	15	22	20	15	11
Total	**57**	**56**	**49**	**40**	**33**
Total regular	1,912	1,665	1,297	1,062	857
Total relief	240	315	271	205	235
Total hourly	2,152	1,980	1,568	1,267	1,092
Total salary	339	330	265	218	182

Notes: The data pertain to 31 December, except in 1994 when the data pertain to 31 January. Local 1 comprises the paper machine, stock preparation, and clothing crew departments; the much bigger Local 76 includes the other departments. Relief crew members are in Local 1 or in 76. Jobs are further designated by the BC Standard Labour Agreement.
Source: Hayter and Holmes 1994: 18, 20.

Between 1980 and 1984, job loss among hourly (and relief) workers driven by technological change and rationalization has affected most departments (Table 8.8). For example, the old groundwood mills that featured hand-fed grinders and labour-intensive, physically demanding work began to be replaced (starting in the 1970s) by capital-intensive thermo-mechanical-pulp (TMP) and chemi-thermo-mechanical pulp (CTMP) processes. Thus, employment in the groundwood department dropped from 248 in 1974 to 28 at 31 January 1994. TMP/CTMP lines only require two operators each. In addition, the virtual elimination of the groundwood process led to job loss in the warehouse, railway, and shipping departments. In the paper-machine department, the job decline from 228 to 87 workers between 1974 and 1994 is related to the rationalization in the number of paper machines from the eight that operated in 1974 to three in 1994. Although the three machines still operating in 1994 are much bigger than the older machines, direct employment requirements per machine are the same: each machine is run by six operators. In addition, technological change is responsible for reducing jobs in the finishing department because one system now handles the paper rolls created by three paper machines, whereas previously each paper machine would have its own finishing line. Instead of six people handling each machine's output, four workers per shift handle the output from the three remaining paper machines.

Maintenance has been equally affected. Some trades, such as blacksmiths, have disappeared, and others, such as laggers (pipe insulators), steam millwrights (millwrights who work in steam plants), and roll grindermen (who maintain a crown – a slight convex slope – on the paper machines), are unlikely to survive, at least as separate groups. The main surviving trades are electricians, millwrights, and pipefitters; welders and instrument mechanics also remain important.

Flattening Management

Similarly, managers are fewer and reorganized into a flatter decision-making structure. In 1980, below the vice president and general manager were seven managers, below each of which were groups of superintendents, supervisors, different categories of engineers, group leaders, coordinators, and various assistants (Hayter and Holmes 1994). This structure has become much leaner. In 1994, for example, five managers remained, and the positions of general superintendent and assistant superintendent had been eliminated. Some functions were combined; for example, the fire, security, and emergency evacuation responsibilities, once in separate departments with their own supervisors, were consolidated into a "protection" group and moved to the human resources department. The management/crew ratio was also increased from 1:8 to 1:16, a change that suggests greater decision-making responsibility among workers, less supervision than in the past, and the impact of technology.

The discretion enjoyed by MB in flattening its management, however, stands in sharp contrast to the shop floor where workers are protected by collective bargaining and flexibility had to be negotiated. These negotiations have not been easy.

Quid Pro Quo Bargaining Over Flexibility

Precisely when MB first decided to achieve labour flexibility at Powell River is not clear. The recession of the early 1980s clearly provoked a rethinking of labour relations, and the success of Chemainus was well known. At Powell

River, the battle lines over flexibility were finally made explicit in a 1988 court case. Since then, management has mainly sought to reduce job demarcation, increase contracting out, and modify seniority while an attempt at teamwork in the sawmill failed. (The mill closed in 1998.)

Despite the extent of job loss in the early 1980s, union concessions towards flexibility required specific quid pro quo. Thus, following a wildcat strike in 1988 over contracting out, MB took the union to court and argued successfully for its right to contract out in this particular case. The union was required to pay the company over $4 million for the cost of the illegal strike, a fine MB offered to waive in return for flexibility concessions on the shop floor. In 1991, the union agreed. The union conceded a second set of shop-floor flexibility demands in 1992 in return for an early retirement package. This package offered workers who were fifty-eight years old full pension rights with no penalties and an enhanced severance package. Severance depended on years of service, and in many cases amounted to an award of $21,000, which was untaxed if placed in a registered retirement savings plan. Workers are allowed to get other jobs, although if they return to the industry, they can only work eighty hours per month before they forfeit pension rights for that month. Apparently, 80 percent of the eligible workers accepted this offer, and in Local 76, of the 158 positions lost in the mill at this time, 120 involved early retirements.

Six months following this early retirement agreement, however, another round of layoffs occurred, and jobs that the union thought they had saved were eliminated. For the union, agreement to early retirement and increased job flexibility was traded for job security. For management, job flexibility was traded for an early retirement package. Whatever the understanding in principle, job security has not occurred in practice. The 1991 and 1992 concessions made by the unions on flexibility primarily involved broadening job descriptions among particular groups of workers. In late 1993, in another example of quid pro quo bargaining, management made an offer to Local 1 (but not to Local 76) that offered another early retirement plan in return for a much wider concept of "total flexibility." This offer was rejected.

Management's criticism of traditional job demarcation centres on the loss of productivity resulting from stopping work on a job until a qualified worker arrives, even if the work can be handled by available workers. For management, traditional job demarcation translates into productivity losses from "waiting times" and the slow completion of jobs. The 1991 and 1992 concessions permit the different trades to perform each other's jobs if they are capable. A related concession, agreed to as part of the 1991 retirement package, involves production-line workers in the paper-machine department (Local 1) providing assistance to tradespeople (Local 76). That is, work flexibility could occur across union local lines, so that, for example, when a paper-machine valve or pump needs maintenance, "spare" production workers can be used. In addition, in a salary-relief agreement, which applies to several departments but is not millwide, if a supervisor is absent (whether sick or on vacation), the senior hourly worker will step in and perform the supervisor's duties on a temporary basis. Consequently, costs are reduced because expensive extra supervisors are replaced by workers. Finally, Local 1 agreed to restrict the seniority rights of workers wishing to realign across paper machines, that is, to move from one paper machine to another, to once every two years. From management's perspective, the advantage of machine-based seniority is that it reduces the extent of job bumping across the mill and the associated need for workers to relearn a machine – a process that may take from three to eighteen months for an established worker, depending on position.

Management also wishes to contract out ongoing functions (such as security and first aid) and temporary functions (such as mill cleanup, aspects of machine repair, and engineering expertise) that are not specific to paper making and can be more efficiently provided by outside firms. Following MB's spinoff of Powell River in 1998, the shift towards worker flexibility remained a goal, and workers were given the opportunity to buy shares in the new company.

Difficulties in the Search for Flexible Work Practices

There is no secret about management desires for a smaller, more flexible workforce operating a smaller, more specialized mill. Union members have been taken on trips to mills, including Bear Island, Virginia, Longview, Washington, and Whitecourt, Alberta, to see operations where flexible work practices underpin highly efficient operations. Such trips demonstrate the competitive challenge posed by new mills and provide examples of "appropriate" labour practices. To further demonstrate the need for change, management has allowed the union to inspect the financial accounts of the mill since the early 1980s.

Yet, the search for flexibility has proven difficult and emotionally harrowing. Flexibility has attacked established principles of job demarcation and seniority, and workers have no legal obligation to change these principles and every legal right to defend them. Indeed, in Powell River, the legal contracts secured by unions are embedded in community and family cultures. Although bargaining is designed to be adversarial, until the 1980s collective bargains reflected compromise (and both sides won something); recent quid pro quo bargaining has been regarded as unfair by unions because they do not involve mutual compromise.

If recession has stimulated demands for flexibility, the relentless downsizing of employment at Powell River undermined worker morale and trust, so essential for new work arrangements, especially those based on cooperation. Without job security, flexibility is seen as threatening rather than job enriching, and flexibility itself is seen as a cause of job loss. Moreover, if seniority means downsizing is not personal, it still implied extensive job bumping in the mill where surviving workers are downgraded to less senior (lower paying, less skilled) positions. Downsizing by seniority, even if modified by early retirement, has also meant that the surviving workforce has become older, perhaps even more wedded to established forms of work organization. At the same time, this older, highly experienced workforce has had to articulate its interests for a new group of more flexible, but younger and less experienced managers.

Ironically, worker training at Powell River has been compromised by the shift towards a smaller, flexible workforce. Thus, the apprenticeship program has been terminated because skilled workers have been laid off; the team concept in the sawmill experienced problems because of difficulties in training everybody to the level necessary to practise job rotation; and extensive job bumping has disrupted traditional on-the-job training. Moreover, the articulation between the new managers and the workforce poses problems for effective, interactive learning; for the workers, new technology has also raised the spectre of testing. For senior management, training and education is costly: it takes people away from their jobs. As a result, training is selective and a potential disappointment for people not chosen. Consequently, even though management and unions have expressed mutual commitment to improving skills, training is not a magic wand that can easily facilitate a shift to flexibility. Rather, training itself involves significant costs, uncertainties, and negotiation.

Is Powell River Shifting Towards High Performance?

The shift to flexibility at Powell River from the perspective of the high-performance model is difficult to assess. Flexibility has been negotiated ad hoc; flexibility concessions have been put into practice, but their extent is unclear; flexibility demands are incomplete; and downsizing occurred again in 1998. Yet, at least until 1997, the high-performance model appears to have been the basis for the search for a flexible operating culture at Powell River. Management recognized that competitiveness with high wages depends on product innovation and quality; new, higher-value papers were innovated (Chapter 4); the mill introduced more participatory decision-making structures, various forms of benchmarking, and more teamwork; it also provided training as necessary. Higher qualifications for job entrants have been established should they ever need to be used.

As of 1997, the Powell River mill could not be classified as high performance. Indeed, more job loss occurred in 1998, further complicating the drive for cooperation. In this context, trust is not important simply for its own sake but to ensure that flexible work arrangements are fair and job enriching as well as efficient. Functional flexibility is not a mechanical concept but requires definition. Workers have to believe that their interests are fairly represented. In 1997, after ten years of seeking a flexible operation, job stability was still threatening the high-performance ideal at Powell River.

Perhaps, the sale of the mill to new owners has helped it move decisively and finally in this direction.

Labour Flexibility in Wood Remanufacturing

An important implication of the shift to flexibility among the large mills of the corporate sector is to increase variability among mills, even if unionized, a trend likely to increase if management is successful in replacing industrywide bargaining with mill-by-mill bargaining, as it has recently tried to do in the pulp-and-paper sector. In the wood-remanufacturing industries, in which SMEs dominate, flexibility and variability in employment relations are inherent characteristics (Rees and Hayter 1996).

Wages, Job Rotation, and Training

Employment in wood remanufacturing grew from 1,800 workers in 1984 to 3,500 full-time workers in 1993, with the growth in new firms rather than in existing firms (McWilliams 1991; Forintek Canada 1993). The average number of full-time workers in these firms is about twenty-three people, although the use of temporary production workers to compensate for supply inconsistencies and demand variations causes employment levels to fluctuate. Among remanufacturing firms, employment characteristics (such as level of employment, degree of unionization, wage rates, nonwage benefits, and forms of flexibility) vary considerably. This variation is captured by the firms examined in Chapter 6 (Table 8.9).

The four nontenured companies, Firms A to D, are all SMEs and nonunion. The variation in wage and nonwage benefits provided by these firms is influenced to some extent by their roles in the production system. Thus, highest entry-level wages are paid by Firm D, a specialist subcontractor and the most technologically sophisticated of the four, while the lowest entry-level wages are paid by Firm B, a contractor that performs low-value work in-house. Indeed, the entry-level wages of Firm D are slightly higher than those paid in Firm E, a union mill. Firm C's entry-level wages are higher than might be expected, given its role as a capacity subcontractor specializing in unsophisticated processes.

Among the firms, job rotation, a key feature of functional flexibility, is an important characteristic. In only one case, the nonunion Firm C, is strict job demarcation practised. Even within the unionized Firm E, a subagreement between the union and management permits job rotation and shift scheduling. The context within which job rotation occurs, however, varies. On the one hand, job rotation in Firms A, D, and E is associated with learning new skills and supported by commitments for training employees; wage scales are based on a pay-for-knowledge system. That is, these firms are pursuing functional flexibility. In Firm B, however, job rotation is practised among unskilled jobs and involves multi-tasking rather than multi-skilling. Moreover, Firm B, which contracts out its more difficult work, has less need

Table 8.9

Employment characteristics of five value-added operations

Firm	Employment 1981	Employment 1986	Employment 1991	Union	Entry wages per hour	Nonwage benefits	Flexibility features
A	13 (3)	28 (10)	40 (10)	No	$12.75	Full medical, dental, and pension	Job rotation among skilled tasks; training, preference for qualified graders
B	6	10	15	No	$8.00	Full medical, 80 percent dental, no pension	Job rotation but jobs unskilled; no qualifications
C	16 (20)	10 (25)	11 (22)	No	$14.50	Full medical, 50 percent dental, no pension	Strong job demarcation, limited training
D	74	72	54	No	$17.81	Full medical, 80 percent dental, pension	Job rotation among skilled tasks, much training
E	–	–	24	Yes	$17.00	Full medical, dental, and pension	Job rotation through sub-agreement, much training

Note: Figures in parentheses refer to temporary employees.
Source: Rees and Hayter 1996: 215. Reprinted with permission of Canadian Association of Geographers.

for a skilled workforce, no qualifications for entry, and a highly unstable workforce.

In nonunion Firms A and D and union Firm E, a core group of functionally flexible workers who are well paid, highly skilled, and stable has been developed. In Firm D, the reduction of full-time employees between 1981 and 1991 from seventy-four to fifty-four occurred because of technological change; job loss was implemented mainly by not replacing retiring workers. In fact, low turnover is important to Firm D because of its investment in employee training. For a production worker to operate machinery without supervision requires a two- or three-year training period, while tradespeople need at least five years' experience and appropriate tickets. Each production worker is trained to know an average of four tasks within the firm (40 percent of total tasks in production), with incremental wage increases awarded according to ability, knowledge, and enthusiasm. As the manager states, "Everybody can do a broad range of things ... We make a crew list every week ... The next week the same people might be on different machines working on different tasks because the order file is pertinent to that machine and there might not be an order file for that machine so we would move everybody" (Rees and Hayter 1996: 216).

Moreover, in times of economic slowdown, unemployment is apportioned evenly throughout the production plant through a job-sharing system operated by management. As the manager describes, "It is not certain kinds of people that get laid off. When you tell a worker, 'Okay you go home for three months,' they never come back. To keep our job base secure and have the people with the knowledge here, we job share. It doesn't matter if you have been here for one year or thirty years, everybody takes a fair kick. In this system if there is going to be five weeks of layoffs everybody gets a week of it" (ibid: 216).

In this way, Firm D retains the long-term investment in training within the firm while achieving numerical flexibility without having to add and drop temporary employees. Firm D can also respond to market dynamics through its extensive subcontracting linkages. Firm A, which, like Firm D, has developed a core group of skilled, stable employees, also uses a pool of temporary workers rather than subcontracting to provide additional numerical and financial flexibility. Temporary workers perform low-skill tasks at times of high demand, and they may only stay for a month.

The Variable Meaning of Flexibility

If Firms A, D, and E illustrate an emphasis on functionally flexible core groups of workers, albeit with varying characteristics, Firms B and F emphasize using a peripheral workforce. Firm F, as noted, relies entirely on subcontractors. Firm B's practice of job rotation is simply multi-tasking that produces efficiency gains but not a core group of workers. Rather, this firm prefers to use a poorly paid, numerically and financially flexible workforce that is constantly turning over. Firm B's full-time workforce is effectively temporary. As the manager noted, "We either need somebody full-time or we don't. Having people work on a part-time basis is not something that we have really thought of. When the business picks up, we take people on to be full-time, permanent employees" (ibid: 216). Yet low wages and no training encourage high quit rates. As the manager points out, "While it is not our intent to hire temporary people, in effect it does happen that way ... Our wage scale basically dictates the type of help that we get, which is inexperienced. That is okay, we are getting what we pay for" (ibid: 216). Even so, as a result of the transient nature of employment, the firm requires production workers to have completed six months of full-time employment with the firm before they receive nonwage benefits. Although unusual, the firm's employment strategy is consistent with its market strategy of contracting out high-value work and retaining low-skilled work in-house. Almost certainly, Firm B's subcontractors have a better-paid, higher-skilled core group of workers.

Firm C's employment strategy is distinct because its workers are both core and peripheral. Thus, Firm C has created a stable group of long-serving employees who are reasonably well paid but who receive minimal training and whose tasks are strongly demarcated, though nonunionized. These characteristics have a logical rationale. Thus, the manager attributes job demarcation to two factors. First, employment is fairly stable within the firm and workers rarely cover for absentees. Second, production processes (chopping, resawing, and sorting) are relatively standardized and require very little worker training. As the manager noted, "You have to have [strict job demarcation], otherwise you would have to train more people and we are too small to really spend time training people. We have very little turnover, the same people might be doing the same job for two or three years" (ibid: 217). That is, Firm C has created an inflexible workforce comprising a stable group of workers who are not well trained. Firm C, however, achieves numerical and financial flexibility by hiring temporary workers to increase output in periods of high demand. For example, twenty-two temporary workers were employed in 1991, although generally no more than ten were used at one time. In contrast to the entry-level wage of $14.50 for full-time permanent employees, temporary employees, who tend to be unskilled and either young or old, are paid $12 an hour and receive no benefits.

Differences in employment strategies are not simply a function of tenured and nontenured status. Indeed, the two tenured case studies pursue totally different employment strategies. As noted, Firm E has developed a functionally flexible workforce that is well paid, highly trained, stable, and unionized. There are no part-time or temporary employees. Given an entry-level wage of $17 per hour, graders start at $19 and the one tradesperson, a millwright, earns $24 an hour, each with a full union benefit package. In addition, production workers are trained on-the-job and by classroom instruction for quality control of a variety of tasks, while flexibility was formally negotiated as a subagreement in the 1991 union-management collective bargain. On entry into the plant, new workers are trained for the majority of jobs at the most basic machine centre before being moved to different machine centres. Promotion is indexed in accordance with ability to perform tasks.

According to management, labour and management have a less confrontational attitude than in many union plants. As the production and sales coordinator noted, "They understand the problem. If they are not flexible then it doesn't work for anybody and you are looking at the closure of the plant, nobody wants to lose their job" (ibid: 217). In this firm at least, flexibility is not associated with intensification. As the union representative acknowledges, "There is no threat of abuse or intensification because the management doesn't treat the workers like machines. There is a real feeling of mutual trust and respect."

In contrast to Firm E, which produces virtually all it needs in-house, Firm F relies exclusively on other firms for labour. Consequently, employment within the specialty products division is limited to one part-time and two full-time office workers performing administrative and coordinating tasks. All production is carried out on a contract basis with local remanufacturers, forming what the firm terms "partnerships" to allow the firm to learn about the business with no investment in plant and equipment. That is, Firm F achieves flexibility externally through the local production system. In this regard, large corporations could also benefit, more than is the case now, from the inherent flexibility of SMEs through subcontracting arrangements.

Conclusion: The Uneasy Search for a Learning Culture

Labour market flexibility is an ambiguous imperative. Theorists distinguish a preferred high-performance model from market-driven, neo-Taylorist alternatives. In reality, distinctions are less clear. In BC's forest economy, the high-performance model is an imperfect template whenever employment downsizing prevails. The model assumes high wages *and* stability. But when bargaining for flexibility occurs in situ at a time of job loss, as has been the case in BC's large corporate sector, high performance lacks appeal as a preferred alternative. Meanwhile, in SME-dominated industries, which have shown some growth, until very recently at least, labour market flexibility has always ruled. Among SMEs, however, the meaning of flexibility is more varied than among large firms.

The juggernaut of labour flexibility poses dilemmas for firms, unions, and governments (Hayter and Barnes 1997). The high-performance model offers broad guidelines for shaping but not dictating labour markets. High performance offers a preferred model for BC's forest economy because it links competitiveness, innovation, and human potential. While market flexibility emphasizes cost reduction and the hiring and firing of workers as

Table 8.10

High-performance strategies: Conflicting impulses

	Advantages	Problems
Firms	Innovation, productivity, adaptability	Cost of programs, loss of apprentices, uncertainty (e.g., raiding of skilled workers)
Unions	Job satisfaction, adaptability, security, high wages, public support	Trainability of members, testing fear of multi-tasking, competition among workers
Government	Acceptability of supply-side policies, logic of promoting higher, more equitable incomes	Conflicting impulses hard to reconcile; potential loss of union support

Source: Hayter and Barnes 1997: 220. Reprinted with permission of Western Geographical Press, University of Victoria.

needed, high-performance flexibility involves manufacturing high-value, quality products by employing high-wage, high-skilled labour that thrives on education, training, and forms of work experience that emphasize cooperative efforts, initiative, and creativity. Devising strategies that steer the BC forest sector in the high-performance direction will not be easy, however. Firms, labour, and government each face conflicting impulses. It is by no means certain that they will make the right choices (Table 8.10).

For firms, core groups of functionally flexible, well-educated, and highly skilled workers are sources of innovation, productivity, and adaptability. Yet, firms face problems in developing such a workforce, especially when changes have to be made in situ. Indeed, the difficulties of persuading existing workers to accept flexibility will likely be compounded by pressures from international forces of competition and pressures for lower costs of production, most immediately by labour market flexibility. The exigencies of recessions and the ability to poach skilled employees as needed also rationalize underinvestment in training (Streeck 1989).

For unions, training and skill potentially provide for greater job satisfaction, adaptability to change, employment security, and high wages. On the other hand, the selective nature of training policies, the trainability of existing members, and multi-skilling are all problematic, while work systems based on skill and qualifications potentially undermine the principles of seniority and job demarcation that unions have deemed essential to take competition out of the workplace. In Streeck's (1992: 264-6) view, unions need to give priority to training, education, and qualifications, and their full support to a knowledge-based economy, a strategy Streeck claims would be effective in garnering public support and allow unions to participate in the direction of change rather than simply reacting to it (see also Marshall and Tucker 1992). This view may seem to neglect the less able, but strong unions ultimately need to be associated with the strong part of the economy. After all, unions in Fordism helped segment labour markets into privileged and less privileged ones.

Finally, there is government. Both provincial and local governments have promoted the high-performance variant of flexibility by increasing funding for skill training and job upgrading, and by enriching locally available educational and training opportunities, not only temporarily to deal with specific problems of adjustment following large-scale layoffs, but also permanently within the school and postsecondary system. Such initiatives herald a potentially significant change in attitude within resource communities. These same initiatives, however, face considerable problems related to cost and trust as well as the form this training should take or the nature of the jobs (if any) when training is completed. These difficulties are especially evident among SMEs that need to be encouraged to train without being overregulated.

In principle, the government, unions, and business should share a strong common interest in education, skill formation, and training. This mutual interest should be the basis for a partnership that is socially beneficial. As Streeck (1989) advises, firms need to shift from organizations of production to organizations of learning. Similarly, communities – where families, businesses, local government, and various voluntary alliances, as well as schools and formal institutions of higher education, form reinforcing networks that encourage a culture of learning – must heed the same advice.

9
The Diversification of Forest-Based Communities: Local Development as an Unruly Process

For many specialized, forest-based communities in British Columbia, the recession of the early 1980s created an immediate social crisis and a longer-term problem of economic development. During the boom years of Fordism, many forest towns enjoyed prosperity and relative stability. For these communities, local development was equated with the strategies of the dominant employers typically headquartered elsewhere. Locally, the promotion of industrial development was a nonissue, and local government dealt primarily with managing basic community services. The underlying vulnerability of forest towns, however, was starkly revealed by the sudden, substantial employment downsizing of dominant employers that began in the early 1980s. The downsizing was geographically uneven. But for those places that experienced job loss on a scale not seen since the 1930s, the problem of local development took on entirely new dimensions.

To use Harvey's (1989) terms, local development in BC's forest towns in the 1980s necessitated a shift from an emphasis on managerialism to an emphasis on entrepreneurialism. Under managerialism, local governance during Fordism in BC's forest towns was preoccupied with administering basic community services. Forest policy and investment strategies were left to outside decision makers in the provincial government and corporate head offices. In the flexible world of the 1980s, however, forest towns have had to confront the task of diversification and how to attract new investment. That is, forest towns had to contemplate becoming entrepreneurial and try to create ideas for development from within the community. As Sjoholt (1987) argues from Norwegian experience, entrepreneurially based local development is unruly. For him, locally based entrepreneurialism, whether in the public or private sector, is highly varied in nature, outcome, and social benefit – as well as hard to predict. For BC's forest towns, unruly development potentially offers an option that differs radically from the structured forms of local development orchestrated by MNCs and senior levels of government under Fordism.

This chapter focuses on issues facing local development for forest resource towns in BC. First, forest towns are conceptualized within the context of the core-periphery model of BC's economy; the discussion considers alternative views of how the Fordist elaboration of this model during the prosperous years of Fordism has been challenged by recent instabilities. Second, the problems created by job loss for families and individuals in forest towns are outlined. The theme of diversification, a widely touted solution to the vulnerabilities of specialized communities everywhere, is addressed in the last section. Diversification, however, is an ambiguous concept, not straightforward to measure or promote. Among forest communities, strategies of diversification are themselves diversified.

Core-Periphery Dynamics in the Transition to Flexibility

The industrialization of BC's forests after the 1880s established Vancouver as the province's principal forest-product centre and principal city. By the early 1900s, Vancouver was BC's largest sawmilling centre, the major port, the major manufacturing centre where forward, backward, and final demand linkages were concentrated, and the home for many of the forest industry's decision makers (Hardwick 1963). Other forest towns were smaller, more specialized, and isolated. The massive forest-product expansions of Fordism elaborated BC's core-periphery structure.

Beyond the boundaries of metropolitan Vancouver and Victoria, hinterland BC largely evolved around numerous, highly specialized, relatively small and dispersed resource towns (Marchak 1983; Edgell 1987; Halseth 1999 and forthcoming). According to a federal government estimate (Canada 1979), there were about 100 single-industry communities in BC (excluding agricultural, fishing, hunting, trapping, tourist, and market-town communities) that included forty forest-resource towns. More recent provincial estimates suggest that in the 1980s there were 200 forest-dependent communities in BC, including twenty-six with populations of more than 1,000 (Travers 1993: 209). The extent of forest dependence was considerable, even among the larger towns. On the coast, in Powell River and Port Alberni, with 1981 populations of around 14,000 and 18,000, respectively, the forest industry accounts for over half the workforce. In the interior, similar degrees of dependence can be found in such communities as Prince George, Golden, Mackenzie, Quesnel, Revelstoke, and Castlegar. Many smaller communities can be listed. In some places, such as Kitimat and Williams Lake, forest products combine with other resource sectors to dominate the economic base.

BC's forest towns vary in their degree of isolation and specialization as well as in terms of ownership structures, population, and labour force characteristics (Barnes and Hayter 1994; Randall and Ironside 1996). Yet, the fortunes of forest towns in BC have been collectively shaped by close ties to the forest industry and typically by the strategies and structures of one or a few dominant companies. As a group, BC's forest towns had much in common in terms of their economic and social context (Marchak 1983; Edgell 1987). Moreover, during Fordism, powerful forces of standardization shaped this context. BC's forest towns comprised the dispersed outposts of an international division of labour whose primary function was to supply low-value commodities to distant markets. Workers were subject to the same principles of specialization and job demarcation and were represented by the same unions. As Marchak stresses (1983, 1990), forest towns are dependent places, vulnerable to forces beyond their control.

During the Fordist years, glimpses of this vulnerability were provided by the failure of small operations and tiny communities. In this period, however, such losses were readily absorbed by the successes of bigger mills and communities whose economic problems were typically temporary and as much self-inflicted (by strikes and lockouts) as resulting from outside business cycle forces. But even among larger forest towns, deeper sources of vulnerability were present. If ultimately rooted in specialization, this vulnerability in part reflected the influence of distant head offices mediating the priorities of different communities in BC (and beyond), in part the influence of global dynamics even beyond the reach of corporations, and in part the nature of resource dynamics. Moreover, after thirty golden years, the vulnerability of BC's forest communities was again exposed, beginning with the job losses of the early 1980s. These job losses are the basis for community crisis.

The Spatial Division of Labour under Fordism

At the heart of the core-periphery structure underlying BC's forest economy during Fordism is the spatial division of labour, organized within large, vertically and horizontally integrated corporations. Typically, if not invariably, the head-office functions, or technostructures to use Galbraith's (1966) term, were increasingly concentrated in the central business district (CBD) of Vancouver. In BC, as elsewhere, increased corporate concentration was associated with the geographic concentration of control functions, notably head offices (Hayter 1978b; Hutton and Ley 1987; Ley and Hutton 1987). In addition to directly generating jobs, many of which were high income, head offices generated myriad localized demands for business or producer services from among an expanding population of legal, financial, engineering, advertising, management, cleaning, security, and personal service firms. Typically, head offices and professional services engaged in an intricate and sophisticated web of information exchange that depended on face-to-face contacts, best facilitated by proximity. Directly and indirectly, Vancouver's forest-product technostructures helped diversify the city's white- (and pink-) collar occupational structure.

In contrast, branch plants in the resource communities, their strategic decision making provided for them by Vancouver head offices, employed largely blue-collar workforces, supplemented by middle management functions needed to supervise production and articulate with the head office. The information flows underlying this articulation, however, were mainly of a standardized, controlled nature, centred on operational performance. If the resource communities had responsibility for maintaining this performance, Vancouver decision makers were more concerned with longer-term problem solving and in ensuring coherence in the corporate systems as a whole.

During Fordism, Vancouver's core status in the BC forest economy became stronger and institutionalized as more and more branch plants in resource communities were integrated within large corporations whose head offices in Vancouver became bigger. In addition, the city and surrounding suburbs retained a significant manufacturing base in forest products and linked industries and expanded transportation and wholesaling functions. If key government functions remained centred in Victoria, they were not far from Vancouver, and for Hardwick (1974), within an expanding and increasingly well-connected Georgia Strait urban region. In general terms, the diversification and autonomy of the core became closely and functionally related to specialization and dependence in the periphery. Core-periphery relations were decidedly asymmetrical in influence, potential, and opportunity. Even so, during Fordism, core-periphery relationships in BC's forest economy implied mutual benefits.

The Autonomy of the Vancouver Technostructure

According to Kerr (1966), by the mid-1960s, Vancouver was verging on metropolitan status; that is, the city was beginning to influence social and economic life beyond its immediate provincial hinterland. In the economic sphere, this view is supported by developments in the forest-product sector, especially by the structures and strategies of leading corporations (Hayter 1978a). MacMillan Bloedel (MB) is the exemplar (Chapter 4). Established as the largest forest-product corporation in BC by the 1950s, MB built its new head office in 1968, then the tallest building in Vancouver's CBD, symbolizing the strength of industrial capital and Vancouver's control of BC's hinterland, while anticipating MB's international growth and Vancouver's growing metropolitanism.

By the mid-1970s, most of BC's main forest-product corporations had a Vancouver head office (Table 9.1). Among these, MB was easily the largest with over 1,000 employees. The internationalization of its activities in the 1960s spearheaded the economic basis of Vancouver's emerging metropolitan status. Decisions made by MB on Georgia Street affected operations in several continents, and the increasing size of MB's head office (and R&D functions) directly reflected the responsibilities of this growing international reach. Other BC-based firms (such as Canfor) that chose to limit their investments to BC and nearby places were nevertheless autonomous

Table 9.1

Head offices of integrated forest-product corporations in British Columbia, c. 1975

Corporation	BC head office[1]	Organizational status	Parent head offices	Scope of manufacturing operations
BC Forest Products	Vancouver (250)	Subsidiary	Toronto, Dayton	BC and (1978) Quebec
Canadian Forest Products	Vancouver (150)	Private corporation	Vancouver	BC and Alberta
Crestbrook Forest Products	Cranbrook (100)	Joint venture[2]	Cranbrook, Tokyo	BC and Alberta
Canadian Cellulose	Vancouver (72)	81% BC Government owned	Vancouver	BC
Crown Zellerbach Canada	Vancouver (200)	Subsidiary[2]	San Francisco	BC
Eurocan	Kitimat (50)	Joint enterprise	Helsinki	BC
Finlay Forest Products	Vancouver (3)	Joint venture	Vancouver, Tokyo	BC
MacMillan Bloedel	Vancouver (1,000)	Incorporated in Canada	Vancouver	Multinational
Northwood Pulp and Paper	Prince George (150)	Joint venture	Toronto, Dayton	BC
Intercontinental Pulp	Prince George (50)[4]	Joint venture	Vancouver, London, Dusseldorf	BC
Rayonier Canada	Vancouver (150)	Subsidiary	New York	BC[3]
Tahsis	Vancouver (150)	Joint venture	Montreal, Copenhagen	BC
Weldwood Canada	Vancouver (330)	Subsidiary[2]	New York	Canada
Weyerhaeuser Canada	Kamloops (100)	Subsidiary	Tacoma	BC and Ontario

1 Approximate employment totals given in parentheses.
2 Incorporated in Canada with shares issued on Canadian stock exchanges and annual reports published.
3 Rayonier's Quebec operations are administered by a separate company.
4 Head-office staff also administers Prince George Pulp and Paper, another joint venture.
Source: Hayter 1978a.

in shaping investment strategy. On the other hand, the activities of the foreign-owned (and central Canada-owned) subsidiaries in BC were ultimately controlled by the head offices of parent companies located elsewhere. High levels of foreign (and external) ownership sharply limited the extent of Vancouver's head-office autonomy (Chapter 5).

Typically, foreign subsidiaries in BC located their head offices in Vancouver, their size reflecting the size of the operations they controlled. Thus, Crown Zellerbach Canada (CZC) and British Columbia Forest Products (BCFP) were smaller only than MB in BC, and they had relatively large head offices in Vancouver. In contrast, the BC operations of Eurocan, Weyerhaeuser, and Crestbrook Forest Products were more specialized and retained smaller local management staffs, split between communities where their manufacturing operations were concentrated and Vancouver. For these subsidiaries, strategic-planning mandates were limited to BC, sometimes to Canada, but not beyond. BCFP did acquire a paper company in the US, but this acquisition reflected unusual circumstances (Chapter 5). All investment proposals by subsidiaries were themselves subject to parent company control elsewhere. In turn, within BC, forest-product plants had limited autonomy to invest without first seeking approval from Vancouver.

During Fordism, Vancouver's role as the control centre of the forest industry in the province, and modestly beyond, as defined by the location of corporate head offices, was much enlarged by the similar concentration of decision makers in producer services, equipment manufacturing, and transportation activities. The Vancouver area also remained the province's dominant forest-product manufacturing centre.

Resource Towns Enjoy Stability

The spatial division of labour that evolved between the core and periphery needs to be seen as a matter of degree and tendency rather than in absolute terms. If the Vancouver region retained major manufacturing activities, managerial functions were not absent from the periphery. In addition to the few subsidiary companies whose local offices were located in resource communities, entrepreneurially run logging and contracting companies could be found throughout BC along with some locally controlled medium-sized sawmills. Even local subsidiaries of corporations based in Vancouver had substantial middle-management staffs. MB's Powell River mill in the 1970s, for example, had over 300 managerial staff.

Moreover, traditional characterizations of resource towns as ephemeral places in which economic instability is compounded by various social problems – sharp class divisions, unbalanced gender structures, limited recreational options, male-dominated job opportunities, lack of higher-education facilities, and company-union-dominated politics – were tempered, if not eliminated, by the prosperity of the Fordist decades. Thus, according to Lucas's (1971) classic model of the evolution of Canadian mill towns, the four developmental phases of resource towns involve the changing and difficult stages of construction, recruitment, and transition, and ultimately culminate in the stage of maturity featuring stable population levels, work relations, social relations, and linkages with the outside world. In this model, families rather than young males provided the social fabric of communities, company power was countered by union power, and high incomes provided the basis for a growing range of community services. The Instant Town Policy of 1965 sought to quicken the process of social maturation by requiring the residential areas of new towns, such as Kitimat, Gold River, and Mackenzie, to be designed as a Vancouver suburb (Bradbury 1978).

The Lucas model accurately depicted the situation of many, if not all, forest towns of any size in BC by the 1970s. Economic problems existed even among the most stable towns. As Marchak (1983) demonstrates, job opportunities for women were extremely limited in number and kind, while even for males, occupational and industry specialization was high (and job prospects accordingly circumscribed). Yet, socially, forestry towns offered advantages in relation to metropolitan Vancouver, a point often missed by the critics of resource-town life. Moreover, by the 1970s, stability appeared to be the norm rather than instability. For those individuals who wished to stay in their communities, the mill offered secure, well-paid jobs, and the only condition was that entry was limited to the lowest-level positions. The mill looked after local youth either by promising full-time jobs or, for those who wished to leave the community, by providing the funds for education elsewhere, after a stint of readily available part-time and seasonal employment (Behrisch 1995).

Typically, forest resource towns in BC were union towns and high-income towns, a situation underpinned by the collective bargaining process initiated in the 1950s. Port Alberni, for example, was consistently ranked within the top ten Canadian communities in terms of per capita income (Hay 1993). Indeed, Marshall and Tucker's (1992: 8) depiction of Fordist communities in the US could also apply to forest towns in BC: "Workers with no more than an eighth grade education and little in the way of technical skills could end up drawing paychecks that enabled them to have two cars, a vacation cottage as well as a principal residence and maybe a boat for fishing and waterskiing. The system worked for everyone."

But the system did not work forever, or even for long after the publication of Lucas's (1971) model.

The Spatial Division of Labour under Flexibility

During Fordism, the resource sector, especially the forest industries, dominated BC's core-periphery structure. The command-and-control hierarchies of horizontally and vertically integrated forest-product corporations, mostly centred in downtown Vancouver, were the defining feature of this structure. Since the recession of the early 1980s, however, there has been growing intimations of change in BC's core-periphery relations and the hitherto dominant role played by the forest industries (Ley and Hutton 1987; Davis and Hutton 1989; Davis 1993; Hutton 1997). Davis and Hutton (1989), for example, spoke of two economies, and they argued that the economy of Vancouver was now on a different trajectory from the economy of the rest of BC, a difference that is sharper if the Victoria region is not included with the rest of BC (Davis 1993). Indeed, Davis (1993) explicitly raises the question of the Vancouver metropolitan area decoupling from the rest of the province.

Although not precisely defined, the decoupling hypothesis implies that the economy of the Vancouver core is becoming more autonomous, less dependent on the provincial hinterland, and more integrated within a global system of cities. According to this hypothesis, the vibrancy and growth potentials of the Vancouver core can be contrasted with the vulnerability and even stagnation of the periphery. Thus, the hypothesis implies a top-down view of development in which growth in the periphery ultimately depends on connections with the core; if these connections weaken, the periphery has a problem but the core does not. That is, the decoupling hypothesis defines a complex, albeit related set of processes (and outcomes), themselves not easy to measure (or uncouple!).

Explorations of the decoupling thesis have ambiguous results. In the 1980s, for example, there was no significant difference in overall employment stability between the Vancouver core and the rest of BC (Davis and Hutton 1989: 6). Moreover, the clear (and growing) differences in economic structure between the two economies, notably the much greater importance of producer services in Vancouver metro and primary activities in the rest of BC, are *not* prima facie evidence for decoupling because these (and other) differences are inherent to the idea of cores and peripheries. Indeed, in a recent, comprehensive analysis of core-periphery relations in BC, Hutton (1997: 235) emphasizes the continuing strength of such relations and concludes that arguments about Vancouver decoupling from its resource hinterland are premature. However, Ley and Hutton note some "loosening of the bonds" between core and periphery in BC, for example, in the form of the growth of producer-service exports by Vancouver firms (1987). They also argue that core-periphery relations in BC are being reconfigured. Developments in the forest sector support the idea of reconfiguration, if not decoupling.

The Decoupling Hypothesis and the Forest Sector

There is no doubt that the command-and-control hierarchies of the forest industries developed during Fordism are now less important to core-periphery relations in BC as a whole, and they are different in structure and organization. The extent of the changes is encapsulated by reference to MB, specifically employment in its Vancouver head office. Thus, following the recession of the early 1980s, MB sold its head office and moved two blocks away into smaller, rental premises; its head-office staff was reduced from over 1,200 to 560 by 1985 (Hayter 1987: 223). By the mid-1990s, MB's head-office staff had been further reduced to 230. Its restructuring plans announced in early 1998 will see head-office jobs decline to about 100, and these jobs will likely be shifted to smaller, cheaper rental premises in an inner suburb. Yet, MB's sales in 1996 were over twice what they were in 1981. Smaller production workforces, rapid improvements in communications technology, polyvalence, and outsourcing underpin this shift to a smaller, flexible, and flatter command-and-control centre. This flexible flattening of the head office has been duplicated for middle-management hierarchies in the periphery such as at Powell River (Chapter 8).

The downsizing of MB's head office in Vancouver represents a general trend among forest-product corporations. Indeed, in the early 1990s (that is, before more recent cuts), the total number of forest industry employees in Vancouver head offices was estimated at just 1,200 (that is, about the number MB employed in 1981) (Hutton 1997: 252). Forest-product technostructures now account for less than 1 percent of downtown Vancouver office workers. Mergers are unlikely to have been important in this overall downsizing. Big firms such as BCFP and CZC (see Table 9.1) became part of Fletcher Challenge Canada whose head-office employment declined to just 100 jobs by 1996. Overall corporate concentration, however, has not changed, and the underlying factors towards smaller head offices relate to technology change, cost cutting, and flexibility. For Scott Paper, before its acquisition by Kimberly Clark, its head office had been relocated to Toronto. Japanese firms have few employees in their Vancouver offices.

Along with head-office downsizing, several large sawmills and plywood mills have been closed in Vancouver metro since the 1980s. Yet, the forest industries remain extremely important to the metropolitan economy. A recent report estimated that in 1996, one in six jobs in Vancouver metro can be attributed, directly or indirectly through various multiplier effects, including for transportation, wholesaling, services, and government activities, to the forest sector (Chancellor Partners 1997). In addition, the industry in this region contributed $9 billion to provincial GDP, half of the share of the provincial forest industry as a whole. While these estimates will need to be revised downward for 1997 and 1998, head offices and large (commodity) mills still exist in Vancouver metro, and their employment has been supplemented by the value-added sector.

If Vancouver continues to be the command-and-control centre for the sector, the significantly downsized head-office staff, as well as middle management elsewhere, underscores that the nature of communication between the head office and the subsidiary has changed; it is now less hierarchical, possibly less personal. Another potential implication is greater decentralization of decision-making autonomy towards the periphery. There is some anecdotal evidence, for example, that several MB sawmills on Vancouver Island are more actively engaged in the marketing process than they once were (Hayter and Edgington 1997). In the present discussion, such a trend, if it exists, is significant because it points to more proactive roles within the periphery, even among branch plants.

Another development reconfiguring old Fordist-based core-periphery structures in the forest sector is the growing role of the SMEs. Compared to corporate models, command and control within the SMEs is inherently more flexible, idiosyncratic, and local. How this trend is affecting the relative balance of decision-making power between core and periphery in BC is hard to state. Within the manufacturing sector, important agglomerations of SMEs have occurred in Vancouver metro (Chapter 6). For many SMEs in wood-based manufacturing, a Vancouver metro location allows economical access to fibre from the periphery while facilitating interaction with suppliers and services and close access to markets. At the same time, smaller groups of SMEs have developed in various hinterland communities, notably in the Okanagan, the Prince George area, and on Vancouver Island. These same developments point to an important imperative of flexibility, largely neglected in discussions of the decoupling hypothesis, namely, the differentiation of productive activities, not only in the core but also in the periphery.

Volatility and Differentiation among BC's Forest Towns in the Periphery

With increasing force, the stable, structured planned resource towns of Fordism have been undermined by technological change, rationalization, and even closure of once dominant employers. Even as the forest has changed through massive exploitation, land claims and environmentalism, as well as pleas for privatization, have posed new questions about the ownership, control, and use of the forest which provided the raison d'être of forest towns. In an age of flexibility, forest resource towns have become volatile. But they have not simply downsized. Rather, forest towns are becoming different kinds of places. Under Fordism, corporate structures, Taylorism, and urban planning philosophy were powerful forces of standardization. Flexibility imperatives, however, seek to offset volatility through differentiation. Inevitably, differentiation must occur in situ.

In the transition from Fordism, flexible alternatives comprise a multifaceted search for options appropriate to specific circumstances. That is, the flip side of the flexibility coin is local variability or differentiation. Indeed, in the high-cost BC forest economy, there is a growing realization that differentiation, however nuanced, is vital to competitive advantage. In Vancouver metro, many of the conventional sawmills and plywood mills have been closed, and the shift from commodities to a highly diversified range of forest-product manufacturing activities is well advanced, organized by both the SMEs and the branch plants of large organizations.

Similar trends towards differentiation, if not to the same degree, can be discerned in the periphery. In particular, the emergence of SMEs (for example, remanufacturers in Prince George, log-home builders in Salmon Arm, planers, kiln dryers in Chemainus, and more isolated developments such as Sarita Furniture in Port Alberni) has helped internally differentiate the forest-product periphery in the 1990s. In addition, the introduction of flexible mass production in the corporate sector provides a degree of differentiation between resource communities. MB, for example, has distinguished its whitewood mills in Nanaimo, Chemainus, and Port Alberni by market and fibre characteristics (tree species and size), while the Powell River and Port Alberni paper mills respectively concentrated on printing papers and telephone directory papers. (Recently, Port Alberni has added lightweight coated papers.) In general, the forces favouring differentiation are the same in the periphery as they are in the core. High-income markets are more fragmented than commodity markets and so require more specialized suppliers. Moreover, in seeking out and promoting market niches, firms throughout the interior have access to fibre supplies that vary in species mix, quality, and quantity. Such supply-side variation in turn provides a basis for market differentiation. As noted, the pressures for differentiation are stronger in the core than the periphery. In time, however, increasing land costs, taxes, congestion, and planning regulations may stimulate relocation of some activities from core to periphery. Relocation may also be directed south of the border, even to Mexico.

Although they do not overlap exactly, the parallel shifts towards product-market differentiation and labour market flexibility are not coincidental. Firms are seeking to cope with more uncertain market and supply conditions by employing more flexible labour; the forms of labour flexibility are varying within the corporate sector as well as among SMEs. Over time, varying bargains over flexibility may themselves be expected to contribute to changing resource-town cultures and the internal differentiation of the periphery.

There are also forces of differentiation among forest towns beyond the forest sector. Three broad observations may be briefly mentioned in this context. First, resource towns themselves, with varying approaches and degrees of success, are seeking to diversify, as the last part of this chapter discusses. Second, in many towns, population dynamics, through in- and out-migration, are a major source of change, even if overall population levels remain more or less constant. In the last 20 years, for example, Youbou has seen virtually all its sawmill workers and their families replaced by others, many of whom commute elsewhere to work (Grass 1990). Towns such as Port Alberni, Powell River, and Chemainus, have attracted newcomers, including retirees, who brought new attitudes, as well as skills and income, sometimes creating a "culture clash" with long-time residents, especially with respect to the benefits of large-scale resource development (see Blahna

1990). If towns in the coastal region are most favourably situated to attract newcomers, population change is a feature of interior forest towns as well (Halseth forthcoming).

Third, as resource towns seek to survive, environmental and Aboriginal interests can no longer be ignored as they were under Fordism. For residents of metropolitan Vancouver, environmentalism and aboriginalism may appear as relatively abstract forces, seen only via the news media. It is in resource towns where not only direct confrontation occurs but where solutions must be forged, through various forms of cooperation that will potentially help shape business practice, work habits, and the nature of communities.

Within the context of the debate on decoupling, the role of the forest-product industries in shaping core-periphery relations in BC is clearly different than in 1980. Within the dynamism of Vancouver metro, the depth of these changes is hidden; the problems of job loss, for example, are obscured by the permanent effects of relentless urbanization and by intensifying Pacific Rim connections. In the forest towns of the periphery, on the other hand, the problems of transformation, whether conceptualized as decoupling or as reconfiguration mediated by differentiation, are transparent.

Crisis among Forest Towns

Job losses in BC's forest industries since 1980 have been geographically uneven, focusing on the coast. The winds of change are also blowing through interior communities (Halseth 1999 and forthcoming). The local impacts have been especially obvious in communities dominated by the forest industry. Port Alberni is a case in point (Hay 1993). Since 1861, when European settlement began, the economic growth of the once separate communities of mainly residential Alberni and mainly industrial Port Alberni (formally amalgamated in 1967) has been dominated by the forest industries. Port Alberni grew impressively during Fordism to become one of the highest-income communities in Canada. It was a classic company town: relatively remote with a male- and union-dominated work culture and with around 50 percent of its employment base tied directly to the forest sector, vulnerable to change (Egan and Klausen 1998/9).

Indeed, since 1980, Port Alberni's stability has been threatened by the relentless downsizing of MB's operations (Table 9.2). Job losses occurred primarily in the early 1980s, again in the early 1990s, and once more in early 1998. By 1997, MB had laid off about 2,600 or 50 percent of its 1980 workforce in the Port Alberni region. Closures of paperboard, kraft pulp, and groundwood mills, several paper machines, the plywood mill, and shingle-and-shake operations, combined with technological change and the introduction of flexible working practices are associated with this downsizing. All the workers were unionized; the layoffs were organized by seniority and, apart from the plywood mill that employed a large contingent of females, mostly involved males (Egan and Klausen 1998).

Table 9.2

MacMillan Bloedel's Port Alberni forest-product complex: Employment levels, 1980, 1986, 1991, 1996, and 1998[1]

	1980	1986	1991	1996	1998
Woodlands	1,700[2]	1,090	1,060	835	
Somass sawmill	1,064	588	509	450	341
APD sawmill	650	533	476	541	397
Plywood mill	450	377	0	0	0
Pulp and paper	1,522	1,316	1,340	958	868

1 Employment figures are year-end levels and do not include part-time workers.
2 Estimate.
Note: In 1998, the pulp-and-paper operations were controlled by Pacifica.
Source: MacMillan Bloedel, interviews with various managers and company representatives.

MB's downsizing left Port Alberni with a long-term crisis (Table 9.3). Unemployment rates for males and females increased significantly in the 1980s – the rate for females in 1986 was extraordinarily high – and these high rates have continued in the 1990s, remaining much higher than the provincial average. In 1981, labour participation rates illustrate the resource-town work culture: higher than the provincial average for males and lower for females. Since then, the labour force participation rate for males has declined consistently, whereas the participation rate increase for females appears to have levelled off by 1996 and is still below the provincial average. Male incomes in 1996 were still twice as high as female incomes, but male incomes in the 1990s actually declined, even in nominal terms (Egan and Klausen 1998/9). Port Alberni's population has declined only slightly since 1981, but the community has changed. Families and individuals have experienced much distress. Even if rates of unemployment are higher among females, job loss is dominated in absolute terms by higher-waged males.

Table 9.3

Unemployment and labour force participation rates, BC and Port Alberni, 1981, 1986, 1991, and 1996

	1981		1986		1991		1996	
	Females	Males	Females	Males	Females	Males	Females	Males
Unemployment rates								
Port Alberni	13.7	7.3	21.5	14.8	15.6	13.5	14.1	13.7
BC	7.7	5.6	13.4	12.9	10.5	10.1	8.7	9.0
Participation rates								
Port Alberni	47.8	83.3	41.5	74.2	50.4	74.5	49.8	67.2
BC	52.7	78.3	55.5	76.2	59.9	75.6	58.6	72.6

Source: Egan and Klausen 1998/9: 11, 31.

Table 9.4

Selected characteristics of a sample of (99) workers laid off by MacMillan Bloedel in Port Alberni, 1982-4

Age	*N*	Formal education	*N*	Marital status	*N*
< 25 years	29	< Grade 9	7	Single	26
25-34 years	39	< Grade 12	40	Married	62
35-44 years	17	Grade 12	42	Other	11
45-54 years	10	Some secondary	8		
> 54 years	4	University degree	2		
Total	**99**		**99**		**99**

Source: Hay 1993.

Thus, income losses to families throughout the community are substantial. The men and women who found jobs outside the forest sector in Port Alberni almost certainly earned lower incomes or did not receive incomes as high as traditionally achieved in the forest sector. Since 1980, Port Alberni has been unable to supply new jobs to offset attrition in the forest sector.

Port Alberni: Impacts of Job Loss

In Port Alberni, the downsizing of MB's workforce imposed considerable pain on individuals, families, and the community. Specifically in relation to the recession of the early 1980s, the extent of these problems and how laidoff workers and their spouses tried to cope with their situation is revealed by Hay (1993), who interviewed ninety-nine laidoff workers (out of a total of 2,000 or so workers), including eleven who were salaried, and more than fifty spouses. While this sample was organized by the snowball technique of finding respondents by information provided, and therefore cannot be considered random, Hay's results provide interesting insights into the personal response to job loss and the creation of high long-term unemployment rates that would have been even higher in the absence of out-migration.

The Indignity of Job Loss

In Port Alberni, job loss created uncertainty, stress, and indignity for individuals and their families. Of the ninety-nine laidoff workers Hay interviewed, at the time of layoff (typically in 1981-2) 62 percent were married, most had incomes in excess of $20,000, 70 percent were at least twenty-five years of age, and about half had a Grade 12 education or higher (Table 9.4). The recession of the early 1980s was of a sufficient scale to bite deeply into the mature, inherently stable part of the workforce.

The majority of Hay's respondents emphasized the emotional impact as the greatest problem of job loss. Several workers compared job loss to a death in the family. For many employees, job loss was a blow to self-esteem and an indication of their lack of control. Hay's respondents frequently made comments such as "It gave me a total sense of helplessness," "It was the worst time in my life," and "It was truly a mind-shattering experience and one from which none of us has yet completely recovered." Several respondents lost self-discipline. As one respondent stated, "I turned to alcohol and eventually became an alcoholic and then to drugs." Another, "When I didn't work I drank more, got bitchier, fights happened." A third noted, "We all became depressed and sick." A female spouse stated: "You can't imagine the desperate feelings, the helplessness, the anxiety of each day. Hoping and praying that each day maybe, just maybe your husband would get called to work, and each day you get more and more frustrated. Then the bickering about money starts. It never gives up and it doesn't get better. You start to wonder if this hell hole of a town, or jobs you try to get is worth all the disappointments. You wonder how on earth you can keep going, trying to be in a somewhat decent frame of mind, if not for yourself for your kids" (Hay 1993: 97).

Several of Hay's respondents noted how their children became emotionally affected by their parents' trauma, while four of six respondents who divorced laid the blame on job loss. In addition, Hay (1993: 99-100) offers interesting albeit anecdotal comments on the implications of job loss for domestic violence. In particular, she reports that occupancy rates at the transition house for battered women went down during the period of the layoffs and increased following recall. Such unexpected behaviour, Hay suggests, can be explained either because a laidoff and home-bound husband makes it more difficult for wives to leave and/or because wives were more willing to excuse violent behaviour in such stressful times.

For the blue-collar workforce, seniority at least provided a mechanical ("nonpersonal") framework for layoffs and a structure for callback, should any job possibilities occur. Callback, however, introduced another element of uncertainty into the lives of laidoff workers. Many respondents, for example, indicated they had stayed in Port Alberni because of callback possibilities. Yet, for some workers the possibilities never arose; for others, the waiting period lasted five years. Moreover, uncertainty was compounded by the severance formula. To receive severance benefits, workers had to be laid off for eighteen consecutive months, and acceptance of severance implied loss of seniority on the recall list. One of Hay's (1993: 95) respondents told of being recalled after a seventeen-and-three-quarter-month layoff; his dilemma was that he might only work one day before being laid off again, then the possibility of severance would be deferred for another eighteen months! While MB provided no guarantee of security, this worker took his chance with recall.

For managers, the indignity of layoff was compounded because it was selective and not influenced in any way by seniority. Some managers who were laid off were in their thirties; others were near retirement. All salaried managers that Hay interviewed expressed dissatisfaction almost ten years after the event, even bitterness, with how they had been laid off. All were called from their offices without prior notice, given the news of termination, and either prohibited from returning to their office or escorted to their office to collect personal belongings. They were then driven immediately to a counselling office elsewhere in Port Alberni. One manager felt he was forced to take early retirement. The spouse of another noted (Hay 1993: 93): "It was devastating ... the manner in which it was done ... was inhuman. After twenty-five years of service, my husband, a respected employee was called into the personnel office and told his job was redundant, to leave all keys, told not to return to his office, to leave the plant immediately and go to a counselling office set up at another location and headed by strangers [two terminated employees from MB's head office] – the ultimate humiliation."

Although such conduct is hard to excuse, redundancy inevitably imposes indignity that is deeply felt. Moreover, Port Alberni had not felt such a crisis for fifty years – and fifty years ago the community was tiny. Nobody in Port Alberni was prepared to cope with large-scale, permanent layoffs. Without warning, workers and managers, and their families, had to cope with no job and no clear alternative, in a community then lacking mechanisms of adjustment.

The Search for Work

In addition to emotional scars, job loss imposed considerable financial burdens on individuals and families. These problems had an immediate impact on housing. One bank in Port Alberni repossessed 40 percent of the mortgages initiated in the first six months of 1981 (Hay 1993: 101). The courts were so backlogged that repossession took eighteen months for processing; banks soon owned so many houses that they became reluctant to evict because they did not want more vacant dwellings. Many homeowners simply left and forfeited their equity. In Hay's sample, which does not include workers who permanently left Port Alberni, 30 percent were forced to move because of layoff, in most cases to lower-income towns. On the other hand, homeownership likely dissuaded many from leaving Port Alberni.

For virtually all (86 percent) of Hay's respondents, unemployment insurance mitigated some of the financial penalty imposed by layoff. After the exhaustion of UI benefits, 21 percent remained out of work, including 10 percent on welfare and 2 percent who retired (Figure 9.1). The majority who found work took nonunion jobs that typically were low paying. Many respondents changed jobs frequently (Figure 9.2). Indeed, from time of layoff until 1990, a period of about seven to eight years in most cases, 9 percent of the respondents changed jobs more than six times and 48 percent at least three times. By 1990, however, 73 percent of the workers returned to MB and only 2 percent remained unemployed (Figure 9.3). The late 1980s were, of course, boom years. Unfortunately, the early 1990s saw the onset of another severe recession and more cutbacks at Port Alberni. Although

Figure 9.1

Employment status after UIC of a sample of workers laid off by MacMillan Bloedel in Port Alberni, 1981-2

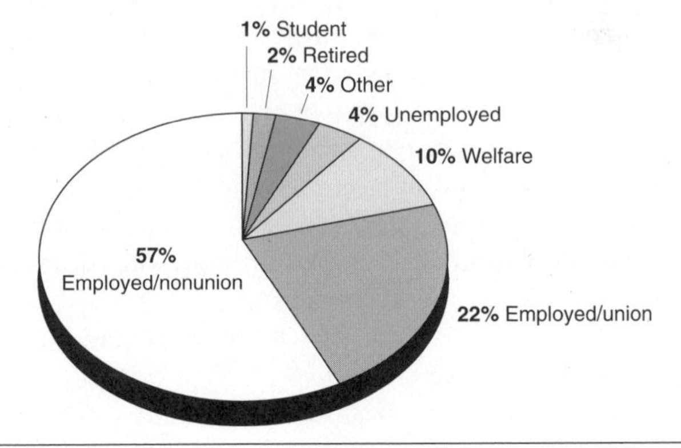

Source: Hay 1993: 107.

information is not available, many workers who waited several years to regain a position within MB have probably been laid off again.

Few laidoff workers became full-time students. About 40 percent of Hay's respondents attended some kind of educational institution to upgrade formal education qualifications (Grade 12 certification or technical skills offered by vocational programs), while almost 60 percent did not. In the early 1980s, there was little experience in dealing with large-scale layoffs, and existing training and education programs were not especially geared for laidoff forestry workers, even if their needs could have been effectively defined. On reflection, workers did not believe that enhancement of skills or qualifications aided their job search.

Figure 9.2

The number of jobs held by a sample of workers laid off by MacMillan Bloedel in Port Alberni, 1981-2

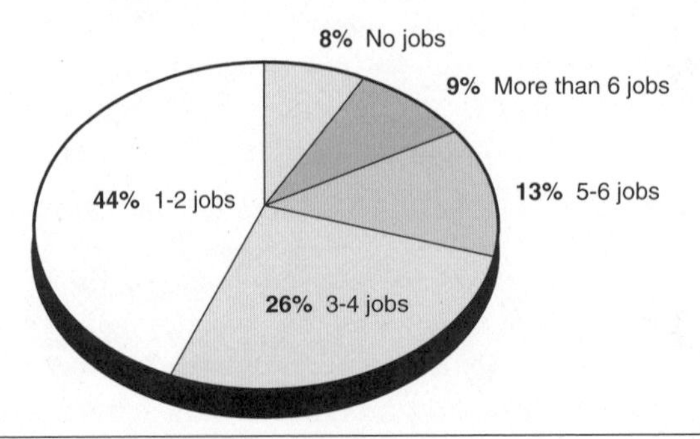

Source: Hay 1993: 108.

Figure 9.3

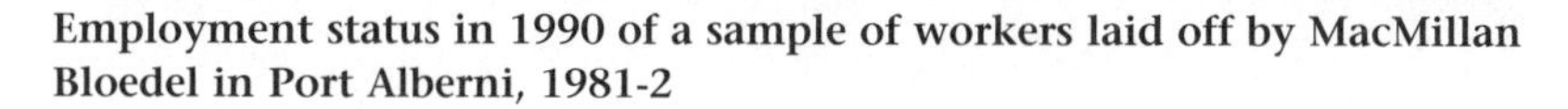

Employment status in 1990 of a sample of workers laid off by MacMillan Bloedel in Port Alberni, 1981-2

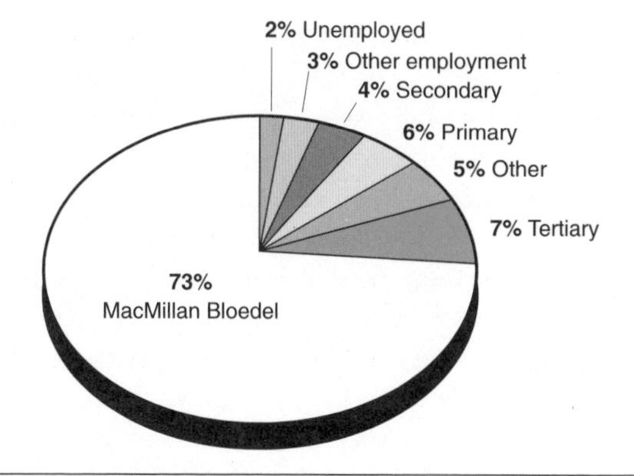

Source: Hay 1993: 109.

In fact, the retraining of redundant forestry workers in Port Alberni and throughout BC is an extremely difficult, if not intractable problem. In the 1990s, communities are now better prepared to advise and counsel laidoff workers, to identify potentially appropriate forms of training, and to implement training programs that target laidoff workers. Yet, for the middle-aged factory worker or ex-logger, there is little convincing evidence that such training actually works. The ongoing problem is the failure to link training to specific and realistic job opportunities. For workers who have families, and left school decades ago, education and training is not an easy prospect to contemplate, and frustration is likely to be considerable if, after such an investment, jobs are not forthcoming. Youth are inherently more flexible, but the problem of job opportunities for young people, and especially school leavers, is no less pressing. After all, unemployment levels are highest among the young.

Unruly Diversification: Towns That Did, Towns That Are Trying

A striking feature of local development among forest communities is the diversity of strategies and outcomes (Barnes and Hayter 1994). These differences relate to community abilities and preferences, not simply to geographic structure. Youbou, for example, opted for a passive approach to local development: as the local mill downsized, the community largely became residential in function; and as most of the former and surviving workers at the local sawmill left, other families chose to move in and even build homes there. On the southern end of Vancouver Island, with good accessibility to the major cities in the region (principally, Victoria and Nanaimo) and with a mild climate, Youbou serves as both a bedroom community for commuters and a retirement destination. So far, the residents have expressed little desire to develop beyond this residential function, or to become part of the larger municipality of North Cowichan, partly because of fears about lack of representation in a larger jurisdiction (Grass 1987; Barnes and Hayter 1994).

Other communities have actively addressed diversification, opting for different approaches and enjoying varying degrees of success. In some instances, such as Chemainus (Barnes and Hayter 1992), these plans have involved charismatic entrepreneurs; in other cases, such as Powell River, Port Alberni, and Revelstoke, local development has been rooted more broadly in the community through various agencies and committees. These differences broadly reflect Reed and Gill's (1997: 266) distinctions between community economic development (CED) and local economic development (LED). CED development is more participatory and democratic or inclusive, whereas LED focuses on more narrowly defined business interests that are not necessarily congruent with community interests. In practice, this distinction is blurred: CED is likely to be shaped by power struggles among groups and agendas, whereas LED entrepreneurship in resource towns generates social benefits (Sjoholt 1987; Trist 1979). Nevertheless, the distinction between CED and LED reinforces the idea that diversification efforts – in process and outcome – among forest communities are highly varied.

Chemainus and Port Alberni illustrate these efforts. Both towns grew rapidly during Fordism, especially Port Alberni, the bigger of the two. Thus, Port Alberni's population expanded from 6,391 in 1941 to 16,000 in 1961 and 19,892 in 1981. Since then, Port Alberni has struggled to maintain its population, which in 1991 was 18,523 and then slightly higher in 1996. The population of Chemainus stood at 2,069 in 1981, barely changed until after 1986 when the population increased to over 4,000 in 1996.

Chemainus: A Town That Did

The local development response of Chemainus to the restructuring of its mill has been proactive, featuring strong roles for entrepreneurial individuals, albeit in the public as well as private sector (see Barnes and Hayter 1992). Thus, Chemainus has benefited from entrepreneurial activity, primarily exemplified by the activity of a long-time local resident Karl Schutz and Mayor Graham Bruce, who began to organize, even before the old sawmill closure, the revitalization of Chemainus's main street, including painting a series of giant outdoor murals by artists from throughout North America. Schutz's idea for murals resulted from a holiday in Romania where he had seen frescoes at several monasteries; the implementation of the mural plan and the revamping of Chemainus's main street was stimulated by the building of a new shopping mall on the main highway, several miles away. In 1980, a provincial grant was obtained to begin the revitalization project. The mill closure in 1982 gave greater urgency to these plans, and in that year a second provincial government grant was obtained to fund painting the first five murals. By 1991, about thirty murals had been painted, largely based on Olsen's (1963) local history. In turn, the murals and revitalization stimulated new tourism-related businesses. By 1986, Chemainus was attracting 250,000 tourists a year, and by 1993, this number had increased to 350,000.

Impressive as these numbers are, as Schutz recognized, the length of stay by most tourists in Chemainus was limited, usually no more than three hours. To encourage tourists to stay longer (and spend more), Schutz developed the idea of the Pacific Rim Artisan Village. The original concept for this village called for a hotel, restaurants, cafes, gift shops, studios, galleries, and most importantly, a complex of open workshops of artists and craftspeople. This plan, budgeted at $50 million, has not materialized. However, the site is used for an annual festival in August, and there is still a plan for a hotel to be built, if financing can be arranged. A more modest but successful initiative in tourism occurred in 1993 when two entrepreneurs, one with strong local connections, opened a new dinner-and-live-theatre facility that specializes in plays for senior citizens. The other entrepreneur is a director of a dinner-theatre-and-theatre school, the Rosebud Theatre, which has been running for twenty years in Drumheller, Alberta. Eventually, they hope to establish a west coast campus for Rosebud.

While tourism has added a new dimension to the local economy, the forest industry is still the town's economic base. The new Chemainus sawmill, although employing fewer people, is extremely profitable, and entrepreneurial activity has sparked the development of Chemainus's industrial park where several new remanufacturing firms are located, including Plenk's Wood Centre and Paulcan, the latter operated by a former worker of the old Chemainus sawmill and now subcontracting with the new sawmill. These jobs are mainly unionized and well paid, more so than the jobs created by tourism, although job stability has recently been a problem for remanufacturing firms.

Local development in Chemainus has faced difficulties. The mural project was slow to recognize the role of Japanese families who were interned during World War II, and native consternation over a mural for its lifelike portrayal of their ancestors was resolved by a community-wide consecration ceremony. There was also concern expressed about the impact of the murals on parking, the arrival of non-local entrepreneurs, and over the town's copyrighted slogan ("The little town that did") which greets visitors as they enter Chemainus. (An extra "e" is occasionally spraypainted between the "i" and the "d"). An unresolved problem is that the first significant burst of entrepreneurialism in Chemainus, notably the mural and downtown revitalization projects, depended on external government funding. As that funding has all but disappeared, opportunities for entrepreneurialism have been less evident.

Nevertheless, the Chemainus economy has diversified since the 1980s. Located between Vancouver Island's two largest cities (Victoria and Nanaimo) and the ferry terminals, and on the coastal island highway, Chemainus is ideally located to capture tourist traffic and the retirement community. An indication of its success is that there is now a lack of sites for commercial development within the community. If Chemainus has not solved all its economic problems, its population has grown in the 1990s. The town also recognizes a broadly based responsibility for local development as a continuing challenge. This acknowledgment differs substantially from Fordist times when Chemainus was a company and union town.

Port Alberni: A Town That Is Trying

In Chemainus, Karl Schutz and Graham Bruce anticipated the downsizing of the early 1980s and allowed Chemainus to more or less instantaneously respond to crisis, following the protest made to MB (and the provincial government) to save the mill it had closed. Such anticipatory action also gave Chemainus a head start in negotiating financial support from the provincial government. Similar prescience did not exist in Port Alberni, a much bigger town than Chemainus: from the perspective of local development, the crisis was completely unanticipated. In coping with the crisis generated by the early 1980s recession, the problem facing Port Alberni was not simply a lack of agencies and resources; the community also had to deal with its in-grown attitudes – or "valley vision."

Valley Vision

Port Alberni, politically and socially, as well as economically, constitutes a prime example of the resource town culture analyzed by Marchak (1983). Port Alberni residents refer to then prevailing attitudes as "valley vision," implying a relatively closed society with an insular viewpoint dominated by the politics of labour relations. Over the years, these politics had increasingly hardened into uncompromising attitudes on both sides of the labour bargain. Throughout MB's operations, grievance procedures were frequently taken to Step 5, the last and most expensive step involving formal hearings and arbitration. The dividing line between management and workers was strong and culturally embedded (Hay 1993). In 1990, the refusal by the union (IWA-85) to take a $2 hourly pay cut in return for MB keeping the plywood mill in operation may well have been justified – MB might have asked for further cuts the following year. At the same time, the failure of the two sides to come to agreement reflected the uncompromising attitudes of valley vision, ensuring the closure of the mill and the layoff of most of MB's female production workers.

Valley vision implied that grievance, demarcation, and adversarial relations within MB contributed to social tensions and to political rigidity throughout the community, at least until recently. Thus, the union was unwilling to support any venture encouraging businesses that paid less than MB (Hay 1993: 151). This view was publicly stated by Dave Haggard, then vice-president of IWA-85 and now president of the IWA, in a speech to the Tin Wis conference in late 1990. To Haggard, the key to Port Alberni's diversification was in attracting large-scale unionized industry, not in more entrepreneurial forms of development. Indeed, workers in MB who decided to leave MB in the late 1980s were reportedly ostracized by fellow workers, a behaviour that may have dissuaded others from making similar choices (Hay 1993: 150). MB, too, was not ready to change past attitudes and practices. Thus, MB management initially did not participate much in local organizations seeking diversification and help for laidoff workers (for example, in the form of counselling), and job-search advice did not occur until the 1990s.

Valley vision implied considerable inertia throughout the community, a comprehensive lack of interest within the private and public sector in initiating change. By 1980, the retail sector, for example, simply relied on the business of a captive market, marketing and advertising efforts were minimal, and storefronts had changed little in twenty to thirty years. But as retail sales declined with MB's layoffs, not to mention competition from Nanaimo, increasingly accessible through road improvements, change became associated with survival. Indeed, throughout the community, regardless of MB or union wishes, various initiatives were undertaken.

The layoffs of the early 1980s unsettled the valley vision, but it was not until the second half of the decade that initiatives to resolve the community's economic problems emerged. These initiatives were given a fillip by the 1988 provincial summer games, which not only created business but more importantly kindled considerable community spirit. Improvements in the highway system, especially from Port Alberni to Nanaimo, and after 1992, the CORE process facilitated union-management cooperation in regional planning and land use. Moreover, in contrast to Chemainus, a more collectivist approach is evident in Port Alberni.

The Diversity of Community Responses in Port Alberni

There are two overriding related characteristics of community responses to crisis in Port Alberni in the 1980s and 1990s: the creation of a wide range of organizations (Table 9.5) and the contemplation of numerous initiatives of development. Following the early 1980s crisis, the only developmental organization in place was the Economic Development Commission (EDC). Others quickly emerged following MB's 1981 layoffs (see Hay 1993: 120-39). Existing church organizations also responded. One Roman Catholic priest organized workshops on poverty and invited radical speakers, but when he was replaced, the same church became less vocal (Hay 1993: 127). Another church invited speakers from MB; in one speech, a reference to the "elimination of deadwood" was apparently not appreciated by the congregation. The newly created organizations, however, represented more focused responses to crisis.

Table 9.5

Community initiatives in support of diversification in Port Alberni

Organization	Mandate and funding
Economic Development Commission, 1978	Promote and coordinate economic development; initially funded jointly by federal and provincial governments, now largely municipally funded (80 percent)
Community Futures Committee, 1987	Federally funded, its seven-year term was renewed in 1998; develop economic programs and strategies
Alberni-Clayoquot Development Society, 1987	Operating arm of Community Futures; loans and grants to help businesses startup
Alberni Valley Cottage Industry Society, 1986	A voluntary society to help those on social assistance and unemployment insurance get temporary job experience
Port Alberni Adjustment Committee, 1991	Began as a joint venture between IWA and MB; received funding from MB and then from federal government under ILAP program to train and provide assistance to laidoff workers
Port Alberni Strong Community Program, 1990-1	A voluntary group established to prepare a vision statement for the community
Port Alberni Commercial Enhancement Society, 1991	Develop and implement plans to renovate commercial areas of the city; funding of a part-time director from a local merchant tax and province

Note: Port Alberni is a municipality with a mayor and council. Other business organizations include the Chamber of Commerce and the Harbour Commission. The Port Alberni Historical Association is directing renovation of the McLean Mill, a nationally designated heritage site.
Source: Hay 1993.

Mainstream Organizations

With the EDC already in existence, the initial mainstream government-sponsored organizations that responded to Port Alberni's economic crisis sought to provide immediate help. Thus, the Port Alberni Development Society, an offshoot of the Chamber of Commerce, was set up to acquire grants from the provincial government for short-term work projects, such as trail clearing or tree thinning, especially to provide individuals whose unemployment insurance benefits had expired with enough work to requalify. Typically, work was allocated according to need. The society closed in 1984 when the government withdrew its funding. The Port Alberni Adjustment Committee (PAAC) lasted longer. PAAC, started in 1983, was the first organization in western Canada to access funds from the federal government's Industry and Labour Adjustment Program (ILAP) which helped communities attract new manufacturing activities. PAAC used $5 million of the $14 million available to it. Some funds were used for training (for example, in the use of power saws and intensive forestry), although these projects were opposed by the unions because of low wages. A German entrepreneur received $230,000 to start a pewter factory, but this project never materialized, and he was subsequently jailed for fraud. MB was more successful in using almost $3 million to renovate part of its Somass Division. ILAP was terminated in 1994.

The Alberni-Clayoquot Development Society (ACDS) was formed in 1984 on the basis of funding from the federal government, initially through the Local Employment Assistance Development Program (LEAD) and then by the Community Futures Program. ACDS, primarily concerned with stimulating jobs within the community, launched its first major project as the Alberni Enterprise Project (AEP). AEP was created in 1986, a result of discussions between the Innovations Secretariat in Vancouver and the West Coast Research and Information Cooperative (WCRIC). It was funded by the National Labour Market Innovations Program as a three-year project. In addition to ACDS, AEP was sponsored by the EDC, WCRIC, Port Alberni, and the Port Alberni Harbour Commission, the latter a federal agency.

AEP sought to stimulate entrepreneurship in the Alberni Valley through four main initiatives, namely, the Entrepreneur Training Project (ETP), a small business incubator, a seed capital fund, and a management and networking component. AEP was seen as "an experiment in community economic development in a single-industry community grappling with economic instability" (Clague and Flavell 1989). Its design was based on successful projects elsewhere, notably in New Zealand and the US (especially Hawaii and Illinois). Although expectations were high, the preliminary assessment of AEP was negative (Clague and Flavell 1989). The incubator project never really took off, although a building was provided; the seed capital collected from the community for local ventures was embarrassingly small; and the networking or coordinating component was never realized. The ETP did graduate twenty-one people at a cost of $156,000 and twelve created some kind of business. One of these firms is Sarita Furniture (Chapter 6), but few long-term jobs appeared to have been generated.

AEP coordinators attributed the program's failure to inappropriate models, too much ambition, and a lack of support in Port Alberni. The models were primarily based on the experience of large American cities, they were not modified to fit a community of Port Alberni's size. Neither MB nor the unions supported AEP. In the late 1980s, the long crisis in Port Alberni was still not recognized. On the other hand, AEP contributed to the community's learning curve for promoting development. Indeed, the 1990s witnessed a substantial range of initiatives, many of which have involved the facilitating and sponsoring efforts of the EDC and Community Futures. One economic development commissioner estimates that by early 1999, she had participated in 1,600 projects over ten years. Because provincial and federal funds were largely withdrawn, local taxpayers chose to maintain the EDC, an indication of a commitment to shift from managerialism to entrepreneurialism.

The Range of Recent Initiatives

The range of proposals that have received serious study, and in several cases were implemented, cover all sectors of the economy – resource, transportation, secondary manufacturing, tourism, education and training, health, retail, and infrastructure. Much effort is made to include Aboriginal peoples. The projects vary considerably in size, funding, actual or potential community impacts, and organization. As with virtually all other forest towns, Port Alberni has been active in promoting tourism, efforts that are based on the fact that about 600,000 visitors a year pass by Port Alberni on the way to the west coast of Vancouver Island. The town has long attracted a salmon sports fishery, rivalling Campbell River as BC's chief sports-salmon centre. There have been attempts to extend this activity by establishing a secluded resource hotel at the mouth of Alberni Inlet and promoting freshwater lake fishing. The biggest recent tourism project is the 1997 opening of the Mount Arrowsmith ski area. Planning studies predict good snow conditions, and the ski hill is scheduled for further expansion. In the centre of Port Alberni, the improved quay area provides viewscapes, catering services, and other facilities for visitors along with a forestry centre offering tours of MB facilities. Nearby, a steam engine railroad is open in the summer, and the plan is to extend the operations of this railroad to the McLean Mill, now under renovation as a national historic monument.

In the context of tourism, the McLean Mill, eleven kilometres from the town centre, is Port Alberni's signature development. Situated next to a poplar tree farm owned by MB, it hearkens back to the age of steam and small family-run sawmills. This project illustrates key features of the local development process; in particular local individuals have played the key creative roles, numerous organizations and levels of government are involved, and financing has been obtained from a variety of sources, especially from the government. The project has required considerable planning and development. People within the community have been thinking about the project since at least 1980, and David Lowe, a long-time volunteer and a founding member of the Western Vancouver Island Industrial Heritage Society in 1984, joined the staff of the McLean Mill project in 1991. The project was designated a national historic site in 1994, when John MacFarlane became general manager, and renovation was still ongoing in 1998. Expertise within the community has been marshalled to reconstruct old structures as originally designed and to make the steam mill operational. The organizers hope to maintain the site by user fees.

In terms of job creation, the McLean Mill is a small development. The management plan anticipated the mill will create seventeen jobs (8.63 full-time equivalents) by the year 2000, including four full-time positions (a general manager, a general works officer, a cultural resource officer, and a marketing officer). The remaining jobs will be seasonal, full-time and part-time, include site and machine interpreters, a machine operator, and a steam engineer. The organizers hope that by 2001 the mill will attract 200,000 visitors.

The McLean Mill project underscores the size of the task facing towns like Port Alberni that are trying to replace as many as 3,000 well-paid jobs. After almost a decade of escalating effort, the project's viability remains an uncertain prospect. At the same time, the mill symbolizes the extent of local efforts, ingenuity, and growing community participation in local development, representing a completely different attitude to valley vision. It also symbolizes broader commitments to fundamentally reconfigure the economic base of Port Alberni. Indeed, the Port Alberni Strong Community Program was mandated in 1990 to prepare a vision statement for the long-term future of the community. In this context, the completion of an airport, the construction of a hospital for injured sawmill workers, and investments in education and training facilities are developments with long-term implications.

The need to cope with the immediate problems of crisis while developing longer-run plans is evident in education. In the late 1980s, the Port Alberni Adjustment Committee sought to provide help to unemployed workers through counselling and basic skills training, and to determine training and education expectations. If this initiative was a belated response to crisis, that crisis was a new experience for Port Alberni. Moreover, the new campus of North Island College is helping to increase the long-term adaptability of the workforce, by offering educational upgrading and a variety of skills-related courses. For example, the college helped train workers in welding techniques to allow a local firm, in association with a local Native band, to win a contract on the provincial government's fast-ferry project. An important criticism of training programs, as noted, is the usual lack of jobs at the end of the program. If job guarantees are not possible, however, skill-development options are available and choices can then be made.

Nonmainstream Organizations

In addition to organizations with some kind of official sanction to help diversify Port Alberni's economy, other nonmainstream organizations sought to make a contribution. These contributions further testify to the range and complexity of local processes of diversification. There are three examples of nonmainstream organizations: the Organization of Unemployed Workers (OUW), the Alberni Valley Cottage Industry Society (AVCIS), and the West

Coast Research and Information Cooperative (WCRIC). OUW was formed in 1982 by politically active laidoff workers with connections to the Communist Party of Canada and the Maoist Party of Canada. In part, OUW sought a radical form for political change, providing speakers and films on political struggles around the world, and in part, OUW sought to provide immediate assistance to laidoff workers. Eventually, the political activists recognized that the interests of laidoff workers were the immediate needs of food and shelter, not political reform. When the activists left town in the mid-1980s, OUW collapsed.

WCRIC was a workers' cooperative also linked with labour. In contrast to OUW and the IWA's official position, however, WCRIC favoured entrepreneurial developments, contributing to the Alberni Enterprise Project as contractors for the Entrepreneur Training Program. Based in Port Alberni, WCRIC was involved in consulting and research projects across Canada.

AVCIS was the brainchild of Cecile McKinley, a resident of Port Alberni since the late 1950s, having moved there from Winnipeg. An active church and school-board member, McKinley was upset by Port Alberni's social problems, and in 1985-6, she formed AVCIS to provide unemployed youth with job skills and experience. Lacking a business background, Cecile and a few friends nonetheless obtained premises by buying an old firehall from the municipality for $1 a year and by acquiring some modest funding (after learning how to write applications) from federal and provincial job-sponsoring programs, notably, Job Trac. The funding provided temporary, relatively low-wage employment for individuals, usually young, females and males, including Aboriginal people. For about 200 individuals between 1987 and 1996, AVCIS provided work experience, skill development, and help towards self-discipline, such as coming to work on time and in good shape. The technical skills taught included carpentry, ditch digging, painting, trail making, (small) bridge building, and sewing and embroidery. Teaching expertise was usually provided free by skilled workers, including IWA members; MB also supplied some materials, such as wood supplies for stream bridges.

Although the jobs generated by AVCIS may have been menial and low paying, the organization provided help to people that went beyond the reach of more conventional, bureaucratic forms of job stimulation. Its activities also affected the community, notably projects that involved cleaning up Port Alberni before the 1998 provincial summer games, painting old people's homes, and trail making (including bridge construction) in and around Port Alberni for walkers, bikers, and horse riders. Nor was AVCIS short of ideas. For example, it tried, without success, to obtain funds for trimming trees and weeds alongside roads and BC Hydro's rights of ways and for manufacturing plastics from recycled materials. As funding sources dried up for AVCIS, and with the closure of the firehall, its movement into new premises shifted its emphasis towards making and selling garments and accessories (handbags and bingo bags) from old materials. Three to five people were engaged in this activity for a couple of years before AVCIS closed in 1996.

If nonmainstream organizations such as OUW, WCRIC, and AVCIS proved temporary, they nonetheless collectively provided insights into the depth of Port Alberni's crisis and the range of attitudes and initiatives within the community addressing its economic difficulties.

Diversification as a Rallying Cry to Promote Jobs

From the perspective of local development, diversification is invariably applauded because of its implications for widening growth opportunities and enhancing stability. Diversification, however, is problematical and such implications should not be assumed. There is no objective or mechanical definition to measure diversification as a trend or to prescribe what constitutes an appropriately diversified community. Definitions are invariably relative and partial, usually focusing on jobs. Thus, diversification is conventionally defined as a trend towards a more even distribution of jobs across a set of industrial categories.

Defined in this manner, for many forest towns such as Port Alberni, diversification has occurred since 1980 simply because forest-industry jobs have declined and jobs in other industries have grown, remained stable, or not declined as quickly. If so, diversification has implied declining stability and lower incomes. Moreover, in towns such as Port Alberni, after more than a decade of downsizing, the forest industry remains the most important component of the local economy. Any proposals to maintain the health of the forest industry, or to expand it (that is, to reinforce the specialized nature of the community), are probably going to be welcomed. In fact, any such expansion might represent diversification within the forest industry, including by the replacement of an old, declining commodity with a new, expanding product. The partial replacement of newsprint with higher-value papers at Port Alberni illustrates the latter trend.

In terms of the practice of local development, setting aside the ambiguities of its definition, diversification is a rallying cry for BC's economically distressed forest communities to seek growth opportunities wherever they can be found. In the present period of transformation, BC's forest towns know that they can no longer rely on the forest industry for economic sustainability. Other jobs are desperately sought, even if seasonal and low paying, a common criticism of diversification through tourism. Moreover, in this search, literally encompassing every sector of the economy, forest towns are gradually becoming more differentiated from one another. Some activities, such as health-care facilities, have few location options, and efforts to diversify around common themes are often locally distinct. Tourism, for example, is almost universally encouraged, but individual projects reflect local circumstances and initiative. Thus, Chemainus has its murals, dinner theatre, and arts festival; Port Alberni has the McLean Mill, steam-engine railroad, and sports fishing; Duncan offers the totem walk, a forest museum, and an Aboriginal heritage centre; Powell River advertises scuba diving (including opportunities for people with disabilities) and its two yearly music festivals; and Revelstoke is developing a railway museum. Similar observations can be made regarding health and education services. In the health field, for example, Port Alberni promoted the idea of facilities for injured forest-sector workers, while Powell River thought about attracting recovery and convalescent facilities to cater for victims of serious accidents.

This diversification of activities among forest towns is reinforcing the differences created by the differentiation occurring within the forest sector as a result of the geographically uneven implications of flexible mass production, the birth of SMEs, and flexible specialization. The latter may favour urban concentrations, specifically in Vancouver metro, but as indicated by groups of remanufacturers in the Prince George and Okanagan regions and log-house builders in Salmon Arm, smaller places are not excluded. If community forests become a more important form of forest tenure, as many advocate (M'Gonigle and Parfitt 1994), then local diversity in forest-based activities can only be enhanced (Chapter 10).

With few exceptions, such as Chemainus, forest towns have been unable to diversify sufficiently to offset forest-industry job loss. Thus, Port Alberni has failed to diversify manufacturing outside of the forest sector; ideas to promote textiles and a small distillery by foreign interests remain vague; and local entrepreneurial investments are small scale. Fish-processing investments are limited to a single new plant, and a proposed new aluminum smelter seems highly speculative. In addition, investments in social and economic infrastructure and plans to promote tourism remain incomplete and their long-term impacts uncertain.

Obstacles to Local Economic Diversification

The forces constraining the diversification of forest towns run deep. Problems of isolation are entrenched by organizational structures, and by a development psychology and attitudes that collectively militate against diversification. Freudenburg (1992) cogently argues that resource communities throughout North America are "addicted" to resource dependence, a metaphor that complements Watkins's (1963) view that the Canadian periphery is in a staple trap. Under Fordism, the quintessential forest town was a company town controlled by a MNC. But the linkages of company towns are international and within the MNC rather than local and outside the plant. Company towns are union towns whose workers are relatively well paid and specialized, hold benefits that increase with seniority, and may have deeply ingrained antimanagement attitudes. In general, institutional arrangements and local attitudes in forest towns in BC reinforced one another during Fordism to create a culture inimical to entrepreneurship, as measured by the formation of SMEs (Chapter 6). This legacy, combined with geographic marginality, has created massive obstacles to local economic diversification.

The obstacles are not absolute barriers (Randall and Ironside 1996). Forest towns vary in size and situation: Port Alberni, Chemainus, and Powell River are significantly more accessible to the population concentration of Vancouver metro, with the potential to attract retirement spillovers and tourism, than communities such as Gold River or Mackenzie. Forest towns also contain populations that have generated numerous ideas for diversification. Even if many were impractical and frequently failed to materialize, the creation of so many ideas suggests factors beyond those considerations implied by the metaphor of addiction or the staple trap. For BC's forest communities, these additional factors relate to financing and physical planning; as well, a final comment here needs to address the role of provincial policies.

Financing is a major constraint facing forest towns seeking to diversify. Diversification depends on costly investments. The kinds of ideas that forest communities have contemplated over the last decade or so include the creation of industrial parks, airport development, various forms of heritage development, downtown renewal, the creation or enhancement of museums, ski-hill development, habitat renewal or development, the upgrading of high schools, creation of secondary education and training centres of one kind or another, campgrounds and related recreational activities, development of alternative resources, and specialized health facilities. Many of these ideas depend on public funds, either for basic infrastructure (for example, new roads) or to partially offset capital costs and, in some cases, operational costs. In fact, even the annual cost of maintaining an economic development office (which for a community such as Port Alberni is about $250,000) is no small matter. Indeed, it is for financing, whether from public or private sources, that forest communities compete, even if the projects they wish to implement differ. In addition, financing is a general problem for the foundation and survival of new and small firms.

The problem facing forest towns is that the cost of their ideas – for example, the proposal for the Pacific Rim Artisan Centre in Chemainus requires $50 million – far exceeds available financing. The federal government finances various programs to promote local development, including the Western Canadian Diversification Fund. Compared to the 1960s and 1970s, however, regional development is a lower federal priority, the scale of funding is less, the few allocations are seemingly ad hoc, and the federal government has shown little long-term interest in BC's economy, except as a source of revenue (see Cannon and McLoughlin 1990). In the latest expression of disinterest, with the forest industry in a state of crisis in early 1999, the federal government has apparently rejected provincial overtures for financial aid (O'Neil and Beatty 1999). Moreover, hopes that Forest Renewal BC, the provincial agency created as part of the high-stumpage regime, would provide rainy-day funds for local diversification have been dampened, if not quashed entirely. Forest Renewal BC lacks focus, is too ad hoc in its implementation, and falls prone to political manipulation. Now, revenues sources are drying up during some very heavy rainy days.

In addition to financing, physical land use planning is a surprising problem facing forest communities contemplating diversification. Despite isolated locations, many forest towns do not have much space available in terms of a supply of suitably serviced, appropriately located sites of sufficient size. Typically, the best sites – waterfront, view sites on flat land – were chosen by the forest industry. Heavy industry is incompatible with most other uses, and even as mills are restructured, it takes time and money to clean sites after decades of degradation. In addition, even restructured forest-product operations that employ fewer people still consume much land that, for reasons related to noise, traffic, safety, aesthetics, and pollution, repel many other types of land uses. Moreover, many of the older forest towns are fragmented between an older part centred on the mill and a newer part centred on residential subdivisions and shopping areas.

In Powell River, for example, the old residential, institutional, and commercial areas around the paper mill are dilapidated; most residents prefer to live several miles to the south, out of sight of the mill. It is the old part of town that has heritage potential, however. In this case, the cost of realizing this potential is compounded by the shift in the town's centre of gravity. As a result, it is difficult to redevelop this centre so that it is meaningful in the daily lives of local residents (for example, by combining retail and commercial activities). In fact, the town's waterfront redevelopment proposals (marinas, motels, and parks) are centred in the newer part of town, and the funding for this initiative is problematic. If the town could physically integrate heritage redesign with this waterfront plan, then the financial viability of both would likely have been enhanced. Unfortunately, such integration is not possible.

Finally, the provincial government exerts powerful, complex influences on the development of forest towns by its control over land use, including Crown forests, provincial legislation governing municipal powers, tax allocations, scarce funds for development, and adjudication powers over competing proposals in the private and public sectors. There is no doubt that the provincial government seeks to promote viable communities. It may be that provincial policies in support of forest towns are more difficult to construct now than in the boom years of Fordism. Regardless, the high-stumpage regime of the 1990s has sent mixed signals to forest towns. On the one hand, the regime is associated with a philosophical shift favouring devolution and local participation; the CORE process, plans for community forests, the treaty process, and support for SMEs reflect such a philosophy. On the other hand, the implementation of these policies has been slow, contentious, and politicized, and the policies have collectively failed to provide a stable framework within which forest towns can effectively plan. Community forests are a case in point (Chapter 10). The provincial government is only planning to award three new community forests, insufficient to affect forest policy as a whole or to meet demands from communities across BC, and these decisions have been delayed several times. Moreover, in the application process, the provincial government has established its own priorities for what it sees as an appropriate community forest so that towns have been scrambling to submit proposals that meet provincial goals rather than just local goals. Indeed, in the new community forests, as with the CORE process itself, the province will retain overall control. Establishing a stable provincial framework that forest towns can identify and relate to is an important policy priority.

Conclusion: Forest Towns as Different Places

Forest towns, slowly, are becoming different places. The massive forces of standardization associated with Fordist production – especially commodity production and Taylorist labour relations – are breaking down. The nature of forest activities is becoming more differentiated among forest communities. In addition, as traditional forest industries downsize, forest towns across BC have engaged more actively in the process of stimulating diversification. Furthermore, planning for forest-town diversification is itself highly diversified. After all, the essence of bottom-up planning is to facilitate interaction and representation among the diverse groups and organizations interested in local development. In practice, the CED and LED distinctions cover a bewildering variety of organizations within and between the public and private sectors. Forest towns have varied in their commitment to, and organization of, local diversification. In this context, unruly implies a broad definition of entrepreneurialism, connoting a patchwork of bottom-up, entrepreneurial developments, only loosely coordinated, if at all, by broader planning frameworks.

Unruliness is not inherently a pejorative epithet for development. It need not imply forest towns are in cutthroat zero-sum games as each town seeks development at the expense of another (Cox and Mair 1988; Barnes and Hayter 1994). But it also does not automatically signal the arrival of entrepreneurs able and willing to diversify communities (Coffey and Polèse 1985). Moreover, for the most part, contemporary CED is an advocacy movement for citizen participation and sustainability, based on utopian pleas for local cooperation, rather than a fail-safe template for diversification. CED is critical, with justification, of industrial forestry. But how does Port Alberni replace 3,000 high-income jobs that had been stable for fifty years? That many jobs requires a lot of SMEs.

BC's forest towns face profound geographic and institutional constraints to local diversification, and these constraints are compounded by a lack of financing, physical planning problems, and uncertain relations with the provincial government. Fortunately, forest towns are not short of ideas for development. They are also increasingly sharing information. Such collective action is welcome. Four suggestions may be offered for future collective efforts. First, resource towns might usefully inventory their projects, identify the full scope of their financial requirements, and try to strengthen individual projects by revealing complementarities and perhaps synergies. Second, they might usefully compare notes on physical planning problems, including their financial aspects, because a collective effort in this regard may well reveal the scope of these problems to senior governments more effectively than an ad hoc approach. Third, they might usefully collaborate to advise the provincial government on how the latter can more effectively create a stable framework for local development. Fourth, they might usefully focus efforts to address their education and training needs. So far these efforts have typically retrained displaced mature workers, usually without much success. Efforts could be redirected to give more attention to youth and to high-school programs.

As Behrisch (1995) reveals in the context of Powell River, job losses in forest towns have enormous implications for youth, and in the long run, diversification is about addressing their needs. As one laidoff worker in Port Alberni notes, to be "too young to retire, too bloody old to work" is an awful predicament (Egan and Klausen 1998/9: 29). But to be in your early twenties, never having had a proper job, and be part of what Behrisch (1995) calls a "lost generation" – youth whose expectations for work at the local mill proved unrealistic – is equally bad. Local education systems now have to prepare youth for a very different future than the one anticipated in 1980.

10 Environmentalism and the Reregulation of British Columbia's Forests

> The principal concern of both foresters and environmentalists over the next decades will be to help society assess and secure what have come to be loosely categorized as the "non-wood" benefits of the forests.
>
> – Westoby 1989: 196

> Most of the environmental concerns of the present day, with the possible exception of fears about the loss of genetic diversity, were voiced two thousand or more years ago.
>
> – Mather 1990: 33

Since the early 1970s, debate over the use of British Columbia's forests has escalated. Reasoned expressions of concern over wood supply and management (Reed 1978; Pearse 1976) were upstaged by a more emotive "war in the woods" in the 1980s (Blomley 1996), and in the 1990s by a gruelling, unruly search for political solutions to forest-use controversies. These controversies have put into question the fabric of resource (and land) ownership and property rights in the province. Science, emotion, and politics, each with their own legitimacies and inconsistencies, are grappling with one another, but with differing worldviews, models, and languages – and with no clear means of adjudication. This grappling can be frustrating; as one facilitator, after failing to bring together industry, union, environmental, community, and Aboriginal groups in support of a community forest proposal, diplomatically confided: "It's like herding cats." This chapter should be read with this uncertainty in mind.

Arguably, the need for rethinking sustainability as a basis for reregulating BC's forests was recognized twenty-five years ago (Chapter 3). Thus, the new NDP provincial government signalled just such a contemplation in 1973, even if the subsequent Royal Commission (Pearse 1976) and the Forest Act of 1978 are not now considered radical. In 1980, the BC Ministry of Forests also publicly announced current harvest levels were not sustainable and predicted a significant decline (of about one-third) in harvest levels in the future (Percy 1986: 11). Such concerns have since been repeated with growing conviction, even if the AAC has only just begun to modestly decline (see Table 3.9). Moreover, contemporary discussions about sustainability and regulation are no longer focused on the needs of industry, as they were during the Sloan hearings of the mid-1940s. Rather, environmental values, Westoby's (1989: 196) "loosely categorized ... non-wood benefits of the forest," have become paramount. Although many, if not all, of the environmental issues created by forest harvesting have been known since classical times, environmentalism has emerged as a powerful influence shaping forest policy in many parts of the world in the latter part of the twentieth century (Mather 1990; Marchak 1995). In BC, where substantial temperate old-growth rain forest remains, global and local environmental pressures are shaping forest-industry behaviour (Cafferata 1997).

This chapter assesses the influence of environmental concerns and pressure on BC forest-industry policy (Hayter and Soyez 1996). Environmentalism is conceived broadly as an institutional force shaping public policy efforts designed to reduce the human impact on the natural environment. Politically, environmentalism has a Janus-like character, imposing two distinct and contrary tendencies on the reregulation of BC industrial forestry. On the one hand, environmentalism represents a democratic model of forestry regulation that emphasizes transparency in decision making and greater local control of the forests (Wilson 1987/88; M'Gonigle and Parfitt 1994). On the other hand, environmentalism represents an authoritarian model that seeks to impose global imperatives on forest regulation, if necessary, regardless of local consensus-building mechanisms.

In outline, the first part of the chapter discusses the emergence of environmentalism in BC in shaping the transformation of forest policy from its Fordist role of providing large volumes of fibre to large-scale industry to a flexible institution accommodating a wider range of values and forms of control. Second, the chapter explores the problematic implications of environmentalism and environmentally inspired forestry regulations for the corporate sector, specifically for logging activities. The Clayoquot Sound Compromise provides a case study. Finally, the chapter claims that environmentalism has encouraged flexibility and unruliness in BC's forest policy by its promotion of public participation, multiple values, and (more ambiguously) local control. However, for forest policy, unruliness and flexibility define a fine and dangerous line between complementary, mutually beneficial differentiation of forest use on the one hand, and unstable hypercompetition of forest use on the other. From this perspective, the institutional challenge is to ensure that unruliness realizes the benefits of a more diversified forest economy without degenerating into a highly politicized, uncertain, and disruptive approach.

Several prefatory caveats to the chapter should be noted. First, the implications of environmentalism for forest policy in BC are complex and impossible to characterize precisely. Environmentalism has emerged at a time of considerable change in technological, demand, and supply conditions. Untangling the effects of these various dimensions of change is not attempted here. Second, environmental groups are diverse in size, agenda, mandate, and values, ranging from localized environmental coalitions to powerful international environmental lobbies, notably Greenpeace and the Rainforest Action Network (RAN). In this context, characterizing environmentalism as authoritarian or democratic essentially represents poles of a continuum. Third, while environmentalism is often associated with advocacy for local control and Aboriginal rights, these issues should not be conflated, as Aboriginal spokespeople have noted (Nathan 1993; see also Willems-Braun 1997a, 1997b). Fourth, this chapter does not assess the actual or potential role of the forest industries in creating environmental problems, such as soil erosion, air pollution, and climatic change (Kimmins 1992).

Fifth, a striking feature of contemporary forest reregulation is its duration. If the election of the first provincial NDP government in 1973 or the bringing into force of the Forest Act of 1978 is seen as the beginning of this rethinking, discussions have lasted from twenty to over twenty-five years. Even the election of the second NDP government in 1991, which instigated a barrage of forestry legislation, marks eight years of discussion. Moreover, closure of the debate is not in sight. As of mid-1999, environmental opposition to forest practices remains in evidence, the *Delgamuukw* judgment of the Supreme Court of Canada added new perspective to the land claims process, and the Nisga'a treaty is only the first of more treaties, announcements about new community forests are pending, privatization of some forest land has become a real possibility, the US may challenge stumpage reductions, future levels of the AAC are unclear, and just how to divide the forest pie remains contentious.

In contrast, the Sloan hearings and the Forest Act of 1947, which provided the basis for forest regulation, with some modification, for at least thirty years, were relatively brief. Clearly, the current circumstances and goals of forest policy are different from the late 1940s. If stability and homogeneity were prized features of forest regulation in the 1940s, the reregulation of BC's forests is demanding more flexibility and diversity – and these features are sought in the context of an established system. Environmentalism is an important driving force.

Environmentalism as Voices of Localism and Globalism

Expressions of concern over environmental matters, for problems of degradation and depletion, as well as attempts at resource management, are ancient (Thirgood 1981; Meiggs 1982). This concern, however, has escalated rapidly since the early 1970s (Mather 1990; Garner 1996). In many countries in the developed world, the public's environmental consciousness is extremely high. The environment is frequently placed alongside economic growth as a policy priority at national and regional levels, and a spate of environmental legislation has been developed since the late 1960s. In addition, the environmental movement has grown rapidly since the early 1970s in terms of the number of organizations and members. The largest, Greenpeace, has considerable power, resources, and global reach. Greenpeace expanded between 1970 and 1995 from twelve Vancouver-based members to 2.9 million members located in 158 countries. In 1995, Greenpeace received donations totalling $152 million to support its activities and offices in thirty-two countries (Austin 1996: A26). If Greenpeace is committed to nonviolent protest, its "navy" can mount embarrassing "greenboat" diplomacy around the globe.

Society's commitment to environmental values is closely associated with economic development and the onset of postindustrialism (see Figure 1.2). In this advanced stage of economic development, in privately regulated and state-regulated forests, environmental goals emerge to rival, possibly supplant, industrial goals (Mather 1990). It is also commonly supposed that environmental regulation is crisis motivated, occurring after, rather than in anticipation of, depletion and degradation. Among developed economies, the escalation of environmental legislation of forest use since the 1970s (Mather 1990: 120) is supported strongly by public opinion, including Canada (Mather 1990: 268-75). One public opinion survey in Canada, conducted in 1981, found 18 percent of the respondents agreed strongly that "the forest should not be economically exploited at all" (Mather 1990: 269). Public support in turn provides the basis for environmentalism.

The Emergence of Environmentalism in BC

A part of widespread global trends, the environmental critique of forest practices in BC gathered momentum in the 1970s (Wilson 1998). This critique has become progressively more complex. Initially, environmentalism was largely locally based, and primarily expressed the fear that BC was running out of wood. Although not an environmental treatise, Marchak's (1983) *Green Gold* intimated that the forest was being mined and deforestation was an emerging problem. Indeed, in the late 1970s, the falldown effect, which occurs when the massive harvests available from old-growth timber start to decline and second-growth supplies, at best, offer lower-volume alternatives, began to be publicly discussed and accepted in government documents. In effect, acknowledgment of the falldown effect signalled that sustainability was a problem in BC's forests.

Second, the prolonged economic crisis of the early 1980s raised concerns about the economic viability of the forest industry, and it helped stimulate recognition of the nonwood values of the forests for alternative forms of economic development, notably those related to the fishing industry and various forms of tourism and recreation. From this perspective, the deleterious effects of logging and forest-product manufacture, in terms of local air and water pollution, erosion, flooding, and aesthetic considerations, need to be reduced if not eliminated.

The third, and most radical plank of the environmental critique, focused on the issue of biodiversity. This critique is led by international environmentalism, especially Greenpeace Germany and various groups in the US that emphasize biodiversity as a global issue and responsibility. This critique rejects the harvesting of old-growth forests on the basis of the argument that the complex and interdependent diversity of such forests cannot be replaced once harvested, even if reforestation is practised. From this perspective, BC's forests, especially the temperate rain forests, have global value. Thus, BC accounts for over 7 percent of the world's softwood growing stock, a wide range of species, and some of the largest segments of old-growth coastal temperate forest ecosystems in the world. Maser (1990: 46-7) claims that the coastal forests, as part of the great Pacific Northwest rain forest, are unrivalled in the size and longevity of individual trees, the accumulation of biomass in individual forest stands, and the range of conifers present (see Hammond 1991; Kimmins 1992). From the point of view of maintaining the values of the old-growth forest, environmentalists argue for conservation, with no logging or extremely limited forms of logging based on selective tree harvesting (Hammond 1991; M'Gonigle and Parfitt 1994; Maser 1990).

Environmentalists have been extremely effective in raising public concern within BC and making BC a global target for its forestry practices. Use of graphic images of clear-cutting, in newspapers, magazines, television, and the Web, have shaped public perception in BC and elsewhere, especially Europe. Apart from stories about environmental blockades, debate has been sparked by the portrayal of Canada, particularly BC, as the "Brazil – or Amazon of the North" (Maclean's 1991a, 1991b), the satellite image of the massive Bowron Lake clear-cut north of Prince George, Greenpeace Germany's 1993 forest campaign, and the more recent declaration of the Great Bear Rainforest on the BC central coast as an old-growth preserve that should be kept free of any logging.

Traditionally excluded from policy making in BC, environmental groups have sought to change forest policy and industry behaviour with highly public campaigns. These efforts were stimulated in the early 1980s when it became known that the provincial government had practised "sympathetic administration," meaning that regulations were relaxed to allow companies to lower logging costs during the recession. They were also affected by the lumber-trade conflict with the US, which suggested BC's forest policies were a subsidy (Chapter 7). Growing public discontent provided the context for a rising tide of environmental protests and helped elect the provincial NDP in 1991.

In practice, environmental forces in BC are associated with a diverse set of concerns, philosophies, and "eco-tactics" (Blomley 1996; Wilson 1987/88). In many valleys, especially in the coastal region, environmental groups engaged in civil disobedience and illegal activities, including blockades to stop logging, and in some cases, tree spiking was advocated. Other environmental groups emphasized legal opposition to industrial forestry, charging companies with violations of contracts and permits. In recent years, the more radical environmental groups, notably Greenpeace and the Rainforest Action Network (RAN), have shifted the focus of their activities from the prevention of logging to consumer boycotts of BC forest products. In late 1998, various environmental groups initiated an advertising campaign in BC and the US, including the *New York Times,* targeting corporate consumers, specifically by highlighting firms that still clear-cut old-growth forests (Hamilton 1998j).

In part, the shift in approach by environmentalists reflects public concern within BC over tactics that are sometimes illegal and that are insensitive to local values. To some extent, environmentalists have lost the high moral ground they monopolized a decade earlier and now face increased opposition. Internationally, this opposition was underlined when the Canadian government, with support from the BC government and other organizations, organized its own "environmental diplomacy" in 1994 in Europe, emphasizing that BC forest policy had been radically changed in response to environmental concerns (Hayter and Soyez 1996: 149). In addition, Aboriginal groups, who participated in this environmental mission, have expressed concern over environmentalism and have occasionally refused to cooperate with Greenpeace. Industry has taken environmental groups to court for mounting illegal blockades, while union workers have become more vocal in their opposition. Thus, the IWA in 1997 blockaded a Greenpeace vessel in Vancouver's harbour and attempted court action against Greenpeace to recoup wages lost as a result of the blockades. In December 1998, after seeing a Greenpeace-sponsored advertisement in the *New York Times* indicating that Hallmark had refused to buy pulp and paper from BC because of its logging of old-growth forests, the IWA advised its members not to buy cards from Hallmark. As it happened, a Hallmark spokesperson indicated that no such assurance had been given to Greenpeace; the US-owned company was still buying pulp from BC (see Hamilton 1998l and m).

Moreover, as environmental opposition to logging has mounted, corporations have responded by distancing themselves from logging activities, most notably by contracting out and by entering into joint ventures with Aboriginal groups. This strategy has reduced jobs for unionized loggers and has redrawn the battle lines between environmentalists and big business to today's conflict between environmentalists and the broader community, including Aboriginal peoples. In this reconfigured battle, environmentalism has a more difficult task in shaping public opinion. It also underlines the

Janus-like character of environmentalism: it is both a local model promoting democratic values and an authoritarian model imposing global values.

Environmentalism versus Globalism

According to Saurin (1996: 81), heightened environmental consciousness is related to the increasingly globalized scale (and production) of environmental degradation. O'Riordan (1976) agrees, noting that if the concept of environmentalism predates that of globalization, the rising commitment to the former is partially a response to the latter. In this view, the deepening of globalization over the past few decades, led by the increased mobility of financial and investment capital, is threatening global environmental values and stability.

From this perspective, environmentalism is a democratic response to globalism. Indeed, Paehlke (1996: 20) argues that it is not coincidental that environmental protection is most successful in democratic countries because environmentalism depends on the ability of individuals to participate in political action geared to changing established practice and values in local decision making. In turn, "the environmental movement has been self-conscious and single-minded in seeking to involve directly the public in governmental decision making regarding the environment" (Paehlke 1996: 18). Thus, as a democratic model, environmentalism is a force for transparency and participation in public decision making because it advocates acting locally while thinking globally. In addition, as a democratic model, environmentalism stands in sharp contrast to the globally defined hierarchical structures of MNCs that impose decisions and strategies on peripheral regions from a few decision-making cores. The decision-making autonomy of branch plants is inevitably curtailed in some way, and corporate information flows are largely internal and secret. In contrast, the democratic model of environmentalism celebrates local sovereignty.

BC has been a part of the emergence of environmentalism as a democratic model that has been a powerful force in BC's forest economy. Indeed, as Wilson (1987/88) argues, forest policy in BC has shifted from a highly secretive process controlled by the "exploitation alliance" of government, business, and labour to an open-book process in which harvesting plans, stumpage, and other charges are now public documents and subject to public approval. In addition, environmentalists in BC, such as M'Gonigle and Parfitt (1994), are strong advocates of local control of the forest resource as an alternative to top-down hierarchical thinking. The politics of environmentalism, however, are not locally confined.

Environmentalism as Globalization

Environmentalism is not simply a local response to globalization and increasingly mobile capital. The impacts of telecommunications and related technology on distances of space and time, which have facilitated the organization of MNCs, have similarly extended the global reach of environmentalism (Ekins 1992; Princen 1994). Indeed, environmental groups are perhaps the most effective form of nongovernmental organization (NGO) that can think globally and act locally as well as the reverse (Barker and Soyez 1994) in a way unparalleled by any representative government (Hirst and Thompson 1992). According to Princen (1994: 29), environmental NGOs are "independent actors with their own, often unique bargaining assets" that can be used to attract media attention, access funds, provide scientific knowledge, coordinate lobbying activities, and communicate ideas, values, and information to environmental groups around the world. In Rosenau's (1990: 301) terms, information about environmental events can rapidly "cascade" through the networks of environmental groups to help form public pressure on decision makers, wherever they are located.

Princen (1994) justifies the activities of environmental NGOs in linking local action with global responsibilities according to principles of transparency, transnationalism, and legitimacy. Thus, he suggests that environmental NGOs pressure business and governments to be honest and to provide information; transnationalism ensures that policy responses to environmental problems are at an appropriate biogeographical scale; and NGOs derive legitimacy by their principled, uncompromising commitment to a single issue – environmental values – that other organizations find difficult to emulate (Princen 1994: 35-6). In this view, the global reach of environmentalism is a form of countervailing power (Galbraith 1952) to global business and arrogant governments.

As a globally powerful actor, however, environmentalism may threaten local decision making. As Taylor (1996) notes, the more radical forms of environmentalism contain strong elements of authoritarian political thought, and the approaches and beliefs of global environmental actors may not coincide with local interests, even local environmentalism. Taylor (1996) emphasizes that the scientific basis for resource scarcity issues is rarely cut and dried, even among science communities, while within environmentalism, controversy remains over such ideas as anthropocentrism and biocentrism. Although environmental messages are communicated instantaneously across the globe, they may be ambiguous and questionable, adding to the uncertainty of local economies trying to deal with environmental issues.

It is conventionally supposed that the mobility of capital is the basis for its power to exploit labour, play off governments, and shun environmental responsibilities (Harvey 1990). If financial capital does whiz around the globe, often motivated by short-term speculative gains, forest-product MNCs have deep roots in particular places as a result of their access to particular resources, huge capital investments, and the accumulated know-how of workforces. Global environmentalism, on the other hand, is inherently mobile, as ideas, ideology, and information are communicated at little or no cost. Even small "armies" and "navies" (protest groups) can be readily forged, at least for selected targets and a limited time. Indeed, BC has been targeted by much international, as well as local, environmental scrutiny. Possibly, BC, with its cosmopolitan liberalism and tolerance, and reliance on exports, offers a relatively soft target. It has not been easy for already invested forest-product capital in BC to simply run away. New capital spending, of course, always has alternatives.

Reregulation and Corporate Forestry

Environmental legislation has become an increasingly important influence on BC's forest industries. Air and water pollution laws were supplemented in the late 1970s by increased reforestation efforts. By the 1990s, the area planted was more or less equivalent to the area harvested. Bearing in mind that natural regeneration does occur in some areas, forest planting has reached a noteworthy scale that is seeking to renew previous, as well as present, harvests, suggesting that the forest turn-round has occurred in BC (Chapter 3). However, the turn-round is only a minimum requirement for sustainable forestry, a basis for more intensive forest policy, and a broader range of responses to environmental concerns (Binkley 1997a).

At the same time, the reregulation of BC's forests, especially its remarkably rapid evolution in the 1990s, seeks more values for more interest groups representing a more diffuse pattern of control. Thus, the high-stumpage regime, in addition to the change in stumpage formula and stumpage escalator, features many initiatives that seek to realize the nonwood values of forests (see Table 3.10). The principal initiatives are the Commission on Resources and Environment (CORE) (1992), BC's Pulp Mill Effluent Standards (1992), the Protected Areas Strategy (1992), the Timber Supply Review (1992), the Clayoquot Sound Land Use Plan (1993), Forest Renewal BC (1994), and the Forest Practices Code (1994). CORE and its related legislation have a wide-ranging mandate, including dispute mechanisms; in particular, the commissioner has discretionary powers for adjudication when related legislation does not mesh. The provincial government also retains the right to adjudicate.

The Challenge to Corporate Forestry

Environmental values are an explicit theme of both the rhetoric and the practice of the reregulation. They have posed a massive challenge to corporate forestry. Four interrelated points can be made in this context. First, the direct costs of logging Crown land have been greatly increased by the change in the stumpage formula, which delinked stumpage from market prices, and by increases in stumpage. Second, the Forest Practices Code and related environmental regulations indirectly increased costs by imposing more restrictions on logging; for example, near streams and rivers, in areas where aesthetic values are important, at sites where cultural features are evident, in areas vital to wildlife, and on the size and distribution of clear-cuts. In addition, the Code increased the bureaucratic requirements for log-harvesting plans. The Code has added substantially to planning costs, including the time required to develop and gain approval for plans. Indeed, MB claims that over 700 regulations govern harvesting in BC; one government assessment suggests that harvesting costs have increased by 10 percent because of the Forest Practices Code (Gunton 1997). Another estimate is that the Code added between $750 million and $1 billion a year to industry's forest management costs (Cafferata 1997: 57). The plans also have to comply with CORE's comprehensive regional planning process that seeks to ensure that land use dynamics are consistent, economically and environmentally. CORE developed four regional plans by 1995, accounting for almost 25 percent of the province (Vancouver Island, Cariboo-Chilcotin, West Kootenays, and East Kootenays); there are a further twelve land and resource management plans or LRMPs (Owen 1995).

Third, the greater transparency of forest policy (Wilson 1987/88) means that industry must now engage in extensive public debate over forest contracts, licences, and stumpage calculations, which were previously confined to private deliberations. The CORE process is broadly participatory, while the Clayoquot Sound Land Use Plan took unusual steps to ensure the widest possible input from interest groups. Such debates not only involve time and costs, but also create uncertainties about how matters will be resolved. Fourth, the AAC available to industry has been reduced because of the Code, the SBFEP, and the Protected Area Strategy; the latter, for example, has led to the creation of more than 140 wilderness parks since 1992 and has increased BC's protected area network from 6.5 percent to 9.2 percent of the land base, removing about 2.5 million hectares from logging activity (Gunton 1997).

In 1997, the changes required in clear-cutting effectively symbolized the reregulation of BC forestry practices (Hayter and Soyez 1996). Traditionally, clear-cuts were large, while the practice of continuous clear-cutting meant that clear-cuts were adjacent year after year, thus creating extensive, timberless areas, crisscrossed by roads and prone to erosion. To mitigate these problems, the Code reduced the maximum size of clear-cuts (40 hectares on the coast) to below that found in most other forest jurisdictions, including Sweden, and banned continuous clear-cutting. Now, clear-cuts must "green over" before logging occurs in adjacent areas, a process that usually takes about ten years on the coast. The expectation is that smaller, more scattered clear-cuts will reduce aesthetic concerns, retain biodiversity and the interactions of plant and animal life, and reduce erosion.

Yet, by 1999 even the practice of clear-cutting, long-justified in BC according to principles of economics, safety, and where appropriate, sound forestry (Kimmins 1992), was under scrutiny, at least in old-growth areas.

In a surprise announcement in 1998, MB indicated plans to shift from clear-cutting, even as narrowly prescribed by the Code, to a selective ("partial retention") harvesting system (Hogben and Hunter 1998). A rival has announced similar plans (Hamilton 1999b). In the rethinking of the reregulation, industry's ideas are ahead of the provincial government.

The full implications of the high-stumpage regime for corporate forestry (and vice versa) are therefore still being sorted out. But two broad related points can be made. First, for large corporations the cost of cutting timber on Crown lands has become prohibitive. MB lost $26 million on its TFLs in 1998. In response, firms such as MB and Fletcher Challenge are choosing to reduce their commitment to logging on Crown land, in part or even entirely, while increasing their reliance on outside suppliers (and, in MB's case, by increased logging of private lands.) This trend is a major shift in thinking when compared to the vertical integration strategies prevalent during Fordism. One related consequence is the reduced numbers of union loggers, who will also have to embrace flexibility. Second, logging activities are likely to be increasingly controlled by SMEs, communities, and First Nations bands, in MB's case through joint-venture agreements. These groups will have to deal with the implications of forest policy and environmental pressures, including certification and demands for "chain of custody" guarantees.

MB's Sproat Lake Division and the Clayoquot Sound Compromise illustrate the issues facing corporate forestry in the context of the high-stumpage regime.

MacMillan Bloedel's Sproat Lake Division

Changes in operations at MB's Sproat Lake Division, part of TFL 44, illustrate the extent of changes to corporate forestry in recent years (Figure 10.1). TFL 44 was created in the mid-1950s when two TFLs and some private land held by MB were amalgamated (Chapter 4). In the mid-1970s, the AAC for TFL 44 amounted to 3.5 million cubic metres, of which the Sproat Lake division accounted for about 18 percent, including Clayoquot Sound. By 1993, the AAC had been reduced by 23 percent to 2.7 million cubic metres, partly because of timber allocated to the small business forest enterprise program (5 percent of the loss) and partly because of the April 1993 decision regarding Clayoquot Sound (10 percent of the loss). In 1996, the AAC was down to 1.9 million cubic feet, as the effects of the Forest Practices Code and CORE recommendations were felt, a 46 percent decline since the early 1980s. TFL 44 once had four divisions: Franklin, Cameron, Kennedy, and Sproat Lake. Cameron and Kennedy are now closed.

Sproat Lake's harvest has progressively declined from 600,000 cubic metres in 1980 to 440,000 in 1993 and 350,000 cubic metres in 1997. The Sproat Lake Division has not lost area, and the decline in the cut reflects the effects of CORE and the Code. Thus, CORE classifies areas in terms of logging

Figure 10.1

MacMillan Bloedel: Forest tenures in British Columbia, 1997

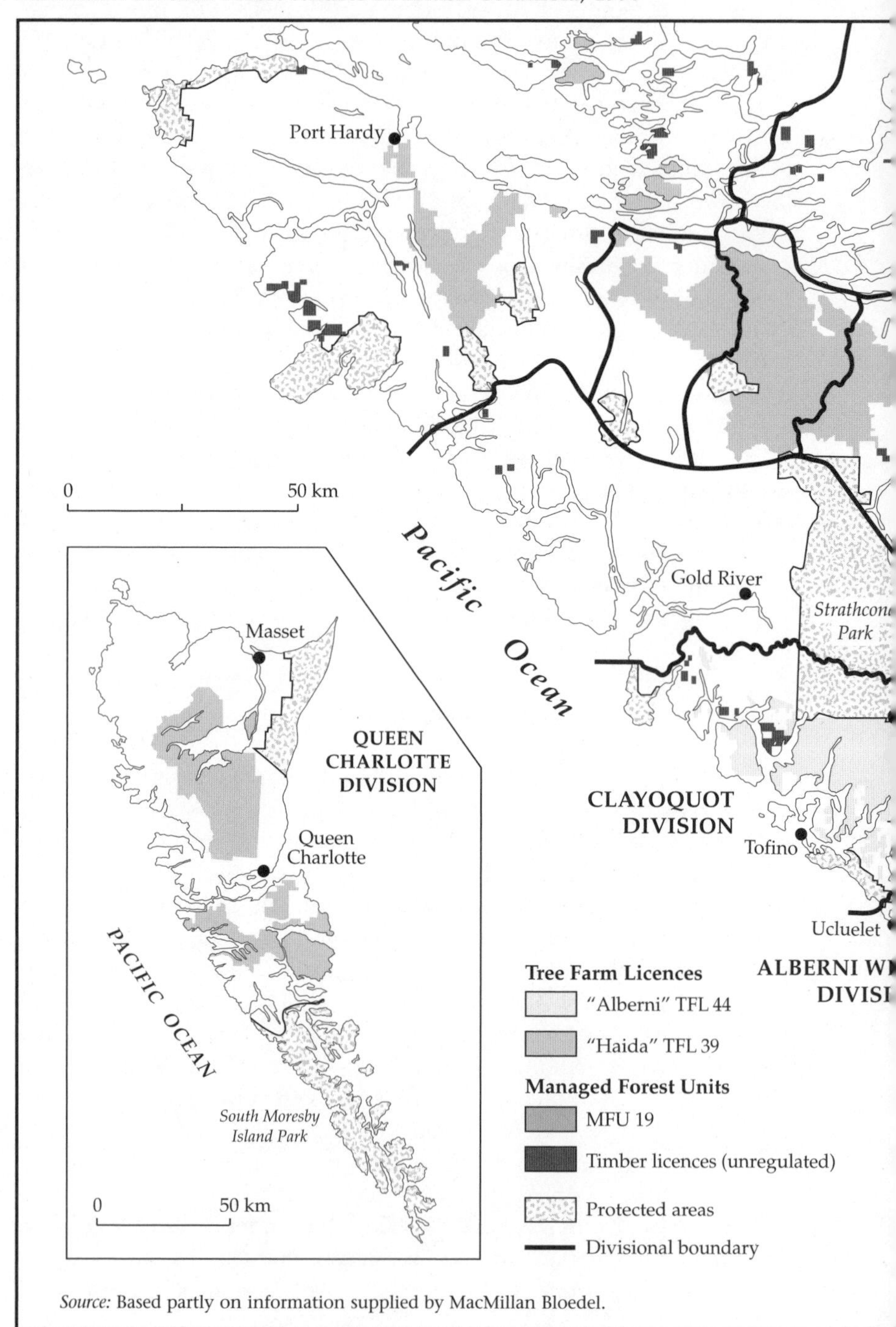

Source: Based partly on information supplied by MacMillan Bloedel.

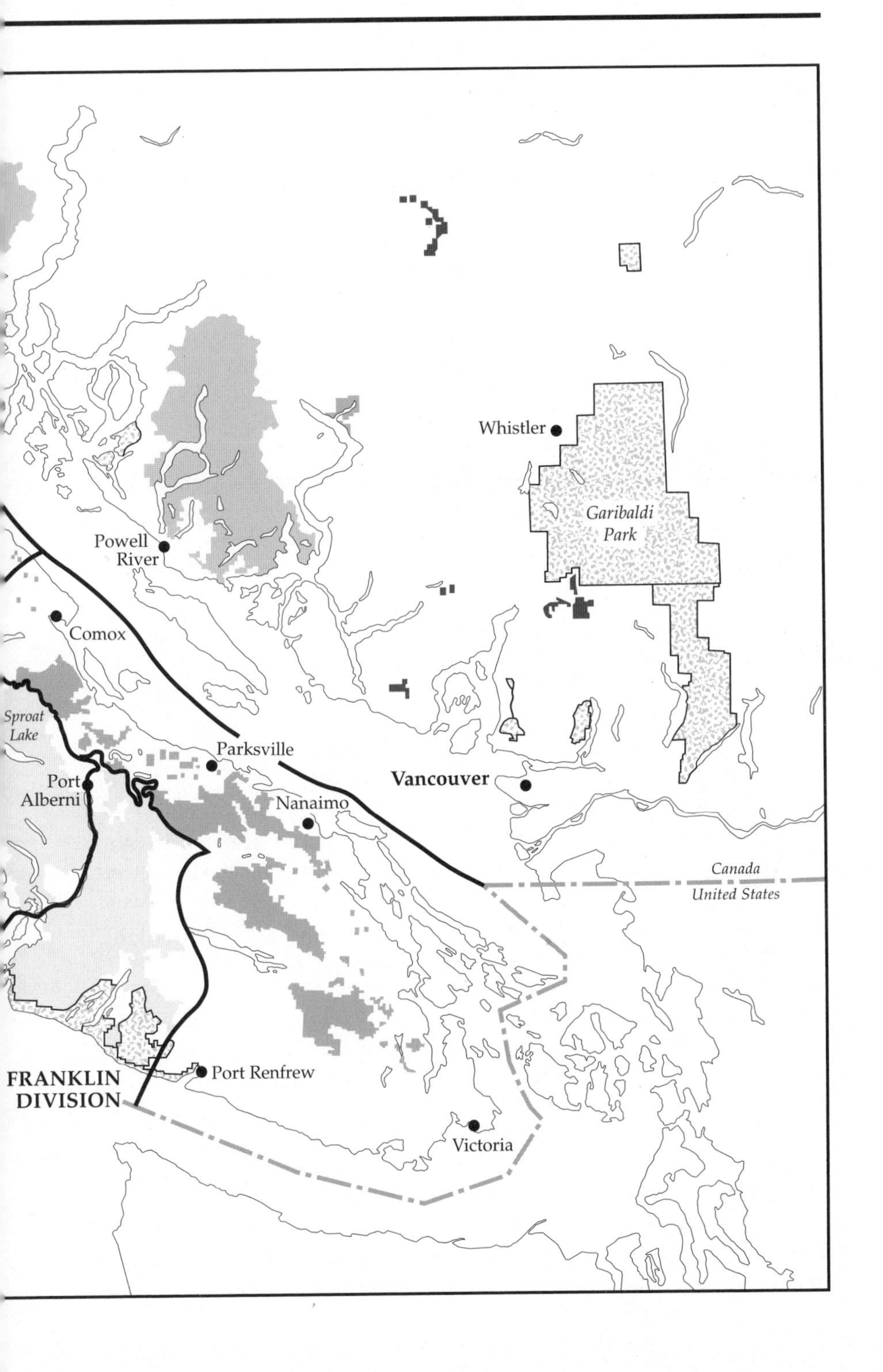
Whistler
Garibaldi Park
Powell River
Comox
Sproat Lake
Parksville
Port Alberni
Vancouver
Nanaimo
Canada
United States
FRANKLIN DIVISION
Port Renfrew
Victoria

intensity required. In Sproat Lake, 45 percent of the land base has been allocated to the special status category in which logging is considerably restricted. In the low intensity area (LIA), the AAC was 100,000 cubic metres and is now 60,000 cubic metres. The LIA is a community watershed, amounting to 35,000 hectares of the division's 100,000 hectares, and so requires special management techniques.

To comply with the new requirements of the Forest Practices Code, 75 percent of the Sproat Lake Division is landscape evaluated. The actual amount harvested depends on viewscapes; if the area is within sight of roads or canals, fewer trees are harvested. Forest ecosystem networks (FENs) also have to be incorporated in harvesting plans to ensure the maintenance of biodiversity corridors that allow for winter ranges for deer and elk and that protect Class A streams and their tributaries. Elsewhere along smaller streams, logging is curtailed a few metres from the bank. Sites of cultural value (for example, sites containing trees used by Natives for barking) are also not available for harvesting. Moreover, the size of clear-cuts has been reduced to an average of 18 hectares, well below the permitted 40 hectares, and continuous clear-cutting is no longer possible. Rather, there has to be visually effected greenup (VEG) before an adjacent clear-cut can be done; VEG occurs when new growth reaches at least three metres in height, or five metres in special scenic areas.

The Sproat Lake Division submits a five-year forest development plan to the Ministry of Forests and the Ministry of Environment, Lands and Parks, as well as for public review (for sixty days) in Port Alberni. Preparing the plan takes the division six to eight months and involves a full-time employee plus consultants who conduct soil, fisheries, and landscape analyses. The ministries take at least one year to review such plans, and the Sproat Lake Division must then respond to comments from the ministries and the public. The plan must also be reviewed by Native bands. The last five-year plan developed by Sproat Lake comprised three volumes. Detailed engineering, logging, and silvicultural plans then have to be developed. If any creeks are located in the area, a fisheries biologist must visit and compile a report. The division must submit its own report to the Forestry Service, including plans for roads and bridges, along with a preharvest silvicultural plan describing how logging will proceed, the nature of reforestation, future commitments of the company to the area, silvicultural prescription defining the height of trees eleven years after planting, and shutdown criteria for wet weather or snow. After submission, these plans take six to eight months for approval.

In the past, logging was done mostly by individual fallers working in clear-cuts. With the onset of second-growth harvests, felling is more mechanized. Helicopter logging has also increased and accounted for 25 percent of the 1997 harvest. Helicopter logging is expensive ($6,000 an hour in 1996 to fly just one helicopter), but it allows access to difficult sites and

Partial retention (corporate) logging near Port Alberni.

preempts the need for roads. Clear-cuts were first reduced greatly in size in accordance with the Code. In 1998, the division shifted to "partial retention systems" on old-growth areas, in which parts of the existing forest are left standing. All harvested areas are reforested, 90 percent by artificial regeneration. Reforestation is mainly done in the spring and in August by MB crews and a Nelson-based contractor. Eleven species are planted, the main ones being fir, hemlock, cedar, and yellow cedar, depending on site conditions (elevation, aspect, and dryness). In 1997, the division planted 4.1 million seedlings, and it has now planted over 95 million.

For the Sproat Lake Division, costs increased so much as to cast doubt on its future. In 1996-7, stumpage levels reached over $60 per cubic metre in some instances. Average costs were over $30 – a high figure taking into account logs from MB's private lands that do not pay the same stumpage rate (although these lands are now paying higher taxes). In one month in 1996, stumpage and tax (not including income tax) amounted to $800,000. In 1997, the division lost $7 million. The direct cost of harvesting amounted to $95 per cubic metre, and with stumpage, the per cubic metre cost was $130. The new chief executive officer of MB demanded that these costs be reduced to $60 direct costs plus stumpage for a total of $95. In February 1998, the union agreed to create a new division; otherwise, the division would have been sold. Jobs have been reduced to fifty-three employees, down from 190 the year before. Staff has declined from thirty-two to twelve, and only two supervisors are left in the bush. Considerable retraining is required to make staff flexible enough to perform the various jobs. In TFL

44, logging is evolving rapidly and eco-certification (and labelling) are now a priority. The Clayoquot Sound Compromise is another illustration of changing times.

Clayoquot Sound: A Compromised "Compromise"?

Clayoquot Sound is a 262,592-hectare land area, 90 percent covered by forests, located on the central west coast of Vancouver Island. There are large tracts of contiguous old-growth forest and other significant natural attributes, including habitat for shorebirds and extensive eelgrass beds. In 1990, harvesting rights were held for about 50 percent to 60 percent of the area's forests, principally by MB, which was logging at an annual rate of about 700,000 cubic metres, or two-thirds of the area's total allowable cut. However, blockades and tree spiking were already a feature of the Clayoquot Sound land use debate, which became a centre of international attention.

Clayoquot Sound poses a dilemma for environmentalism. As a plea for local, environmentally sensitive democracy, environmentalism implies the incorporation of all relevant interest groups within the region in decision making, open and fair discussions in which agendas are not controlled by any particular group, and a search for mutually acceptable compromise. In this view, "compromise" is a valid goal. On the other hand, as an insistence on the maintenance of global ecosystems and biodiversity, environmentalism rejects any further exploitation of old-growth forests in Clayoquot Sound. In this view, "compromise" implies a violation of principle. In practice, in government-sponsored attempts to resolve the Clayoquot Sound dispute, environmentalism first encouraged a local, democratic resolution, but then it rejected this approach in favour of a global imperative that demanded the Clayoquot Sound forests remain intact, subject only to natural forces.

The Failure of Compromise

Initially, environmental representatives, along with Aboriginal peoples, industry, and local community groups, joined a committee set up by the provincial government in 1989. This committee in 1990 became the Clayoquot Sound Sustainable Development Steering Committee whose goal was to develop a land use plan for sustainable development of the area. Internationally recognized principles of sustainable development, specifically those identified by the International Union for the Conservation of Nature, the World Wide Fund for Nature, and the United Nations Environmental Programme, were adopted by the committee and a logging moratorium was placed on twelve of fourteen forest areas that environmental groups wanted to exclude from development. The two active forest areas provided jobs for 250 workers and a much-reduced timber supply. Environmental groups, however, refused the compromise and resigned from the committee. Subsequently, logging blockades were resumed, and Greenpeace Germany and RAN began to lobby for consumer boycotts.

Environmental groups did not rejoin the committee, which still involved thirteen distinct groups representing two large corporations, aquaculture, fishing, three communities, one regional district, small business (general), small business (forest), labour, mining, tourism, and Aboriginal peoples. After three years of study, the committee produced a majority report (signed by eleven of thirteen interests, with Aboriginal representatives in effect abstaining) that formed the basis of the provincial government's Clayoquot Sound Land Use Decision of 1993. In fact, the government modified the majority report to incorporate even stronger environmental provisions. Protected areas were doubled, the commercial forest area halved, and extremely strict regulations imposed on the remaining areas available for forestry, including limiting the size of clear-cuts.

At the same time, the provincial government indicated its interest in having Clayoquot Sound declared a UNESCO Biosphere Reserve. Two further committees were established: the Clayoquot Sound Oversight Committee and a scientific panel. The committee reviews all forestry operations, and the panel provides ongoing refinements to the new standards. For MB, the government's modifications of the Clayoquot Compromise reduced its timber supply by 200,000 cubic metres, imposed strict, regionally specific regulations on logging, and further heightened uncertainty about forestry by introducing the notion of a biosphere reserve. Indeed, MB argued that its large-scale logging operations were no longer viable in the region, and in late 1996, MB announced a moratorium on logging for the immediate future (Hamilton 1997).

A Victory for International Environmentalism

International environmentalism responded positively to the termination of logging in Clayoquot Sound, and Greenpeace reconsidered its anti-MB campaign, as did RAN, although RAN declared the forestry principles developed for Clayoquot Sound should now be applied to the rest of the province. Other groups within the region were disappointed with MB's actions, especially labour whose jobs were forfeited, local businesses that lost opportunities, and First Nations that claimed rights to land and to log in the area.

As of early 1997, global environmentalism had won the latest round in the future of Clayoquot Sound. It had simply refused to compromise its principles. Yet, environmentalism as a local autonomous process was compromised, and new global-local cleavages have emerged in which environmentalism is on the side of globalism and capital is as much local as global. An important new feature of global-local issues in Clayoquot Sound is the cleavage between global environmentalism and Aboriginal peoples, thus ending what appeared to be a close alliance in the 1980s. This cleavage became apparent in 1992 when local bands publicly distanced themselves from environmental protests in the area (Nathan 1993: 156). Local bands were upset with the growing violence of the blockades and concerned about their land claims and rights to develop the economy of their communities. In the government-sponsored tour of Germany to counter the consumer boycotts of Greenpeace, a band leader from the Clayoquot Sound bluntly told Greenpeace members that they had no right telling First Nations what to do with their land. Then again, when Greenpeace Germany tried to fund a blockade of logging in July 1996, a local Native band effectively opposed the plan.

MB has not forfeited its hopes to access logs in Clayoquot Sound. Rather, it has developed an alternative strategy. The company accepted the strictures of environmentalism as a democratic model by its willingness to provide information on its operations, to present its views, and to participate in representative local committees. Indeed, MB became an active promoter of the idea of local control over resources and of building bridges with Aboriginal peoples and local resident associations, again showing a commitment to compromise. Moreover, by refusing to log in Clayoquot Sound, MB brought attention to the fact that it lost money in the region ($7 million in TFL 44 in 1996) and thus redirected environmental opposition towards local interest groups. Indeed, its logging moratorium directly linked environmentalism with job loss, community decline, and violations of Aboriginal rights, and encouraged the provincial government to be more critical of environmentalism as an authoritarian model, especially since it had received zero stumpage from the area in the preceding two years. MB's strategy may be paying off.

A Restructured Compromise and the Growing First Nations Connection

At the end of 1998, MB and Greenpeace announced that they had reached a compromise over logging in Clayoquot Sound (Hamilton 1998k). The compromise involves BC's first attempt at eco-certified forestry, fully supported by Greenpeace. Logging will be under the guidelines of the Clayoquot Sound Scientific Panel and will not exceed 200,000 cubic metres, a much lower level than previously allowed. Logging is not to occur in the region's remaining pristine valleys and will involve relatively untried logging techniques, with helicopters extracting one log at a time. Whether such logging is viable remains to be seen and depends on the ability of MB to receive a premium price for eco-certified wood. Greenpeace claims it will help with marketing.

A key to understanding the unlikely alliance between Greenpeace and MB is the facilitative role played by the region's local Aboriginal peoples. The five bands wanted to resume logging, but they needed conflict to end first. Moreover, logging is to be conducted by Iisaak Forest Resources, a new joint venture between MB and the Nuu-chah-nulth band, based near Port Alberni. The Nuu-chah-nulth will hold 51 percent of the company and MB

49 percent, and Ucluelet has a provision to acquire 10 percent of MB's shares in two years. This kind of arrangement is already part of a trend.

With little fanfare, large corporations have sought ways in recent years to involve Aboriginal peoples in their activities, especially logging. A survey of firms in the private sector, conducted in 1993 with a 36 percent response rate, found that Aboriginal employment in the forest sector amounted to 1,499 people. Individual firms are the main employers, but joint business ventures (JBVs) and contracting arrangements also have a role. Other Natives are employed by the Ministry of Forests, encouraged by the Native Orientation Program. Overall, COFI estimated that just over 5 percent of the industry's workforce, or between 3,520 and 4,840 people, is Aboriginal (COFI 1994: I-3).

Large corporations have shown increased interest in forming JBVs with Aboriginal groups. In part, this interest reflects the fact that forest resources on reserves have been marginal and not well managed. Government programs to rehabilitate these forests have apparently not been successful, and JBVs may provide wood fibre to firms while increasing Aboriginal employment and improving forest practices. In addition, the treaty process will likely shift forest resources to Aboriginal peoples. Industry's perception is that Aboriginal peoples wish to exploit this timber supply for economic benefit. Moreover, by forming JBVs or contracting out logging, forest-product corporations remove themselves from direct conflict with environmentalism. Logging costs may be reduced, although corresponding training costs will arise, and the practice of joint ventures needs to stand the test of time. In the meantime, the forest industry has been silent on the issue of land claims.

Towards an Unruly Forest Policy

Controversy is not new in debates about BC forest policy, whether for tenure allocation, sustainability, appropriate forest management techniques, corporate concentration, or other matters. The Sloan Royal Commissions of the 1940s and 1950s and the Pearse Royal Commission of the 1970s are public records of alternative views on such themes. But these formal proceedings were conducted under accepted rules of debate, and forest policy then essentially meant industrial policy. In M'Gonigle's (1997) terms, forest policy has been dominated by a top-down or hierarchical model of decision making. Decisions about forest communities throughout BC were centralized in a limited number of corporate and government boardrooms in downtown Vancouver and Victoria, often themselves subservient to boardrooms in more distant centres. The hierarchy model implies a standardized template of decision making for the entire province for both forestry and its associated economic activities, which are typically capital intensive and export oriented. Local inputs, at least of a strategic kind, and nonindustrial criteria are excluded, or made secondary.

In the Fordist period, the hierarchical model was conducive (but not necessarily optimal) for organizing rapid and relatively stable rates of growth in BC's forest economy. The recent, tumultuous record, however, reveals this model to be too rigid, not only to cope effectively with contemporary global market dynamics, but to foster local development and meet the demands for prioritizing environmental values. However, as forest policy has become more open to a wider range of influences, it has become more "unruly," complementing trends in local development (Chapter 9).

The Characteristics of an Unruly Forest Policy

In the transformation of BC's forest policy from the hierarchical model to an unruly model, two themes can be emphasized. First, unruliness highlights the contentious nature of current policy making that stems from greater transparency and the influence of diverse interest groups. The value systems of the latter are difficult to reconcile and traditional forms of bargaining by government, business, and labour have been complicated, if not stood on their head, by the intrusion of new power networks, especially those associated with environmentalism and Aboriginal land claims. These new interests have different agendas and have little reverence for established assumptions and procedures. Moreover, forest policy is no longer simply internal to BC; it must now be rationalized to a global as well as a local audience. In this regard, it needs to be emphasized that continuing American pressure on BC to force privatization and log exports is a powerful factor complicating a "made in BC" forest policy (Chapter 7).

Second, unruliness implies a more flexible and differentiated use of the forest. From this perspective, unruliness is consistent with Drushka's (1993) pleas for diversification of forest tenure, and the associated diversification of resource objectives and utilization. In other words, unruliness allows for economic and environmental goals to be met in different ways among different BC communities and locales. Thus, in an unruly approach, M'Gonigle and Parfitt's (1994) plea for "forestopia," which prizes local control of forest resources, Drengsen and Taylor's (1997) various ideas that comprise "ecoforestry," and Binkley's (1997a) suggestion to replace multiple land use planning by land use zones based on dominant values, can all find their place in an unruly BC. Such proposals attempt to achieve Kimmins's (1992: 13) goal of an "appropriate balance between 'preservation' and 'wise use'" of BC's forest resource.

Unruliness is a cause and effect of competing visions of forest policy. Competing values propose alternative ways of using the forest resource, and these alternatives further stimulate debate and perhaps other alternatives. In an unruly BC, territory is contested, and as a result, an unruly map is not easy to draw. During Fordism, forest use in BC could be readily represented as a system of spatially bounded corporate tenures with established rights and responsibilities. For example, TFL 39 and TFL 44 clearly demarcated MB's wood-fibre supply base over which it exercised autonomy within the terms of the lease (see Figure 10.1). These may have been public lands, but MB's control was unequivocal and supported by law to which all acquiesced.

In unruly BC, however, tenures that once seemed to imply "in perpetuity" have been modified. Parks have been created preventing further logging, while wood fibre has been taken away from large corporate use and diverted to SMEs. Moreover, territory is openly contested. Environmental groups and Aboriginal peoples have made claims on forest land that have

Figure 10.2

Contested space: Competing claims on lands and forests on the northcentral coast

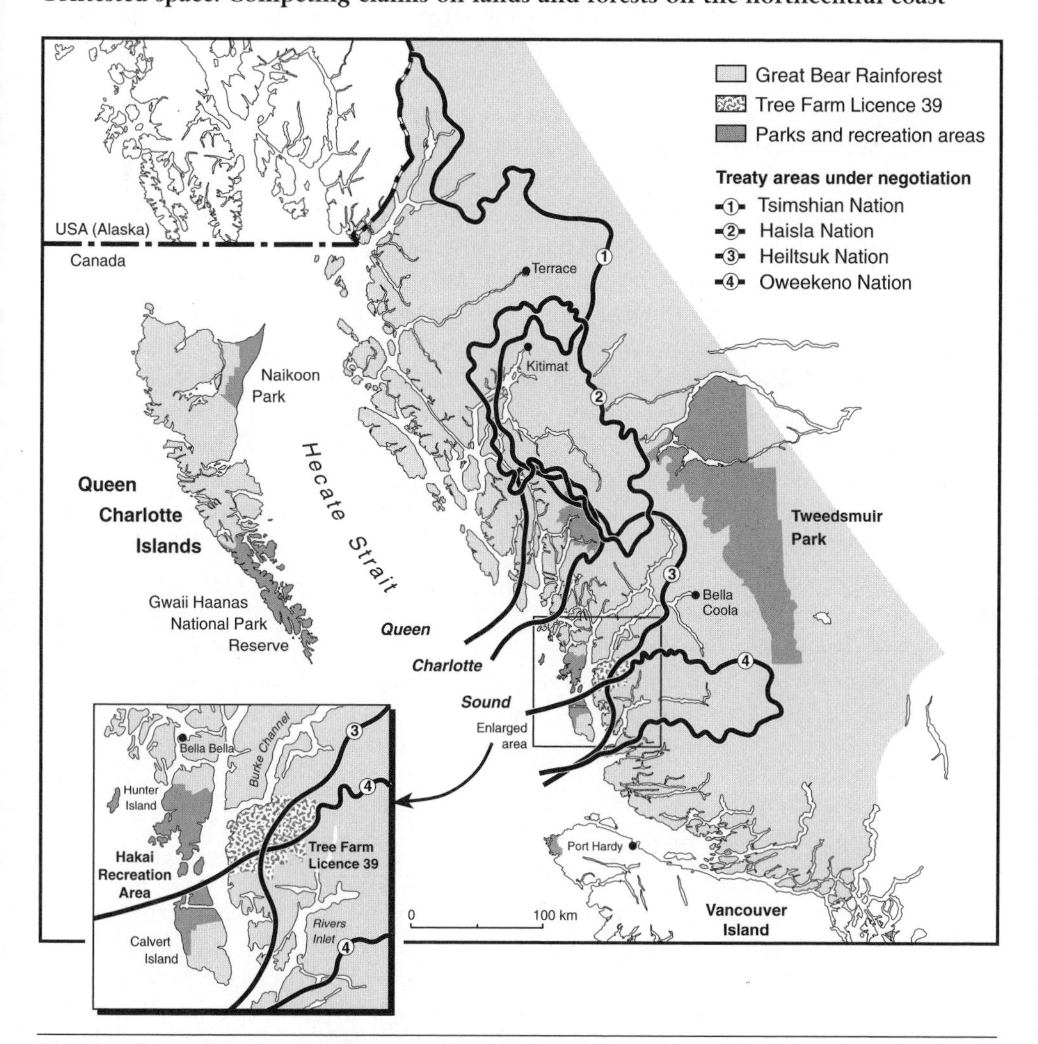

Source: Based on McAllister and McAllister 1997; Federal Treaty Negotiations Office n.d.; and MacMillan Bloedel.

no regard for existing patterns of tenure (Figure 10.2). Thus, Aboriginal land claims overlap throughout BC, and in the northcentral coast, four overlapping claims compete with the so-called Great Bear Rainforest, labelled thus by environmentalists in an attempt to sway public opinion to support the preservation of remaining old-growth forests and remove existing tenure rights (McAllister and McAllister 1997). The area has no legal basis, nor has it been proposed by a formal participatory process, but it is now part of the public's mental map and part of Greenpeace's published literature. Whether such a map is seen as an example of authoritarianism or a defence of nature, it symbolizes that the autonomy once enjoyed by the provincial government and large forest-product corporations is now less clear. Nor is the Great Bear Rainforest the only environmental "land claim." An even larger proposal for an eco-corridor, the Y2Y initiative, cuts through the length of BC as part of a North American wildlife sanctuary (Chancellor Partners 1998).

In unruly BC, sovereignty is an issue. There are more alternatives to consider, and if new claims over the forest resource threaten corporate tenure, they also compete with one another. Moreover, in sorting out forest-resource rights, lines between political and legal jurisdictions are blurred. For MB, even within its existing legal tenures, its autonomy over forest management is no longer assured, as its most recent compromise with Greenpeace in the Clayoquot Sound reveals. Further, *Delgamuukw* establishes Aboriginal rights throughout the province (*Delgamuukw* v. *British Columbia* 1997). Whether forest management can ever be mapped in a way that clearly defines decision-making autonomy, as implied in Binkley (1997a), is questionable.

If forest policy in BC is unruly, as is claimed here, is such a trend desirable? The answer is not straightforward. As a framework that reconciles multiple values, promotes local participation and control of resources, and addresses equity concerns, most notably Aboriginal land claims, unruliness has considerable potential to ensure that the forests are used in society's best interests. However, if unruliness dissolves into a muddling-through approach or, worse, disintegrates into chaos or anarchy, arbitrary choices, and manipulation based on short-term goals, the forest resource is unlikely to be used wisely or in a balanced way. In forestry policy, where the dynamics of change are measured in decades and centuries, as well as years (Kimmins 1992), unruliness can cause problems by simply clouding long-term views.

Sustaining Multiple Values

Stimulated by environmentalism, there is broadly based (if not universally held) agreement throughout the forest sector and the public that the environment be given priority, that nonwood benefits of the forest must be sustained, as well as industrial ones. Whatever the balance of views, provincial forest policy now explicitly recognizes multiple values. In recent years, all serious discussions of forest management, whether by earth scientists such as the pleas by Kimmins (1992) for "balanced forestry," Hammond (1993) for "wholistic forestry," and Drengson and Taylor (1997) for "ecoforestry," or those by social scientists such as Binkley (1997a) for "zonal forestry," and M'Gonigle (1997) for "forestopia," celebrate environmental values. Collectively, even if their differences are recognized, these proposals represent a sea change in forest management attitudes in BC.

Changes in the Valuation of Old Growth

This sea change is best illustrated with reference to old-growth forests, loosely thought of as unmanaged forests that have remained intact since before the large-scale industrialization of BC (for a discussion of the criteria used in defining old-growth forests, see Kimmins 1992: 139-48). Until recently, the prevailing attitude emphasized the rapid depletion of old-growth forests, because they are stable from the perspective of wood volumes. That is, old-growth forests were considered no-growth forests, the cutting and replacement of which would improve forest productivity, defined as annual increments in wood volume. Indeed, in Percy's (1986: 22) analysis, given the presence of substantial "decadent" or "stagnant" stands in mature forests, rather than cutting timber too rapidly, in BC "the main fault may have been in not cutting timber rapidly enough." Economically, by not cutting mature forests before "decadence," income is lost in the short and the long run. From this perspective, a major concern of sustained-yield management is to ensure rotation periods are based on economic rather than biological criteria (Percy 1986: 20-5).

In contrast, in a system of forest management based on multiple values, old-growth forests are not decadent. Rather, old growth has intrinsic aesthetic and spiritual qualities; helps sustain a wide variety of nonwood benefits that include maintenance of wildlife habitat, a gene pool, flood control, and a research laboratory for studying natural processes; and contributes to other industries, such as tourism and fishing. In this view, the idea that old-growth forests have not been cut fast enough is not credible.

Moreover, the extent to which attitudes have changed is not simply a matter of ideology, even if radical environmentalism is a primary causal factor. In this context, Binkley's (1997a) analysis contrasts with Percy's (1986) study published just over ten years previously, both of which are rooted in neoclassical economic analysis, the foundation of forestry economics. Thus, Percy's work focuses on forest-product demand-and-supply issues as they pertain to industrial forestry. Only a fleeting reference is made to the environment, as a cause of removal of land from timber production (Percy 1986: 7); here, as noted, old growth is equated with mature (or "overmature") timber. In contrast, Binkley's analysis begins with reference to the environment, incorporates environmental values in his management schema, and makes no mention of the undercutting of old growth. Rather, Binkley (1997a: 15) notes the "dire consequences of forest depletion" while characterizing old growth not as "mature and decadent" timber, but as "virgin forest" and a "wild estate."

Yet, transparency in unruly BC stimulates rather than closes debates. Indeed, environmental groups are sceptical about the reality of the sea change, especially in light of recent modifications to the Forest Practices Code. Moreover, environmental controversies regarding principles of forest use remain, most notably, for old growth and clear-cutting. The forest-policy initiatives of the early 1990s allow for the continued harvesting of old growth, if not to the extent once anticipated, and for clear-cutting, albeit reduced. Binkley's suggestion for forest zones based on gradations in exploitation, from conservation areas to forestry-intense zones, including clear-cutting, is consistent with this thinking. Greenpeace, however, is against the industrial use of old growth and, despite the restructured Clayoquot Sound Compromise, is especially adamant about clear-cutting old growth. New forestry alternatives tend to be closer to Greenpeace's thinking by limiting forestry in any old-growth forest to alternative silvicultural systems that remove less wood from the harvesting area compared to clear-cutting (Kimmins 1992). The difference between clear-cuts and the alternatives progressively broadens until the "selected tree" harvesting option; the corporate alternatives chosen are more likely to reflect MB's choice of partial retention, and their viability for large-scale forestry still has to be proven.

MacMillan Bloedel's Surprise Commitment to Selective Logging

A recent (1998) announcement by MB to stop clear-cutting old growth in five years in favour of alternative systems promises to significantly modify its practices to more closely resemble the environmental ideal (Hogben and Hunter 1998; Hogben 1998). MB's plan also appears to draw on Binkley's (1997a) zonal forestry model. Thus, for its one million hectares of public and private forests, MB plans an old-growth zone, in which 70 percent of the trees will never be logged, a habitat zone in which 40 percent of trees will never be logged, and a timber zone in which 28 percent of the forest will be retained. Clear-cutting is to be replaced by "variable retention logging," including a mix of shelterwood systems (strip, uniform, and irregular shelterwood systems) and group selection and group retention systems. For MB, the principle underlying the variable retention system is flexibility in "choosing the correct method for the appropriate site" (Hogben 1998: D8). No opening will be greater than one hectare. This plan is expected to increase logging costs by $4 per cubic metre, training and equipment costs will be higher, and the risk of injury to loggers will be increased. MB, however, believes its plan will help stabilize timber supplies and enhance its market potential among increasingly green customers. MB also intends to acquire some, as yet unclear, certificate of sustainability.

Environmentalists immediately applauded MB's initiative, apart from a relatively mild rebuke from the Sierra Club of BC about taking five years to

Single-tree retention on a community forest.

stop clear-cutting and a more threatening reminder from Greenpeace that it was against *any* form of old-growth logging. This latter comment suggests that this latest compromise will again depend on public perception. In the meantime, MB's decision raises implications for provincial policy that traditionally emphasized clear-cutting as essential for economic and safety reasons and as environmentally acceptable.

Several further points can be raised about MB's plan. First, MB is logging about 50,000 cubic metres from second growth (10 percent to 15 percent of the cut) in TFL 44, and this level will rise. MB's plan does not discount clear-cutting second growth. Second, MB's plan does not appear to be exclusively dependent on "selective logging," narrowly defined as the harvesting of one tree among many left untouched. Rather, it includes logging systems that are between clear-cutting and selective logging. Third, MB's plan does not preclude contracting out logging to SMEs or Aboriginal groups, partnerships that MB is already promoting. Fourth, as noted (Chapter 4), MB has doubled its cut from private lands between 1997 and 1998 and increased its level of log exports.

Enhanced Local Control

There are growing pleas for diversification of tenure ownership, specifically to give a greater role to small firms, communities, and Aboriginal peoples; that is, to enhance local control over the forest resource (Marchak 1983; Drushka 1993; M'Gonigle and Parfitt 1994). The basis for these pleas is democracy and local development, and relates closely to the many arguments for bottom-up development (Friedmann and Weaver 1978; Trist 1979; Sjoholt 1987). The main justification for local control of BC's forest resources is the expectation of a strong commitment to the sustainability of forests and the reinforcement of local identities. Harvesting decisions are made according to local priorities rather than the dictates of distant markets. Local control promises more value added, enhanced forest stewardship, and stronger local multiplier effects as cooperation among SMEs substitutes for internal economies of scale orchestrated by outside interests.

The extent to which local control fulfils these promises is likely to vary, and important issues remain for the articulation of local preferences with provincial regulations. Moreover, the idea of local control is ambiguous in terms of the meaning of both "local" and "control." Local could mean a few hectares within a municipality or thousands of hectares throughout a river basin; control may be defined in terms of the greater autonomy of branch-plant operations, the privatization of Crown land, or the granting of "sovereignty" to Aboriginal peoples. Despite these concerns, and the obstacles to in situ change, the diversification of tenure system ownership as a generalized goal, specifically by enhancing local control, potentially offers desirable characteristics in terms of stability, regional development, and multiple values, that is, towards a balanced and wise use of the resource. In practice, the provincial government has acknowledged the potential of local control, specifically for community forests and the treaty process.

The North Cowichan and Mission Community Forests

In 1997, the government created a multi-sector advisory committee to develop community tenure models to specifically provide local communities with more direct control of the forest resource (Hamilton 1997a). The Ministry of Forests also sought proposals from communities to participate in pilot projects to determine the form community forests should take. While the government plans to choose three proposals, over forty communities made applications and well over 100 communities have sought information on community forests (Simpson 1997: B1).

The idea of community forests is not new. The 1976 Pearse Commission advocated community forests by the creation of wood-lot licences. About ten community forests exist in the province, of which those in Mission and North Cowichan are the best known. They also provide slightly different models of community forests. Thus, the Mission community forest of 10,414 hectares is primarily (88 percent) leased from the provincial government (TFL 26), and the North Cowichan community forest of 4,800 hectares is community owned (Figure 10.3). In the latter case, the municipality acquired the land as a result of unpaid taxes in the 1920s. In 1946, the community successfully petitioned for a change in the Municipal Act to allow it to put land in a forest reserve. This reserve remained unmanaged until the 1960s. The Mission lease was obtained in the late 1950s. While the Forest Practices Code guidelines formally apply in Mission, North Cowichan follows them on a voluntary basis.

Figure 10.3

The North Cowichan and Mission community forests

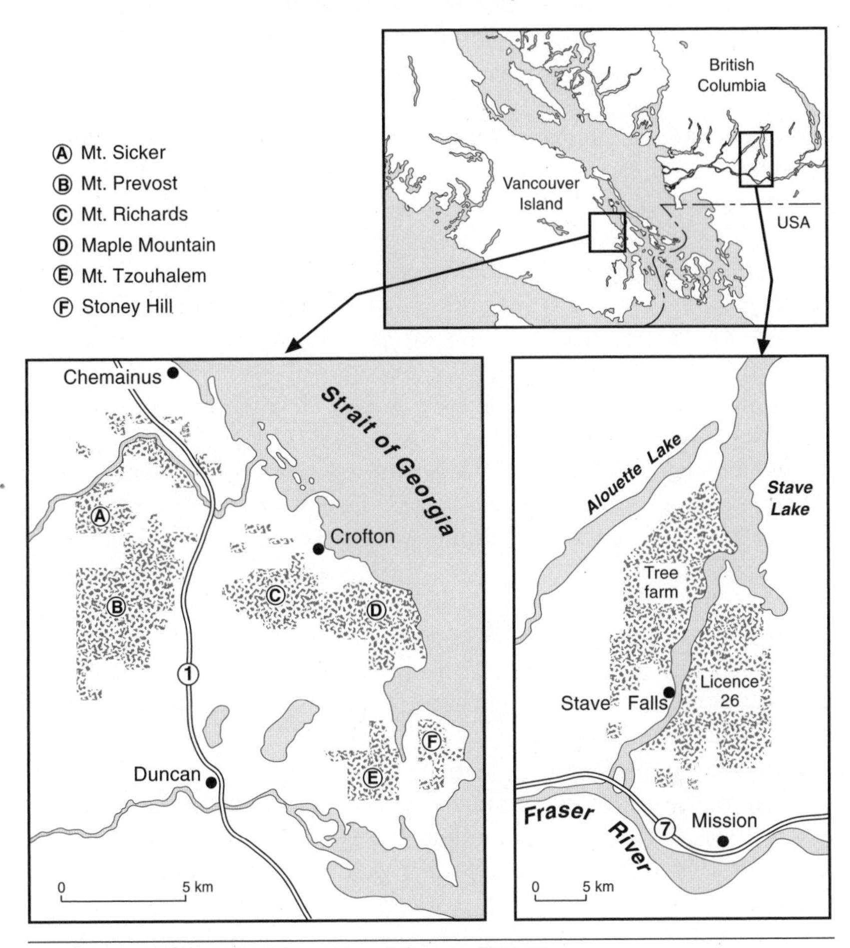

Source: District of North Cowichan n.d. and Simpson 1997: B1.

Management of the North Cowichan forest evolved in three phases. After the first phase of doing nothing, following a consultant's recommendation in the 1960s, ten woodlots were allocated to contractors who selectively logged in return for stumpage payments to the municipality. In practice, selective logging in this case meant high grading with the resulting deterioration of the forest. In the late 1970s, the local mayor considered the financial return to be unsatisfactory. Following recommendations from forest-industry experts, the municipality hired its own forester in 1982, replaced selective logging with small clear-cuts (up to 13 hectares), and introduced a silvicultural program. By 1992, the land reserve (rainy day) bank account was relatively high ($700,000) and used to continue management, even when log sales dropped. A Forest Legacy Fund ($137,000 in 1997) has been created (for special scholarships and other honours awarded by council). Profits and taxes ($1 million in 1997) were also generated for general community use. Although funds for forest management purposes have been obtained from senior governments, the forest always has operated at a profit.

Mission's community forest has been in operation since 1958 and logged according to TFL guidelines. Mission also employs a community forester, and its community forest has not drawn on municipal funds since 1985. Rather, it provides important contributions to local community funds. Recent contributions from 1996 timber sales included $685,000 for a new library, $132,000 for a new fire truck, $170,000 for an ice-rink conversion, $62,000 for the arts, and $1.2 million for budget stabilization (Simpson 1997: B5). As in North Cowichan, all logging is of second-growth timber.

The allowable annual cut in Mission is around 45,000 cubic metres, and in North Cowichan, 20,000 cubic metres. Since 1958, Mission has planted 3.2 million seedlings, and by 1995, North Cowichan had planted over one million seedlings. The forests are managed flexibly under the close supervision of the local councils. In North Cowichan, logging is done by the municipality and contracted out. Special contracts have been awarded for particular species (such as maple) and for salal, a greenery used in floral displays. The community employs eight to twelve people for forestry-related work. In North Cowichan, land can only be taken out of the forest reserve following a long referendum process. In fact, the municipality is considering buying more forest land.

Both Mission and North Cowichan have successful models of community forests. Others are likely to follow. Decisions are still pending regarding almost 100 applications for community forests, and in Revelstoke, residents approved a referendum to invest more than $1 million in a local forest agency.

The Treaty Process

About 1990, the provincial government reestablished the treaty process with Aboriginal peoples as a policy priority. The broad rationale of this initiative is to resolve deep-seated political, social, and economic grievances. A basic thrust of the process is to empower Aboriginal peoples by providing greater autonomy over resources, including forest resources. From a forest-policy perspective, the treaty process is part of a broader effort to diversify and enhance local control so that the forest resource can help Aboriginal peoples escape poverty. The government also argues that the treaty process will reduce uncertainty over resource rights throughout the province and encourage more investment.

In early 1999, the Nisga'a treaty was ratified by the Nisga'a and the provincial government; the treaty has yet to receive federal approval. Another

Figure 10.4

The Nisga'a treaty area as negotiated, 1998

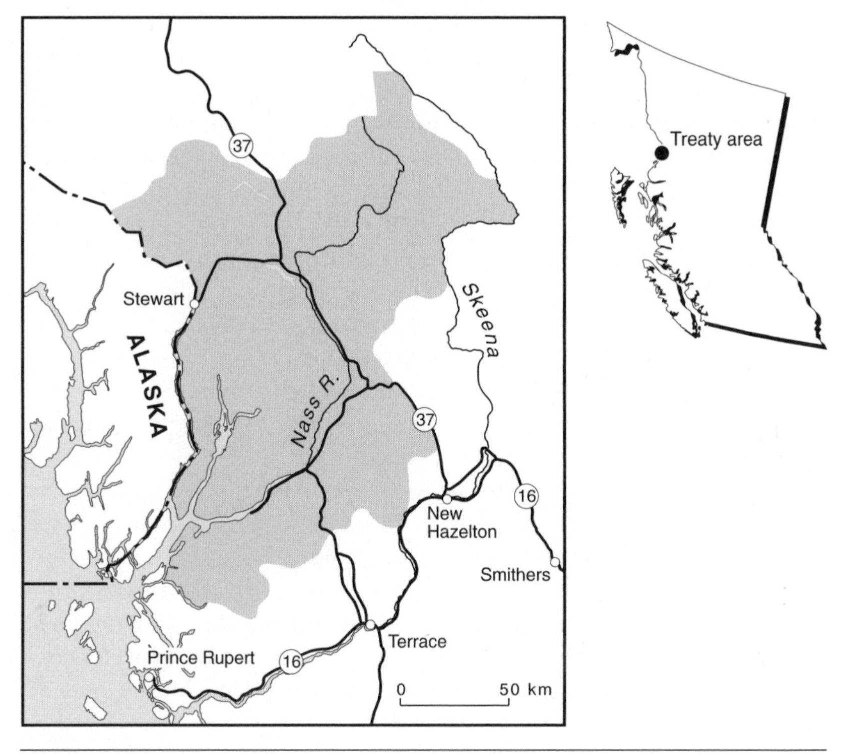

Source: Based on Canada 1996.

fifty claims are in stages three or four of the five-stage approach established by the Treaty Commission Act of 1992. Initially, negotiations primarily involved the federal government, the provincial government, and Aboriginal peoples. In 1994, the province signed an agreement with BC local governments to guarantee their involvement. Moreover, to the extent that treaty negotiations refer to established property rights, other parties have legitimate interests, even if they are not included in the formal process. The forest industry has not been formally included in the process nor has it offered comment.

The Nisga'a treaty, following an agreement-in-principle in February 1996, issued jointly by the federal government, provincial government, and the Nisga'a Tribal Council, was signed by these parties in July 1998. The general provisions of this agreement, which still await federal approval, include acceptance of the Nisga'a as an Aboriginal people and as Canadians with all associated rights and benefits. Significantly, "all parties apparently agree that the final agreement will provide certainty with respect to Nisga'a rights, title and obligations" (Canada 1996). Nisga'a jurisdiction over Nisga's citizens will be phased in, and Canada's Criminal Code and the Canadian Charter of Rights and Freedoms will apply. Nisga'a lands will no longer constitute reserve lands governed by the Indian Act.

The proposed agreement allocates 1,930 square kilometres as Nisga'a communal lands as well as up to 15 square kilometres of fee simple lands (Figure 10.4). On the Nisga'a communal lands, the Nisga'a will own the forests (and subsurface resources) and will manage these resources, following some transitional period for existing licensees to adjust operations. Provincial standards, including the Forest Practices Code, will apply, and stumpage fees will transfer to the Nisga'a. The Nisga'a have limited rights (up to 150,000 cubic metres) to buy an existing forest licence elsewhere. There are numerous other provisions dealing with access, fisheries, wildlife, environmental protection, dispute mechanism, eligibility as a Nisga'a person, taxation, financing (notably, a $190-million settlement as well as infrastructure provision), and the local government comprising a Nisga'a government and four village governments.

Just how the forest resource will be developed will reflect Nisga'a choices. The alternatives include logging to supply established mills located outside the treaty area, logging to supply new wood-processing mills on Nisga'a lands, not doing anything, and developing nonconventional uses of the forest resource. Clearly, these alternatives have implications for the forest industry outside the area, and compensation still has to be reached for companies that have lost timber rights in the treaty and companies whose activities will be taken over by the Nisga'a.

Whether the treaty process will be effective in meeting the needs of the Aboriginal peoples and in creating certainty regarding the framework for forest resource utilization throughout BC is debatable. In the latter context, two broad notes of caution need to be made. First, as noted, Aboriginal communities will have discretion over how any forests assigned to them are used. To the extent that Aboriginal communities seek economic development, such aspirations inevitably will turn towards the BC and indeed the global economy, for markets and for expertise. In turn, such integration introduces new vulnerabilities as well as opportunities, economic and cultural. There is no guarantee that industrial use of the forests will allow Aboriginal groups to escape the poverty trap or to preserve traditional identities. If the forest resource is to be used to supply market needs, then market forces, technological change, and the impacts of boom-and-bust will affect Aboriginal projects like any other. There is also a basic contradiction in linking the goals of economic and social development, and the associated implications of integration and mobility, with the preservation of traditional values and racially based communities that have distinct rights and powers in geographic enclaves. Failure to resolve this contradiction is likely to be a source of grievance.

Second, the provincial government claims that the treaty process will reallocate no more than 5 percent of BC's forests to Aboriginal peoples (in 197 bands comprising 5 percent of BC's population with status Indians about half this percentage). Meanwhile, Aboriginal land claims cover the entire province, and they overlap. Even if the treaty process is completed, it will probably take a considerable amount of time, and it is impossible to anticipate precisely the outcomes of these complex negotiations. In any case, the idea that treaties in the future will provide finality on the question of Aboriginal forest-resource rights is suspect. The recent Supreme Court *Delgamuukw* decision was intended to facilitate Aboriginal claims to rights throughout BC (*Delgamuukw* v. *British Columbia* 1997), a decision Flanagan (1998: 279) argues "fashions aboriginal title in a unique way and subjects it to broad and vague restrictions." Thus the decision allows oral evidence and introduces not easily understood principles, namely "discontinuous continuity" and "shared exclusivity," as a basis for land claims. It also intimates that "pre-existing systems of aboriginal law" may be relevant to Aboriginal title, and that proposed Aboriginal uses of land should not be "irreconcilable" with the nature of traditional attachments to land. Whatever interpretation is made of these principles, Aboriginal power in contesting land use in BC has been enhanced. This power will be exercised, and some Aboriginal groups may see the courts as a better alternative to treaties, although Bell (1998) favours political negotiation. In any case, given *Delgamuukw* and the overlapping nature of land claims, disputes are likely to continue even after treaties have been negotiated.

As the growing trend towards joint ventures and contracting relationships involving Aboriginal peoples indicates, however, there are growing attempts at meaningful cooperation at the grassroots, in the public and private sectors, which are designed to realize mutual long-run benefits. Much may be riding on these forms of cooperation. In the long run, local outcomes regarding Aboriginal use of, and benefit from, the forest resource, are likely to be highly varied.

Rethinking Reregulation?

The government's high-stumpage regime has received increasing criticism. Industry sees it as a source of red tape, excessive regulation, and instability. In response, the Minister of Forests in June 1998 announced that stumpage levels would be reduced (by the year 2000) and the Forest Practices Code modified. Indeed, the Minister of Forests claims that, in addition to lower stumpage (amounting to $5 per cubic metre on the coast), the paperwork required will be cut in half (Hamilton 1998e). The total impact of these changes on industry is claimed to be savings of $300 million in stumpage by 2000 and $300 million in red-tape reduction by 1999. For industry, the changes to the Code are seen as a step in the right direction. They mean reducing the number of plans that have to be submitted (from six to three);

eliminating some review processes; minimizing circumstances in which approved logging plans can be reversed; and allowing foresters to exercise more judgment in the field. For government, these changes are modifications but not reversals of policy. For environmentalists, however, this relaxation of regulations will gut the intent of the Code. Their particular concern is that undermining procedural requirements will modify the proactive approach to environmental problems implied by the Code to a reactive approach in the "new" Forest Practices Code.

Forest policy continues to be in a state of flux. Apart from changes to the Code and to stumpage, the evolution of the Clayoquot Sound Compromise reflects the experimental and incomplete nature of forestry reregulation in BC. Apart from uncertainty over land claims, long-expected decisions over community forests were still to be made, as of May 1999, two years after the first deadline, and the full implications of the results from timber-supply assessments have yet to be implemented (Hunter 1998: A1).

Another source of confusion regarding forest policy was created by big subsidies awarded to the pulp mill at Prince Rupert. The local social and political rationale was clear enough. Yet, this subsidy was contrary to the provincial priority of promoting value-added activities and of realizing higher values from BC's forest resource. Indeed, given these goals and declining timber harvests, a direct implication is that many of the old commodity mills will have to close.

Perhaps the most fundamental challenge to the rethinking of the reregulation of BC's forests has come from industry, specifically MB's announcements that it will stop clear-cutting old growth and its pleas for privatization. These statements are reverberating throughout the industry, apparently catching the government off-guard and turning the tables in the relationship between provincial government and big business (Chapter 4). It is worth recalling that policies diverting timber to SMEs by means of auctions claimed the rationale of the free market which in turn provided a more realistic and higher valuation of BC's timber. In other words, privatization in the early 1990s – which is what timber sales at auctions imply – was seen as a stick to wield against large corporations and their cozy timber monopolies. The government further cited American countervail action as another reason to support higher valuations of BC's timber through auctions for SMEs (Chapter 7).

In fact, there seems to be growing support within the provincial government for some limited form of privatization. Indeed, rather than compensate MB for land lost to parks, the government offered support in principle for MB's next-best proposal to receive land in lieu of cash (see Figure 10.5). MB would have preferred cash, but the government already had a sizable debt (Hunter and McInnes 1999). Yet, in unruly BC, the ability of the government to compensate MB with land is itself in question as a result of vociferous opposition from environmentalists and Aboriginal groups. Indeed,

Figure 10.5

The MacMillan Bloedel land compensation proposal

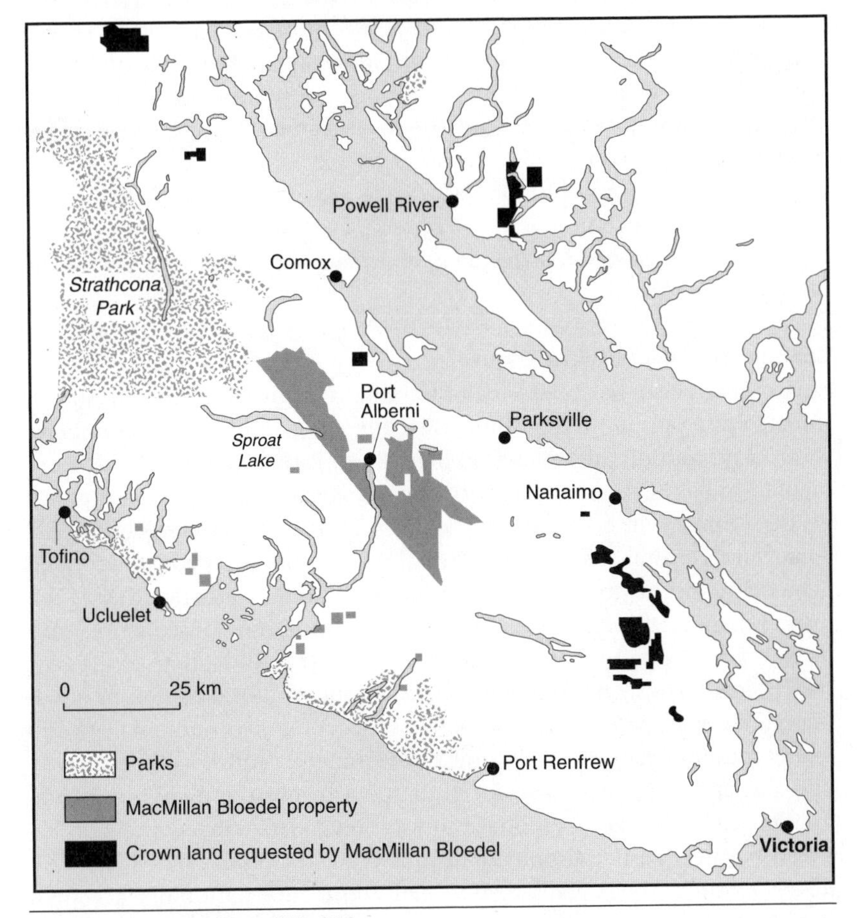

Source: Hunter and McInnes 1999: D12.

and before the government responded, MB recognized this opposition and publicly indicated its willingness to give up more of its timber tenures (and AAC) in return for the privatization of the remainder, or some of the remainder, of its lands.

More generally, privatization casts a giant shadow over the rethinking of the reregulation of BC's forest policy. There are four broadly based motivations underlying recent pleas for privatization. First, privatization is seen as a way to diversify forest tenures in BC and to provide an incentive for better forest management (Drushka 1993). Second, privatization is touted as a way to remove US protectionist pressures that now criticize BC's log export restrictions. In fact, the US government has supported such pleas and it has explicitly linked privatization with the need to remove log export restrictions

(Hamilton 1999f). Third, MNCs see privatization as a way to enhance flexibility in the use of timber supplies, including by log exports. Fourth, financial institutions such as pension funds are likely to increasingly favour privatization to increase their investment options, a trend that is well underway in the US, led by such organizations as the Hancock Timber Resource Group (Binkley et al. 1996). This range of motives is neither mutually exclusive nor mutually compatible. Privatization offers both great benefits and great costs to BC, the latter related to deepening foreign control and the narrowing of industrial (value-added) options. The challenge for policy, if privatization is to be contemplated, is to realize the benefits, local control, improved management, and value-added developments while mitigating the costs.

Conclusion: Diversifying Tenure

Environmentalism has been a major force in BC's forests, advising, cajoling, threatening, and ultimately stimulating the reregulation of the forests towards a profoundly more environmentally sensitive framework. Environmentalism has also demanded, with considerable success, that forest-policy decision making be transparent; traditionally, it had always been opaque. Some voices within environmentalism, however, have grudgingly accepted democratic forums to combine environmental and economic criteria. Yet, if local democratic action is undermined, the long-run consequences may be bad for the environment and for industry, and all the relevant stakeholders. Thus, environmental failure to support local consensus building mechanisms potentially risks public support for environmentalism and renders rational planning by business, governments, and workers extremely difficult, and may encourage social diaspora or even anti-environmental militancy. A potentially important innovation to facilitate cooperation between business and environmentalism is eco-certification, by which businesses commit themselves to nationally and internationally accepted standards of environmental behaviour which are evaluated by recognized third parties (Cabarle et al. 1995; Lyke 1996; Wallis et al. 1997). Eco-certification is not without its dilemmas; for example, it imposes higher costs, especially on SMEs and potentially onerous "chain of custody" guarantees. Nevertheless, both business and environmental groups have cooperated in taking initiatives towards eco-certification, and such cooperation should be encouraged.

Indeed, there are increasing indications that industry and government are seriously committed to eco-certification. Weyerhaeuser is continuing MacMillan Bloedel's initiative to seek certification of its Sproat Lake (and other) operations, according to the International Standards Office (ISO) and Canadian Forestry Service (CFS) standards. Eventually, when international environmental groups have established specific standards for BC, MB will seek this certification. MB is taking steps towards labelling its products consistent with certification standards. Other firms, such as Canfor, are also showing leadership on this issue, and the provincial government has declared its intentions to commit to the certification process, notably in its SBFEP. By investing in certification and green labelling, industry hopes to establish environmentally acceptable practices and offset consumer boycotts. The potential of the certification process, in turn, places pressure on environmental groups to be pragmatic in their attempts to influence consumers.

Similarly, environmental opposition to (a degree of) privatization should be more nuanced. After all, government ownership of forests in BC has not prevented the resource cycle, and in centrally planned economies, forest abuse has probably been worse. Privatization is no magic wand itself, and it poses problems for BC. As Drushka (1993) pointed out, however, some form of privatization, regulated to some degree, can help contribute to more effective forest management. Drushka's (1993: 18) views, citing the work of Peter Sanders – that two-thirds of BC's forest land, split evenly between corporate and small-firm control, should be privatized, while the other third would remain in the public domain – are probably too radical. He also does not mention Aboriginal land claims. Nevertheless, his proposals again underline the need for an explicit framework for forest tenure and ownership, specifically, a framework that is more diversified. Because Drushka also suggests the idea of "earned title" as a way of gradually privatizing lands, he implicitly links forest policy with innovation. This link is to be encouraged, as the next chapter argues. In the absence of innovation, log exports may well increase, with all their negative implications for jobs, value added, and the environment.

11
The BC Forest-Product Innovation System and the (Frustrating) Search for a Knowledge-Based Culture

> Technology is changing at an ever increasing rate, from axe to crosscut saw in 100 years, crosscut saw to bow saw in 45 years, bow saw to chain saw in 30 years and currently chain saws are being displaced by machines with shears or circular-saw felling heads, each period decreasing by one-third.
>
> – Silversides 1984

> We are heading towards a technological backwater in BC.
>
> – Binkley (quoted by Hamilton 1998f)

Innovation is the central challenge facing BC's forest industries. The provincial government's high-stumpage regime (see Figure 3.8), environmentalist pleas for assigning the forest its real worth (M'Gonigle and Parfitt 1994), and corporate viability in a high-cost environment all depend on innovation of products and processes. Innovation provides no iron-clad guarantees of market success, because it is itself a competitive process. Failure to meet this challenge, however, threatens the survival of the forest industry (Hayter 1988; Binkley 1995). The Fordist legacy of mass-commodity production based on cost minimization has clashed with the new imperatives of product differentiation and value maximization. How this clash is resolved will define the future of BC's forest economy.

Since the early 1970s, there has been an escalating litany of pleas for greater commitment to innovation, and related research and development (R&D) processes in BC's and Canada's forest industries (beginning with Smith and Lessard 1970). Despite substantial policy experimentation, it is questionable that BC's forest industry recognizes the need to *create* competitive advantage by applying its know-how. The forest industry evolved around a narrow, conservative approach to innovation, only rarely investing in longer-term R&D. Such ingrained technological conservatism has been hard to change. Indeed, MB's recent (1998) closure of its R&D laboratory eliminated the most important corporate exception to this sorry legacy and underlined the problems facing BC's innovation policy. This chapter reflects on these problems and reiterates the need for such a policy.

The chapter assesses innovation in BC's forest economy in terms of how technological capability is organized. Technological capability is defined as "the ability to solve scientific and technological problems and to follow, assess, and exploit scientific and technological developments" in the marketplace (Britton and Gilmour 1978: 130). A key indicator of this ability is provided by R&D inputs. The first part of the chapter examines the scale and organization of the R&D system in BC's and Canada's forest industries from the 1970s to the mid-1990s. Second, the chapter outlines the technology strategies of BC forest-product firms, and related private-sector organizations, notably equipment suppliers, strategies that link the R&D system to innovation. The final part of the chapter argues that innovation needs to be the priority of BC forest policy, regardless of the forest industry's disinclination for R&D.

This chapter tells an old story. It is well known that R&D activities across the Canadian industrial spectrum are underfunded compared with most advanced industrial economies, and that innovation is a problem. This situation is true for Canadian industry in general (see Marshall et al. 1936; Canada 1972; Britton and Gilmour 1978; Britton 1996), and for the forest industry in particular (Smith and Lessard 1970; Solandt 1979; Hayter 1982a, 1987, 1988; Binkley and Watts, 1992 and 1999; Binkley 1995; Binkley and Forgacs 1997). Thus, BC's situation has to be understood as part of a Canadawide dilemma. However, the forest industries are technologically dynamic (Silversides 1984), and there are significant private and social rewards for innovation.

The R&D System: Engine for Innovation

In contrast to other advanced forest-product regions, in BC (and Canada) technological attitudes and choices are conservative and narrowly based (Hayter 1988). In the midst of the early 1980s recession, when BC's forest industry was in a dire situation, the overriding conclusion of a study commissioned by the Science Council of BC is that the BC forest-product industry had become locked in to technologies and products that yield low average rates of return on investment (for example, construction grade lumber), face declining markets, and, in other cases, have not capitalized on changes taking place in the marketplace (Woodbridge, Reed 1984: 93). This report reveals that for decades BC's technology initiatives were limited, favouring known technology to efficiently manufacture a limited range of commodities. The report further noted that BC equipment suppliers were similarly confined to manufacturing mature technologies (Hayter 1987).

Fifteen years later, another study commissioned by the Science Council of BC, primarily written by individuals representing the major forest-industry institutions, reached similar conclusions. The study recognizes that the forest industry has developed new products and applauds the industry's many technological strengths. Nevertheless, it argues that the "main focus of technology application in the BC forest industry is cost minimization" (Ernst and Young 1998: A-5) and that BC's forest industry is now paying the price for its heavy dependence on highly cyclical lower-value commodity products. In effect, it is "trapped in the commodity box" (1998: A-10). In comparison to leading-edge competitors elsewhere, a "culture of innovation" is lacking (1998: A-7). Indeed, reference to the "commodity box trap" restates the "staple trap" (Chapter 1), with its Catch-22 dilemma. Staple production does not require long-term R&D and innovation, but to diversify from staples, R&D and innovation is essential.

These reports emphasize that the BC forest industries are slow to innovate products in newly emerging, fast-growth areas. A key underlying problem is the lack of long-term R&D necessary for technological leadership in product innovation. In this regard, Parallam and Spacekraft define exceptions rather than the trend (Chapter 4). The failure to anticipate research needs extends to forest management. Thus, in another report commissioned by the Science Council of BC, the authors noted considerable underfunding of forestry research, especially for long-term studies, and suggested: "In the future, research projects will emphasize the management of second-growth (whether planted or naturally established) *about which relatively little is known at present*" (Science Council of BC 1989: 15, italics added). Apparently, although the falldown effect was predictable (and government laboratories have long existed in forestry), little thought was given to anticipating its long-term consequences.

Forest-Sector R&D Budgets in BC (and Canada) since the 1960s

The industry's blinkered approach to technology is directly reflected in R&D investments, the acknowledged engine of innovation. Because forest-sector R&D in BC is an integral part of a national system in which the federal government has historically played a key role, national statistics provide context and insights into the provincial situation.

In Smith and Lessard's (1970) pioneering study, forest-sector research expenditures throughout Canada amounted to $54 million in 1968, with public-sector organizations accounting for just over half (Table 11.1). Because equipment suppliers were excluded from the survey, the actual private-public sector split is likely to have been close to equal. At that time, public-sector R&D was dominated by federal laboratories that funded forestry research throughout Canada, including the Pacific Forest Centre in Victoria and wood-processing R&D in the Eastern and Western Forest Product Laboratories (EFPL and WFPL) in Ottawa and Vancouver, respectively. Limited forest-industry efforts mainly concentrated on pulp-and-paper projects.

Smith and Lessard were concerned by both the level of R&D funding in Canada and its distribution. Thus, they advocated considerable expansion

Table 11.1

Forest-sector research expenditures in Canada by various agencies, 1968, with projections to 1978

	1968 expenditures		1978 projection	
Agency	$M	%	$M	%
Public sector				
Canadian Forestry Service	22.0	40.7	46.2	33.0
Universities	2.7	5.0	8.6	6.2
Provincial bodies	3.3	6.2	6.9	5.0
Private sector				
Forest industries	26.0	48.1	78.0	55.8
Total	**54.0**		**139.7**	

Notes: The CFS data include the budgets of the WFPL and the EFPL. The wood and paper industries data refer to in-house R&D. Equipment suppliers are excluded. The 1978 estimates are based on continuing growth and R&D expenditures account for 3.2 percent of GNP by 1978. Provincial bodies refers to research services and research councils. The totals for the forest industries include the funding of the Pulp and Paper Research Institute of Canada. Data should be regarded as estimates.
Source: Based on Smith and Lessard 1970: 183.

of R&D, noting that if the Science Council of Canada criterion for funding was to be met in the forest sector, annual expenditures would have to grow by 15 percent to result in a total of $139.8 million by 1978 (Table 11.1). At the same time, they suggested that the forest industries should play a relatively bigger role, noting that in the US, industrial R&D in the mid-1960s accounted for 60 percent of forest-sector expenditures, compared to 48.1 percent in Canada. In particular, Smith and Lessard (1970: 184) hoped that industrial contributions would increase by 1978 to 52 percent and 55 percent by 1988.

In practice, despite the availability of incentives for industrial R&D, and the "privatization" of the EFPL and WFPL to create Forintek, and the creation of FERIC (a logging research organization), the relative share of industry R&D declined during the 1970s (Hayter 1988). In a study providing inputs to a federally sponsored forest-sector advisory committee, Solandt (1979) found that private industry's share of forest-sector expenditures had declined to 38.9 percent of the total; his estimates included contributions by equipment suppliers, not calculated by Smith and Lessard (Table 11.2). If industry's share of funding for FERIC and Forintek is included, 50 percent and 25 percent respectively, industry's overall share in Solandt's estimates increases to 41.7 percent, still noticeably less than Smith and Lessard's calculation of a decade earlier. Solandt (1979) further estimated that in 1979 R&D expenditures would have to be $151 million to maintain 1968 levels of activity. The actual figure was $115 million, which is well below Smith and Lessard's desired levels for 1978, estimates that did not take into account inflation.

Table 11.2

Forest-sector research expenditures in Canada by various agencies, 1979

Agency	$M	%
Public sector		
Canadian Forestry Service	30.3	26.4
Other federal	6.8	5.9
Universities	3.7	3.2
Provincial bodies	18.8	16.3
Private sector		
Forest industries	23.0	20.0
Suppliers	11.8	10.3
Paprican	10.0	8.6
Cooperatives		
Forintek	8.8	7.6
FERIC (1978)	2.0	1.7
Total	**115.2**	

Note: Data should be regarded as estimates.
Source: Solandt 1979: 68.

Binkley and Forgacs (1997) have updated these studies and provided indexed estimates of forest-product R&D budgets between 1982 and 1997 (Table 11.3). They found that in real terms, indexed in 1986 dollars, R&D spending increased between 1982 and 1987. However, they regarded this increase as modest in terms of annual growth or international comparisons, and they further note that in 1997, fully 20.8 percent of the entire budget was generated by the BC provincial government whose spending had recently been increased by the formation of Forest Renewal BC. The funding levels of Forest Renewal BC, however, are uncertain.

Significantly, Binkley and Forgacs note that direct industry (in-house) R&D declined between 1982 and 1997, and that by 1997, overall industry contributions, including funds for the laboratories at Paprican, FERIC, and Forintek, to R&D expenditures was 43.9 percent, less than in 1982 and less than in 1968. After thirty years of recurring pleas for more industry R&D, the situation has deteriorated. For in-house R&D, the meagre levels of the 1970s have become pathetic by the late 1990s. The closure of MB's R&D laboratory, the biggest in the country in forest products, and of Canfor's smaller R&D group, in 1998 adds to this trend, specifically in BC.

Indeed, the situation in BC follows national trends. Smith and Lessard (1970: 76) report that in 1956, R&D funding in BC amounted to $1.35 million, or about 0.22 percent of the value of the forest industry in that year. In 1968, they note that forest research amounted to $3.179 million, over two-thirds provided by the federal government and just 11 percent by industry,

Table 11.3

Forest-sector research expenditures in Canada by various agencies, 1982 and 1997

	1982 expenditures		1997 expenditures	
Agency	$1986M	%	$1986M	%
Public sector				
Canadian Forestry Service	62.6	36.9	55.1	25.3
NSERC (universities)	2.0	1.2	8.2	3.8
Provincial - British Columbia	10.6	6.3	45.2	20.8
Provincial - other	11.6	6.9	13.4	6.2
Private sector				
Forest industries	59.2	34.9	56.2	25.9
Cooperative				
All institutes	23.6	13.9	39.1	18.0
Total	**169.6**		**217.2**	

Note: The figures are indexed to 1986 dollar values and are not the "actual" values for either 1982 or 1997. Also, the authors only apportioned industry's contributions to the "All institutes" category; government contributions are incorporated in the appropriate government categories. Suppliers are not included.
Source: Binkley and Forgacs 1997: 23.

even though industry controlled all production, and just 12.4 percent by the provincial government, even though it owned 95 percent of the land (Smith and Lessard 1970: 78). In 1985-6, R&D funding in the BC forest economy amounted to $49.8 million (Table 11.4). Of this total, industry contributions, including shares made to the cooperative laboratories, amounted to 43.2 percent. By 1991-2, according to Binkley and Watts's (1992) estimates, the relative contribution of industry has declined further to around 33 percent (Table 11.5). In 1998, industry's share was still significantly less (Binkley and Watts 1999: 611). Smith and Lessard's (1970) prescriptions for more rather than less industrial R&D have been reversed. Although not as complete – provincial and university expenditure data are absent – estimates for 1984 for federal government, private-sector, and cooperative R&D are consistent with this trend (Hayter 1988: 32-45). Moreover, BC and Canadian R&D funding in the forest sector is low by international standards.

International Comparisons

In comparison to competitors in advanced economies, the intensity of R&D funding throughout the Canadian forest economy, including BC, is relatively low. In 1975, Hanel (1985: 57) noted that for Sweden, Finland, Japan, and the US, shares of world production of pulp and paper were less than shares of OECD forest-product R&D. The reverse is the case for Canada. In 1986, the Canadian pulp-and-paper industry spent just 0.3 percent of sales revenue on R&D, less than one-quarter of the ratio for Sweden and the US

Table 11.4

Forest-sector research expenditures in British Columbia for selected agencies, 1985-6

Agency	$M	%
Public sector		
Pacific Forest Centre (CFS)	7,200	14.5
Other federal	1,254	2.5
Universities	3,877	7.8
BC Forest Service	9,213	18.5
Science Council of BC	1,482	3.0
Private sector		
Forest industries	18,513	37.2
Paprican	800	1.6
Cooperatives		
Forintek	6,000	12.0
FERIC	1,480	3.0
Total	**49,819**	

Notes: The forest industries totals are high for 1985-6 because of the introduction of Parallam into the market by MB. Suppliers are probably not included in this figure. Paprican's activities in BC were expanded in 1986-7 by a new $6M pulp-and-paper centre.
Source: Forest Research Council of British Columbia 1986:6.

Table 11.5

British Columbia's forest sector: Sources of research funds, 1991-2

Source	$M	%
Federal	12.0	17
Provincial	29.4	41
Universities	1.8	3
Research organizations	9.1	13
Industry	18.8	26
Total	**71.1**	

Source: Binkley and Watts 1992: 732. Reprinted with permission of Canadian Institute of Forestry.

and much less than the commitment in Finland (Table 11.6). More recently, Binkley and Forgacs (1997: 7) claim that forest-sector R&D expenditures overall in Canada amounted to just 0.4 percent of the value of industry shipments in 1994, while Binkley and Watts (1992: 733) cite a figure of 0.85 percent based on a Science Council of BC (1989) study. Whatever the precise level, Canadian efforts are low and much less than in the US and Sweden, where R&D expenditures represented 1.5 percent and 1.75 percent of gross sales, respectively.

Table 11.6

Corporate R&D expenditures in the pulp-and-paper industry, 1986

	R&D expenditures (US$M)	R&D as a % of sales
United States	1,224	1.25
Japan	230	0.75
Sweden	86	1.35
Finland	48	0.75
Canada	50	0.30

Source: Science Council of Canada, 1992.

In BC, R&D intensities are even lower than the Canadian average. Thus, in the Binkley and Watts's (1992) study, forest-sector R&D expenditures as a proportion of sales is 0.69 percent compared to the above-cited Canadian average of 0.85 percent. According to this study, R&D intensities in forestry are slightly stronger in BC than Canada, but much lower in forest products. A comparison with the US and Sweden is even less flattering to BC (see Binkley 1993: 295).

Moreover, international comparisons reveal significant differences in how BC and Canadian R&D efforts are allocated within the public and private sectors. Thus, Smith and Lessard (1970: 105) noted the much greater scale of forest-sector R&D in the US than in Canada in the 1960s, especially for industry and university R&D. In 1965, forest-sector R&D expenditures were estimated at US$125 million, several times more than the Canadian total for 1968. Moreover, industry in the US accounted for 60 percent of R&D expenditures compared to 48 percent in Canada; government in Canada plays a larger role than industry.

These differences were underlined in a later comparison of R&D employment (Hayter 1982a). Thus, by the late-1970s, more professional scientists and engineers in Canada and BC were employed in government laboratories than in in-house laboratories, whereas the reverse was true in the US (Table 11.7). In the US, the numbers of private-sector professionals overwhelmed their public-sector counterparts. This pattern of underrepresentation of industry R&D and overrepresentation of government R&D in BC (and Canada) is evident when BC's and Canada's shares of industry and government forest-sector R&D jobs in North America are considered in terms of population, forest-product employment, and wood harvest (Table 11.8). The share of government R&D is invariably much greater than the share of industry R&D, except in BC when population is used as the yardstick.

Since the late 1970s, federal government R&D has declined (see Table 11.3). This decline has not been offset by industrial R&D, especially in-house R&D. Rather, cooperative R&D and, by the 1990s, provincial government funding,

Table 11.7

Professional employment in industrial and federal government R&D in the forest sector of the United States, Canada, and BC, c. 1977

	Industrial R&D		Federal R&D		Total R&D	
	Professional employees	%	Professional employees	%	Professional employees	%
United States	3,461	93.2	978	69.0	4,652	84.9
Canada	258	6.8	441	31.0	833	15.1
(BC	69	1.9	111	7.8	193	3.5)
Total	**3,719**	**100.0**	**1,419**	**100.0**	**5,485**	**100.0**

Notes: Industry R&D refers to in-house R&D by manufacturing firms. Total R&D includes cooperative and industry R&D. BC is part of the Canadian totals.
Source: Based on Hayter 1982b: 257-9.

have become more important. The underrepresentation of industry R&D has remained a feature of Canada's and BC's R&D systems.

Canada's and BC's weak R&D performance extends to forest equipment suppliers. Although comparative data for R&D budgets and employment levels among forest equipment suppliers do not exist, Canadian firms are known to do little R&D, and the biggest and best-known suppliers are foreign, including Beloit (US), Valmet (Finland), Escher Wyss, Stihl and Voith (Germany), and Kamyr (Sweden). These firms, much bigger than Canadian suppliers, all have major R&D centres in their home countries.

Table 11.8

Location quotients for professional R&D forest-sector employment by industry and federal government in Canada and British Columbia with North America as benchmark region, c. 1977

	Population	Forest-product employment	Wood harvest
Industry			
Canada	0.7	0.5	0.3
British Columbia	1.9	0.4	0.2
Government			
Canada	3.2	2.2	1.3
British Columbia	7.8	1.8	0.7

Notes: The location quotient (LQ), as measured, is the ratio of Canada's (or BC's) share of R&D employment in North America (US and Canada combined) to Canada's (or BC's) share of population, forest-product employment, and wood harvest in North America. If the LQ is less than 1.0, then the region is said to have less than its "fair share" of employment according to the yardstick used.
Source: Based on Hayter 1982b: 257-9.

Table 11.9

Location of inventors and owners of patents issued in the Canadian forest-product sector, 1950-75[1]

Location	Inventor	%	Owner	%
British Columbia	37	3.8	20	2.5
Rest of Canada	111	11.5	95	11.8
United States	621	64.2	550	68.4
Europe	179	18.5	124	15.4
Elsewhere	19	2.0	15	1.9

1 Approximately a 10 percent sample.
Source: Hayter 1980.

Patent data, which provide a crude measure of R&D outputs, further confirm the weak position of Canadian forest equipment suppliers. Thus, Hanel's count of patents between 1978 and 1980 reveals that Canadian inventors are most competitive in forestry machinery but progressively less so for wood-processing, pulp-and-paper, and paper-converting technology. For example, Canada accounted for 28.9 percent of the 135 patents in forestry equipment, but only 8.9 percent of the 395 pulp-and-paper patents and 5.8 percent of the 208 paper-converting patents in 1978-80. Another count of patents, this time for 1950-75, reveals similar trends for BC (Table 11.9). Thus, patents originating in BC were relatively few, and based on percentages of industry totals, were mainly in logging and wood processing (Hayter 1980). In fact, BC accounted for most of the Canadian patents in these industries, but for a much smaller share of pulp-and-paper patents, which were predominantly assigned to Americans. Hanel (1985) also felt that Canada's traditional strengths in forestry equipment were being eroded by Swedish competition. Indeed, innovations in logging, silvicultural site preparation, and wood-processing equipment in many European countries are now making inroads into Canadian markets, and "distinctive" Canadian conditions are no longer providing the same protection to local suppliers (Hayter 1988: 72).

The Evolution of the BC Forest-Product R&D System

In BC, as in Canada as a whole, the relative weakness of private-sector R&D and the relative strength of public-sector R&D is a historically distinctive feature of the forest-product industries. Indeed, these two trends are related because various forms of public-sector R&D have sought to compensate for deficient private-sector R&D (Smith and Lessard 1970: 65; Hayter 1982a; see also Marshall et al. 1936).

In-house R&D laboratories, themselves an institutional innovation of the latter part of the nineteenth century in the German chemical and American electrical industries, marked the "professionalization" of inventive and innovative processes (Freeman 1974, 1995). The independent inventor was to be increasingly replaced by the formal, systematic application of scientific principles involving numerous specializations. As Hull (1985) demonstrates, the transformation of the Canadian pulp-and-paper industry from one based on craft principles to one founded on a cellulose-based industry controlled by university-trained chemists, chemical engineers, and other scientists, occurred between 1900 and 1930. The Canadian federal government, however, deemed private-sector efforts to be insufficient and sought to stimulate directly and indirectly an appropriate, professionalized R&D infrastructure. Thus, the R&D system evolved primarily around various public-sector initiatives. Three broad phases in the evolution of these initiatives can be discerned.

First, until the 1970s, the federal government played the key role in creating a national network of forestry and forest-product laboratories controlled by the Canadian Forestry Service (CFS). In BC, the main centres were the WFPL in Vancouver and the Pacific Forestry Research Centre in Victoria. In the 1920s, the federal government also helped set up Paprican in Montreal, which has since been largely voluntarily funded by industry as a classic "association laboratory." In BC, initiatives included a small provincial forestry research effort, the UBC forestry faculty, and a small industry association laboratory in North Vancouver, which was in the 1950s a structural testing facility for plywood financed by the Plywood Manufacturers of British Columbia (PMBC) and then became part of the Council of Forest Industries (COFI). BC members also had access to Paprican in Montreal. A few in-house R&D groups were established, including MB's in 1966, the only group to survive on any scale.

Second, in the 1970s and 1980s, the federal government again took the initiative by trying to privatize some of its R&D operations and by encouraging companies to invest more in R&D. In particular, following the report of a federal-sector advisory panel in the late 1970s, chaired by the CEO of Vancouver-based Crown Zellerbach Canada, a decision was made to "privatize" the WFPL to create Forintek. This decision complemented the privatization of the logging division of the EFPL in the creation of FERIC in 1975 which then had offices in Vancouver and Ottawa. In both cases, the government hoped that FERIC and Forintek would be entirely funded by industry to become proper "association laboratories." In fact, government funding has remained essential to their operation as "cooperative laboratories," mandated to meet industry priorities. In BC, MB's in-house group was joined by the smaller efforts of Canfor.

Third, most recently, the main stimulus to the BC forest-product R&D system has stemmed from the provincial government. Elsewhere in Canada, the Quebec provincial government has been active. In BC, Forest Renewal BC is at the centre of R&D initiatives. Although amorphous in character, one feature of Forest Renewal BC's funding is to encourage the involvement of universities and related postsecondary institutions in the forest industry.

Figure 11.1

The comparative advantages of alternative agencies in the forestry sector R&D processes

Agency type	Comparative advantage				
	Basic research	Applied research	Process development	New product development	Technology transfer
Universities	strong	strong	moderate	weak	weak
Government	moderate	moderate	weak	weak	weak
Research cooperatives	moderate	strong	moderate	moderate	weak
Industry R&D	weak	moderate	moderate	strong	strong
Equipment suppliers	weak	weak	weak	weak	strong

Source: Based on Hayter 1988: 30.

In addition, with provincial government help, Paprican has established a small research group at UBC. In the meantime, industry R&D has all but disappeared in BC, and it is not healthy elsewhere in Canada.

In-house R&D: The Weak Link

In BC's and Canada's forest sectors, R&D is organized in a variety of ways by the public sector, private sector, and cooperative institutions, each of which offers distinctive strengths or comparative advantages in the R&D process (Figure 11.1). There are no rules to resolve "the institutional assignment problem" for the allocation of resources among different R&D organizations (Nelson 1988). The R&D process is too uncertain and the distinctions between R&D processes are often blurred. Nevertheless, experience indicates that in BC's and Canada's forest sector, R&D is not simply underfunded. Rather, it is inappropriately allocated. Specifically, industry in-house R&D is the weak link in the system (Hayter 1982a, 1988).

As Nelson (1988) argues, diversity, a key feature of R&D organization in capitalist countries, has a twofold rationale. First, diversity ensures competition in the how, why, and when of technology development. Although at times wasteful and inefficient, as failures before and after commercialization attest, the offsetting benefits of technological competition relate to the inherent technical and economic uncertainty of the innovation process. As Nelson (1988) notes, the presence of uncertainty undermines potential agreement on the direction of technological change. Indeed, for Nelson (1988), the economically superior performance of capitalist economies, compared to centrally planned economies, results from the extensive privatization of technology, the multiple and often rival sources of new technology, and reliance on market forces to determine success.

Second, diverse R&D organizations fulfil complementary roles in the R&D process. Thus, public- and private-sector R&D reflect the relative merits and demerits of the "public and private facets of technology" (Nelson 1988:

314). Private control of R&D is important partly for its competitive aspects and partly because it links the profit motive, and therefore market opportunities, to innovation. In turn, such privatization depends on the ability of individual firms to appropriate a sufficient proportion of the returns to investment from R&D. Publicly available technology offsets the disadvantages of private-sector R&D in that widespread access to generic knowledge broadens the understanding of technical possibilities. On the other hand, public-sector technology cannot duplicate the advantages of private-sector technology The thrust of in-house R&D is to promote the competitive advantage of individual firms in relation to other firms (Cohen and Mowery 1984). Public-sector R&D is not so well informed nor so biased.

In effect, within the BC forest economy, federal government R&D, cooperative R&D, and now provincial government and university R&D represented successive attempts to not only supply the "public facets" of R&D but to compensate for the missing or limited "private facets" of R&D. Cooperative R&D helped address the criticism that federal laboratories failed to transfer technology to the private sector. But cooperative R&D does not duplicate in-house R&D. Nor does Forest Renewal BC substitute for in-house R&D. The policies to promote federal, cooperative, and provincial government R&D have all failed to appreciate the specific and unique role of in-house R&D.

The various benefits of in-house R&D cannot be derived easily, if at all, from other forms of R&D. Government, university, association, and cooperative laboratories generally have no mandate to supply firm-specific product innovations. While government and university R&D give priority to basic R&D and to extending publicly available knowledge (although military R&D is an important exception to this observation), association and cooperative laboratories are mandated to serve the collective interests of their members, a role that usually implies a focus on industrywide processing problems. Even in the context of processing technology, cooperative R&D may not be able to respond to technological needs when required by a member firm; and external technology suppliers may not be readily available either.

In-house R&D is viable in BC, as MB's thirty-year-plus experience attests, and it plays a distinct role in the overall R&D system (Figure 11.2). Thus, MB's in-house R&D provided firm-specific advantages, most significantly market advantages derived from product innovation, but also process innovations specific to its operations and resource endowments. In addition, the availability of in-house R&D provided MB with advantages for the timing of innovation and the transfer of technology to particular plants (Hayter 1988). MB's in-house R&D generally provided a window on the world of science and technology, and it engaged in technological liaisons with a variety of different organizations, including collaboration with university scientists and other in-house R&D groups. Such collaboration requires that

Figure 11.2

Technological liaisons and in-house R&D: The case of a forest-product firm

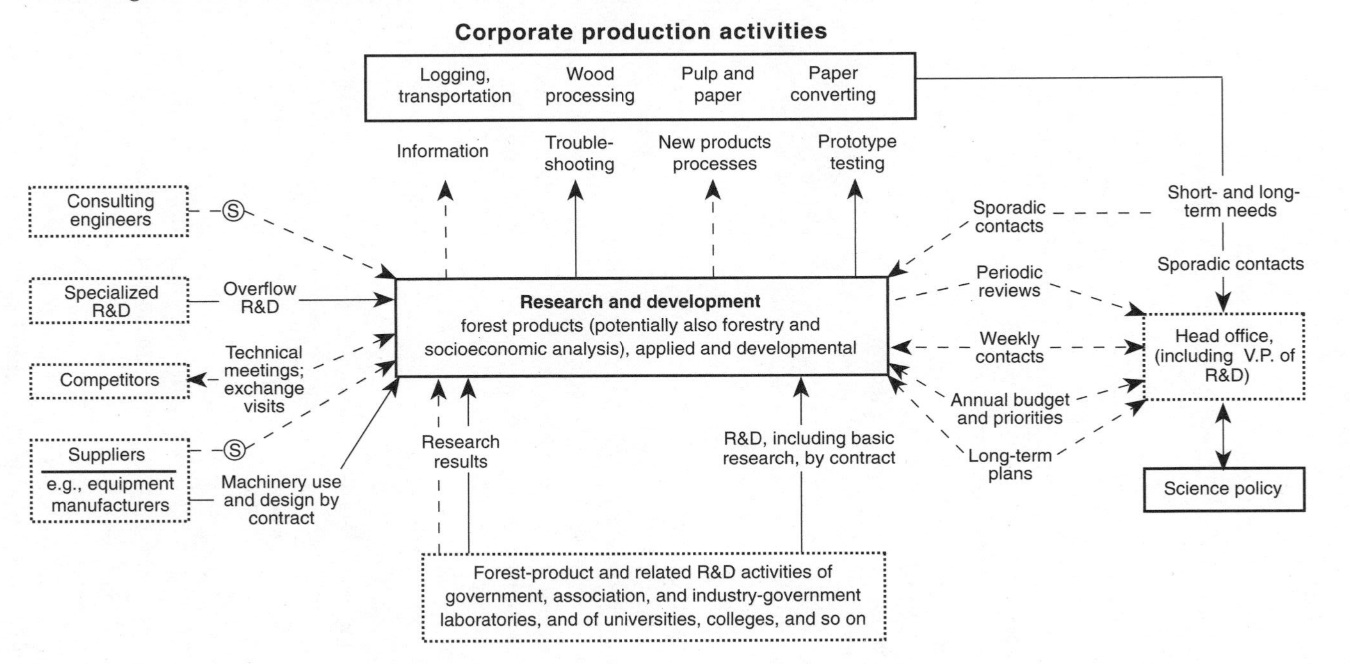

Source: Hayter 1988.

firms understand the nature of the liaison and contribute to it. In this respect, in-house groups can contribute to a firm's knowledge base and its bargaining power.

The presence of in-house R&D groups potentially places sophisticated demands on the rest of the R&D system. Alternatively, in the absence of in-house R&D, industry is likely to make less sophisticated demands on the rest of the R&D system. In other advanced forest-product regions, in-house groups are the cornerstone of the R&D system. In BC, especially with MB's R&D group now closed, the R&D system is based on relationships between the public sector and cooperative organizations that conduct R&D and many companies that do not. In practice, the collective, accumulated impact of public-sector R&D initiatives has been primarily to sustain the private sector's dependent or imitative technology strategies and adaptive technological capability.

Conservatism of Technology Strategies

Technology strategies define how firms access and use technology to meet long-run goals. In Freeman's (1974) terms, these strategies may be "traditional," "dependent," "imitative," "defensive," or "offensive." Traditional firms rely on craft skills and are essentially noninnovative. Dependent firms do not attempt to initiate technical or product change, except at the specific request of customers or parent companies, which may provide the expertise necessary. Imitative strategies involve firms in adaptive R&D that copies industry leaders and serves, in the first instance, to respond to local situations. Defensive and offensive technology strategies depend on in-house R&D to establish global head starts in new product markets or new sources of efficiency (offensive firms) or to respond quickly to global leaders in a unique or differentiated way (defensive firms).

In BC's forest economy, the overwhelming majority of manufacturing firms, forest-product companies, and equipment suppliers have adopted dependent and imitative strategies. Among large corporations, only MB has pursued a defensive/offensive technology strategy; the closure of its R&D facility indicates a choice to more closely follow the industry norm of an imitative strategy. It is hard to think of a single equipment supplier that has sustained a position of global technology leader. Foreign-owned companies have enjoyed some discretion in making technology choices, and they have supported adaptive technological behaviour. But, with one temporary exception (Chapter 5), they have relied on parent companies for in-house R&D. Indeed, foreign subsidiaries have collectively *not* been strongly associated with maintaining state-of-the-art or best-practice technology. Even within the context of imitative behaviour, and in contrast to what is often supposed as a benefit of foreign direct investment, foreign firms are not known for providing technological benchmarking for the rest of the industry.

An important implication for BC firms arising from relying on imitative/defensive strategies is that they imply continuous catching up and a commitment to a cyclical commodity mix. By the early 1980s, however, catching up had become more difficult because of increasing capital intensity and competition. The federal government's pulp-and-paper modernization program, and its $613 million worth of subsidies (1980 dollars), however, excluded BC and focused on Ontario and Quebec (Anderson and Bonsar 1985). Without the technological capability of global leaders, BC's forest firms have been squeezed by the catch-up strategies of competing lower-cost regions. Indeed, by the late 1990s, "most primary mills are no longer on the leading edge of advanced manufacturing and systems technologies ... [and] productivity gains from technologies currently in place have either peaked and[/or] are not always fully achieved" (Ernst and Young 1998: B-21). These productivity problems stem from a preoccupation with cost rather than value, a misplaced priority for a high-cost region.

Proven Equipment and Local Adaptation

At the heart of the conservative approach to innovation – the emphasis on dependent and imitative strategies – is the use of proven equipment, modified or adapted as needed to fit local circumstances (Hayter 1988). Thus, in woodlands, wood-processing and pulp-and-paper investment choices in new equipment almost invariably emphasize proven equipment. Innovation of global-first machinery is rare. Before choosing equipment, BC forest firms, and Canadian forest firms, want to see it in operation. This decision may involve sending teams of people around the world to visit mills. Close consultation with competing equipment suppliers regarding possible machinery choices is also an established practice, but typically the machines are already in operation somewhere. There is very little tradition throughout BC (or Canada) of forest firms working with equipment suppliers in an experimental way, thereby helping suppliers face the costs and uncertainties of long-term R&D. In Scandinavia, on the other hand, forest firms and suppliers typically cooperate at the R&D stage.

At Port Alberni, MB's NexGen (Pacifica) project involved collaboration between MB's R&D group and Valmet, a Finnish paper-machinery manufacturer. However, MB's technology strategy was exceptional, and an innovation solution was deemed essential to the survival of the paper plant. A local BC-based paper-machinery supplier does not exist, and Valmet's cooperation also implied importing a Valmet paper machine. Indeed, it is not unusual for BC mills to modernize by importing significant items of machinery. With some exceptions, BC's forest-product firms are typically unwilling to be the first users of new technology, and they are reticent about sharing the R&D costs of developing new technology with equipment suppliers, committing themselves to buying technology that is in development, or allowing equipment manufacturers to use their factories for experimental purposes. Forest firms in BC prefer to deal with equipment suppliers at arm's length to ensure competitive prices for established technology. No preference is given to local suppliers.

In Scandinavia, in contrast, purchase of local technology is given high priority and normally comprises 90 percent of the equipment purchases in mill modernization schemes. Moreover, collaborative and long-term relations between forest firms and equipment suppliers typically include R&D and the transfer of technology. Both users and producers of Scandinavian forest-product technology have strong commitments to R&D as the basis for offensive technology strategies that work to mutual benefit.

A conservative approach to technology choice is deeply rooted in BC's industries, typically justified as a cost- and risk-reducing strategy. Industry argues that it faces considerable costs and uncertainties in building facilities on the geographic margin of production that will serve distant markets over which it has little or no control. In these circumstances, proven technology is an important source of security, regardless of the technology source. Industry, until 1998 at least, has consistently supported free trade and has long lobbied for the reduction of tariffs to facilitate both its exports and imports of technology. As technology itself has become the basis for competitive advantage, however, the legacy of conservatism has become more problematic.

Technological capability in BC's forest industry, although evolving around the reliance on proven equipment, should not be underestimated. The transfer of technology, including proven equipment, invariably implies some adaptation to meet the needs of local circumstances related to distinct resource and site conditions as well as firm-specific and factory-specific needs and constraints. Moreover, technology has to be integrated. Modern forest-product mills contain thousands of pieces of equipment, and new mills mean all these technologies have to be operational at the same time. Modernization schemes pose the problem of linking new and old technologies. Technology also has to be maintained and repaired if productivity potentials are to be realized. Indeed, the level of engineering know-how in the BC forest industry is deep, in terms of a practical understanding of how large-scale, complex technology works. This engineering know-how is strengthened by a highly experienced workforce. Yet, this engineering and operating know-how supports processing efficiency and technological conservatism rather than product development and technological innovation. As backward linkages to the forest industry, local equipment suppliers and independent engineering consultants serve a similar purpose.

Forest Equipment Manufacturers within BC

The size and structure of the forest equipment supply (or capital-goods) industry reflects its role in adapting proven technology for forest firms. The industry primarily includes logging equipment and wood-processing equipment firms; there has been little development of pulp-and-paper mill machinery suppliers.

A major contributing cause to the development of logging and wood-processing machinery suppliers relates to the distinctive nature of conditions in BC. According to a provincial government publication, probably published in the early 1970s, "economics and the basic characteristics of the rough terrain [in BC] have forced provincial manufacturers to be particularly innovative in design. The result is a highly sophisticated industry with a large degree of automation" (Bedford, n.d.). As this document intimates, the technological strength of equipment suppliers in BC rests on design capability, not in its long-term research capability. Typically, design work has been the responsibility of owner managers, perhaps supported by engineers and technicians. Formal commitment to ongoing R&D programs is minimal, and the sophistication of the industry has been less than leading-edge international competitors in Scandinavia, other European countries, and the US.

In effect, BC's physical geography – variations in climate, landforms, soil, and vegetation – "protected" local equipment suppliers from foreign competition. Examples of imported technology in wood processing and logging that failed because of the technology's ineffectiveness in Canadian conditions further illustrate this point (Hayter 1988; Silversides 1984). In the case of logging, Silversides (1984) has characterized Canadian efforts as a "mousetrap syndrome" – a myriad of local solutions developed for broadly similar problems. The BC forest equipment supply industry comprises such a population of SMEs that are oriented mainly around domestic and American markets. The biggest firms, which probably reached peak size in the 1970s, have rarely employed more than 500 employees. In the 1990s, it is unlikely that any supplier has maintained this level of employment.

On the other hand, pulp-and-paper technology is scientifically based and depends on formal R&D programs. For the most part, pulp-and-paper machinery developed elsewhere can be imported by BC's forest firms and adjusted on-site to meet local circumstances, such as those defined by species mix or specific plant configurations.

Insights into the technological capability of local equipment manufacturers in BC is provided by a survey of thirty-four firms based in the Vancouver area around 1980 (Hayter 1980). These firms produce a range of equipment, not always exclusively for the forest industry, and while they range in the number of employees, they are not very big. Nine of the plants were controlled from outside the province, including seven foreign-controlled branch plants. About an equal number of locally owned plants and branch plants employed twenty-five to 499 workers, although most of the former employed fewer than twenty-five people and branch plants were more likely to be bigger. None of the plants had (1979) sales in excess of $20 million.

The imitative and defensive technology strategies employed by these forest equipment suppliers are directly reflected in their limited and sporadic R&D efforts. Investments in product design are made as needed to ensure the profitability of an established product line. Most of the thirty-four plants surveyed claimed to have design capability and half claimed R&D functions. Half the sample noted that at least three-quarters of their sales derived from locally developed products. For the remaining sales, these firms were more or less equally likely to adapt foreign products as they were to sell entirely foreign-developed technology. All but two plants recognized the role of some local input in product development. Yet, these efforts were extremely limited. Few firms consistently funded R&D or maintained specialized design offices, and only eight of the plants had taken out patents.

Generally, design work involves the practical application of known principles involving few employees who specialize in product development or adaptation. Thus, the thirty-two firms that provided information on this question employed only two scientists, forty-one engineers, and seventy-seven technicians. Admittedly, CanCar, then one of the largest BC forest equipment manufacturers, is not included in this sample. However, given that this well-known (foreign-owned) manufacturer of sawmill equipment employed one scientist and seven engineers plus sixty-nine technicians in 1977, the overall picture is not distorted by its exclusion. Furthermore, the R&D of these suppliers was performed only occasionally and was limited in duration; projects exceeding six months were rare (Hayter 1988).

In the 1990s, the size of the forest equipment industry has declined. Almost half the firms surveyed in the 1980s have closed down, as did CanCar, and in some cases branch-plant operations have been consolidated outside the province. The recession of the 1980s, NAFTA, and declines in the number of wood-processing mills in BC have contributed to this restructuring. Important local equipment suppliers still exist, such as Newnes and Madill, but the collective R&D efforts of these manufacturers are apparently so small that Binkley and Forgacs's (1997) recent inventory of R&D budgets in the Canadian forest sector omitted them entirely. Yet, capital expenditure over the years has revealed a massive demand for technology in the BC forest economy. The small and fragmented nature of the equipment-supply industry in BC creates a substantial loss in terms of jobs, value-added products, R&D, and innovation.

Consulting Engineers

After 1950, directly in response to the forest-product boom in kraft pulp, a consulting engineering community grew rapidly in BC, led by H.A. Simons and H. Sandwells Ltd. The explosion of new mills (and their subsequent expansions) created a substantial demand for engineering expertise to help choose, design, install, and adapt technology at specific sites. Given the sporadic nature of demands for this expertise, most firms chose to contract independent engineering consultants as needed. Engineering consultants do not regard their activities as comprising R&D, although firms such as Beak and Envirocom had laboratory facilities, in both cases to help with environmental impact analyses; Reid Collins owned a nursery and a computer-mapping facility; and all firms had design offices. These firms were all directly involved in technology transfer, and many exported their expertise. However, internationalization has not been supportive of local equipment manufacturers. Indeed, the consultants have valued their independence – their arm's-length relations with equipment suppliers – as an asset in winning contracts.

Precise estimates of the size and importance of this community are difficult to make because the industry is largely privately owned and statistics are not collected on a regular basis. In addition, consulting engineers are rarely confined to one industry. Even so, by the mid-1970s, Vancouver was recognized as a specialist centre in Canada for engineering consulting. By 1974, at least 280 consulting engineering firms operated in BC, employing more than 6,000 people and with billings in excess of $180 million (Hayter 1980). Simons and Sandwell, which went public in 1977 (with billings of $39 million), effectively acted as core companies, principally by acquiring in whole or in part smaller companies and by subcontracting. Since the early 1980s, the importance of the forest sector to the engineering consulting community in BC has probably declined.

Privatization and the Confirmation of Adaptive R&D

During the heyday of Fordism, relationships between federal forestry R&D and private companies were distant and ineffectual in transferring technology (Smith and Lessard 1970). Cooperative R&D has led to closer ties, although R&D for the most part involves technology transfer and some developmental work. Both FERIC and Forintek are committed to meeting industry's needs, as established by their boards, and their activities are consistent with industry's preference for imitative and dependent (or adaptive) technology strategies.

Thus, FERIC is almost exclusively concerned with technology transfer and acting as a technology catalyst by promoting best-practice technology wherever it originates (Hayter 1988: 62-5). FERIC's approach is hands-on; it prefers to visit wood-harvesting sites and identify technological needs in specific circumstances. By suggestion, collaboration, and sometimes financial support, FERIC acts as a catalyst to innovation. An example of FERIC's role is provided by the development and application of high flotation ("wide") tires, a project needed to reduce the environmental damage of logging (Hayter 1988: 63-4). In a program that lasted about six years, FERIC identified the best approach for developing wide tires for logging machines in Canadian conditions, organized meetings among key parties (tire manufacturers, equipment manufacturers, and logging development engineers), purchased a set of tires that had potential to be developed, provided advice, and orchestrated field trials and performance tests.

In Forintek, activities are strongly oriented to providing a technological bridge between forest manager and manufacturers through developmental R&D and technology transfer (Hayter 1988: 65-6). In addition to research, Forintek is heavily involved in codes and standards. Like FERIC, Forintek urges the use of best-practice foreign technology wherever possible. For example, Forintek played an important role in encouraging stellite tipping, a process pioneered in Germany, in BC sawmills beginning in 1982. Another program developed by Forintek was the sawmill improvement program (SIP), designed to allow sawmills to fully realize productivity potentials. Although Forintek will invest in long-term research projects, in some cases (for example, biotechnology) even to do basic research, its overall orientation reflects the immediate technological needs of wood-processing mills.

Historically, Montreal-based Paprican focused on long-term R&D and patenting processes, including the paper machine (Hayter 1988: 57-62). Since the 1980s, Paprican has become more involved in technology transfer. In 1978, Paprican had established a presence at UBC by agreeing to collaborate in postgraduate research programs at the university, with an initial focus on chemical engineering. This liaison with UBC subsequently expanded when the UBC Pulp and Paper Centre was officially opened in 1986, the funding for which has come from Paprican and the provincial and federal governments.

Forest Renewal BC: Shotgun Research

Recent attempts to stimulate R&D and innovation in BC's forest economy have been promoted by the provincial government. In particular, Forest Renewal BC has become the main source of funds for forestry research in BC, $36.6 million in 1996/7 (Binkley and Watts 1999). In practice, funded projects are extremely wide ranging and organized according to five themes (land and resources, environment, value added, communities, and workforce). In these areas, research goals are broadly defined: what is meant by "research" ranges from the funding of endowed chairs at universities to promote basic research to small-scale inventories of community characteristics. In addition, funding is based on proposals that are initiated by a myriad of researchers rather than in response to a clearly focused R&D plan.

Forest Renewal BC has supported academic, government, consultant, industry research, and related information-gathering activities by a wide range of individuals and organizations throughout the province. It has also contributed to a wide variety of community development projects, such as new trails, park facilities, and the McLean Mill historical monument in Port Alberni. It has become octopus-like in its reach on developments in BC so that it is hard to disentangle research from other activities. But funding has also proven to be sporadic. The government almost transferred the funds to general revenues before Forest Renewal started; its research budget for 1998 has been sharply reduced; and $8.5 million of funding for some long-established silvicultural research programs within the Ministry of Forests and at UBC was withdrawn (Binkley and Watts 1999).

In organization, mandate, and activity, Forest Renewal BC differs from any other research organization in the BC forest industry. Its shotgun approach to research will undoubtedly generate beneficial results. At the same time, Forest Renewal BC's research mandate is cloudy and its long-term impact on the industry's ability to innovate is hard to assess. Forest Renewal BC again raises questions about how well BC and Canada are solving the "institutional assignment" problem for R&D in the forest sector.

Weak In-house R&D: The Ultimate Expression of the Staple Trap

Historically, forest-product firms in BC have been diffident about investing in in-house R&D. One common argument is that there is no incentive to invest in in-house R&D because the industry is "open" with respect to technology and most new technology can be readily purchased from equipment suppliers, especially foreign suppliers (Hayter, 1982b). Other common views stress the lack of incentive to invest in R&D during booms and the lack of funds during busts. Most firms, until recently at least, have regarded themselves as commodity producers with little need to invest in technology. (How do you improve a two-by-four?) Given that competitive advantage was firmly based in nature's bounty, the implicit view of in-house R&D is as an unnecessary cost whose benefits are unlikely to be appropriated.

Antagonistic attitudes towards in-house R&D are deeply entrenched. Such attitudes have been shaped by high dependence on commodities and exports, and location in an "empty land" on the geographic margin. Commodities narrowed industry's concern to cost minimization rather than value maximization, the search for exports was seen as challenge enough without the additional uncertainties of developing technology, and a free trade mentality reinforced the "common sense" of exchanging resources for imported expertise. Over time, these attitudes have been institutionalized by high levels of foreign ownership and tariffs. Thus, the tariff policies of BC's major markets in the US, Europe, and Japan traditionally imposed limited or no barriers on commodity exports but much higher tariffs on higher-value products. Vestiges of these policies are still in effect. High levels of foreign ownership served both to reinforce a commitment to commodity production and to preempt and even replace indigenous R&D (Chapter 5). Virtually all major foreign firms in BC have invested in R&D in their home countries but not in BC. Invariably, subsidiary companies pay charges to maintain parent company R&D, but they have no mandate for R&D themselves. For foreign firms, government incentives for R&D are virtually irrelevant.

The president of Weyerhaeuser Canada recently stated that forest-product companies in Canada are not big enough to conduct R&D. There is some substance to this view (Hayter 1982a). R&D is dominated by MNCs in the forest industry, as it is in many others. Yet, Weyerhaeuser Canada is a large subsidiary of a giant parent company, Weyerhaeuser of Tacoma Weyerhaeuser (US), which has major R&D facilities in Washington but not in BC. A high level of foreign ownership means many foreign subsidiaries have no interest or ability to invest in R&D in BC, and their presence limits the potential of local firms to reach a sufficient size to justify R&D. It is not coincidental that the two most important in-house R&D programs in BC were maintained by MB and Canfor, two BC-based firms.

The fact that MB and Canfor have recently closed their R&D centres suggests foreign ownership (and corporate size) is not the sole factor affecting levels of in-house investment. Indeed, the effect of traditional anti-R&D attitudes, foreign ownership, commodity dependency, and tariff structures on in-house R&D investments has been compounded by a market crisis and uncertainty about timber supply in BC. Such uncertainty is not conducive to encouraging the long-term thinking necessary for R&D. Because technological know-how is cumulative, however, it becomes harder to develop new R&D-based technology strategies.

The lack of in-house R&D compromises the industry's ability to reduce its commodity reliance and diversify into faster-growing value-added market segments. Indeed, the open-to-technology-transfer thesis begs the question why so many foreign firms invest so much in R&D. The thesis vastly oversimplifies the situation by ignoring the costs, timing, and uncertainties involved in technology transfer, and overlooking the competitive advantages of lower costs of production and market diversification and penetration that can be achieved by aggressive technology strategies. In fact, deficiencies in funding in-house R&D has meant that its potential impact on the rest of the R&D system (for example, through various kinds of technological exchanges or liaisons) is very much reduced. Indeed, the forest-product R&D system in Canada is fragmented, and cooperation among organizations is more limited compared to Sweden, Finland, and the US.

The costs and missed opportunities stemming from a low level of private-sector R&D are substantial. First, high-income jobs directly associated with R&D are lost. Second, imported technology in the form of equipment, licences, and R&D services from parent corporations contribute to deficits in both visible and invisible trade. For equipment and machinery manufacture, it is well established that in-house R&D is important for export success and the growth of firms; the meagre levels of R&D by BC firms in this regard are contributing to the firms' disappointing and declining export performance (Hanel 1985; Hayter 1981). Canada as a whole has a balance of payments deficit in forest-product equipment trade.

Third, because of the continued emphasis on proven technology to manufacture a narrow range of standardized commodities, BC's forest-product firms have lost market share and position to their competitors. Worldwide, Canada's share of the value of forest-product export markets has declined from 25.7 percent in 1961 to 19.4 percent in 1989. In contrast, the US increased its share of forest-product exports from 8.7 to 13.1 percent during the same period. Particularly in paper products, the US, Finland, and Sweden have invested more strongly in R&D and have been more innovative in shifting towards new and faster-growing market segments (Croon 1988: 271). In contrast, major BC firms have been slower to shift from low- and medium-value commodities such as kraft pulp and standard newsprint (Woodbridge 1988: 83). The exceptions to this generalization simply confirm that more can be done to develop technological leadership.

The staple thesis prediction of an underinvestment in in-house R&D has unfortunately become increasingly evident in the BC forest economy, and an important stimulus for innovation has virtually disappeared. Cooperative, association, government, and university R&D have compensated to a degree for missing in-house laboratories, but their mandates limit their interest in firm-specific product-market innovations. SMEs also contribute to innovation without investing in ongoing R&D programs, although historically many factors have dampened the size of the SME population in BC. It is questionable whether the speed and extent of the transition to value-added activities by BC's forest industries are sufficient to realize available job and income potentials.

Towards a Smart Forest Sector: Innovation Imperatives

Since Smith and Lessard (1970), many studies have argued that the forest industry in Canada needs to be more innovative. In BC, this plea is the theme of consultant reports (Woodbridge, Reed 1984; Ernst and Young 1998) and numerous academic studies (Hayter 1987, 1988; Binkley 1993, 1995, 1997a; Binkley and Watts 1992, 1998; Binkley and Forgacs 1997). These studies have called for much greater levels of investment in R&D, an argument consistent with a general literature that emphasizes the central role of innovation in economic development, and the close ties between R&D and innovation (Freeman 1987). While caveats can be offered, variations in patterns of innovation among firms, industries, or regions, are closely and positively paralleled by variations in patterns of R&D.

Yet, the forest industry is unimpressed by this evidence, and the global imperative it implies. Or if industry recognizes the need for R&D and innovation, it seems unable to effect a solution. The industry's situation in relation to innovation has not changed much since the 1980s and may have worsened. Increases in overall R&D budgets in BC largely stem from the ambiguous, wide-ranging activities of Forest Renewal BC. If in-house R&D is seen as the core of the forest-product innovation system (Hayter 1988), the situation has become desperate, because no important company laboratory remains in the province.

An Innovation Culture: The Celebration of Smartness

To use Ernst and Young's (1998) phrase, the BC forest sector needs to develop an "innovation culture." Such a culture may be broadly defined as a population of firms that seek to continuously innovate as a way of life, pursue higher value as a production philosophy, and aspire to global leadership in product development. From this perspective, innovation is not simply about major leaps forward, whether thought as large-scale machinery or new products that stem from formal R&D centres. More generally, an innovation culture implies a holistic commitment to innovation by all groups in the industry – workers, managers, and scientists – in support of innovative activities that include the smallest incremental steps as well as major developments. Alternatively put, an innovation culture depends on smart firms, smart factories, smart production systems, smart workers, and, ultimately, a smart forest policy.

There is no shortage of specific, thoughtful recommendations for stimulating an innovation culture for BC's forest sector (Binkley 1995, 1997a; Ernst and Young 1998). The remainder of this chapter, therefore, advocates elevating innovation to the dominant theme in forest policy and forest utilization.

Towards an Innovation-Based Forest Policy

Remarkably, the basic policy documents for forest policy in BC since the 1940s have given little attention to R&D and innovation. From the Sloan Royal Commission of the 1940s and 1950s to the Pearse Commission of the 1970s, the Peel Report of the 1990s, and the high-stumpage regime (see Table 3.10), the theme of innovation appears tangential. In the most important public policy discussions about BC's forests, R&D and innovation are sideshows. Thus, the high-stumpage regime explicitly focuses on the level and calculation of stumpage and environmental considerations. Admittedly, value added is the key ("magic wand") to the effective realization of both the economic and the environmental goals of this battery of policies. But value is in turn created by innovation. Yet, incentives for innovation by the private sector are secondary to this reregulation of BC's forests. Big increases in stumpage may just as easily undermine as promote innovative activities.

A start to the promotion of an innovation culture in BC's forest economy is to give priority to innovation in broad statements of forest policy. The casual treatment of R&D and innovation evident in all Royal Commissions since Sloan needs to be reversed: innovation must be established as a central theme of forest policy, especially as it relates to industry. Forest policy needs to stimulate and reward smart factories and firms.

Given a commitment to innovation, one mechanism to ensure its implementation is to assign timber rights according to the innovativeness of proposals and the implication for industrial investment to promote technological capability. Big corporations with big tenures should be expected to make big contributions to innovation in BC by means of their own manufacturing and marketing operations, stronger relationships with equipment suppliers than in the past, and R&D investments. Small firms can develop innovative proposals for the use of specific sites and tree species to serve specific market niches. Alliances among firms and other interest groups, including Aboriginal peoples, could be fostered around imaginative ideas. In general, efforts should be made to promote the technological strengths of equipment suppliers in BC, presently a fragmented group of small firms. If the provincial government can subsidize fast-ferry technology, support for forest equipment manufacture makes more sense. Local demand for forest-product technology is massive.

Stability and Flexibility

Stability of forest ownership and regulation and flexibility are vital requirements for an innovative forest policy. Achievement of these apparently paradoxical features constitutes a major challenge for policy, not rendered easy by the fact that BC's forests are already allocated in some form.

Stability is essential to justify long-term attitudes in the forest industry and investments in uncertain projects, programs, and training schemes. In the late 1990s, a downturn in investment in BC's forest industries was not helped by uncertainties over forest policy. Yet, modernization and innovation depends on investment. Without stable terms of reference for wood supply, even investment in worker training is questionable. The government needs to ensure that the long-term goals, basic approaches to resource allocation, and the rules of the game for forest policy are transparent. At the same time, forest policy has to be flexible in realizing the diverse values of the forest and in rewarding innovators. Whether innovative firms and factories are rewarded by tax breaks, lower stumpage, and more timber, or some other incentive, may be debated. Innovators at least should not be penalized by stumpage and regulatory policies, which happens under the present high-stumpage regime. Moreover, if the results from an American survey of manufacturing firms apply in BC, and the a priori reasoning is similar, firms that are technologically innovative are also likely to be innovative in labour relations and environmental practice (Florida 1996). Thus, support for innovators might be expected to have widespread social benefits.

As an overall framework within which to promote stability and flexibility, Drushka's (1993) argument to diversify tenure is important. Privatizing one-third of BC's working forests primarily for large-scale corporate forestry and dividing the remainder among various forms of community ownership and provincial ownership would achieve an overall framework to enhance forest use flexibility by diversifying ownership structures and therefore the motivations and ideas for utilization. If the provincial government defined a clear plan to allocate forests along these lines, an important requirement for overall stability in forest allocation would also be met. Such a plan, of course, presupposes that forest values are defined, and wood-fibre supplies and appropriate management regimes are clear.

In BC, tenure diversification generally implies a shift in timber allocation from corporations to other organizations, notably communities, various types of SMEs, and Aboriginal bands. Privatization generally implies the conversion of Crown forests to some form of privately held forests and/or the creation of open markets for raw logs. If implemented on a substantive scale, diversification and privatization require massive changes in provincial forest policy that presuppose an overall plan and principles of adjudication.

Leaving aside the question of practicality, both diversification and privatization are potentially important to an innovation-based forest policy. Thus, as a general idea, tenure diversification has intrinsic merit from an innovation perspective because more varied ownership implies more varied use of the forest as well as the application and generation of more varied knowledge. It would increase the number of decision makers – potential innovators – influencing the direction of the forest economy. The benefits of privatization for innovation are ultimately rooted in providing strong incentives to individuals or organizations to invest and innovate. If forest users were also forest owners, they would be more willing to invest and innovate because they would be able to gain from their efforts.

Moreover, if diversification and privatization of tenure are implemented together along the lines of Drushka's plan, then potential conflicts between the two are mitigated. In particular, if privatization is implemented without the broad constraints of a diversified tenure structure, privatization could just as easily be associated with concentration of tenure. Moreover, if privatization means more log exports, innovation potentials will be reduced. In addition, if privatization is associated with higher levels of foreign ownership, R&D and innovation are unlikely to result. Private-forest owners may be better forest managers, but not invariably. Privatization might therefore be restricted to Canadian owners or at least largely to Canadian owners.

Profile of a Smart Forest-Product Sector

It is unlikely that the achievement of an innovation culture in the BC forest sector can now result from the stimulation of in-house R&D. Industry is impressively reluctant to commit to in-house R&D. Unfortunately, cooperative and Forest Renewal BC's shotgun approach does not provide an adequate substitute. If R&D and innovation are profoundly related, this linkage has never been appreciated in BC's forest economy, especially in the sector that counts – the private sector. An alternative perspective to promoting innovation has to be contemplated.

The starting point for an alternative approach is recognizing that an innovation culture is a holistic experience and that the link between R&D and innovation is not a linear process beginning with basic R&D and ending with technology transfer and manufacturing (Britton 1991). Innovation can be incremental and originate on the factory floor, where innovation is shaped by factors such as work organization, skill levels, ability to translate new information into practical results, and an understanding of consumer needs. From this perspective, the key operating unit of a smart forest sector is the smart factory. In turn, smart factories require smart firms that develop explicit technology strategies that provide long-run guidelines defining competitiveness in a high-cost environment. Smart factories also employ smart unions. But there is no single type (or size) of smart factory and firm; smart production systems are predicated on facilitating diverse organizational forms.

Smart Factories (and Harvesting Operations)

Smart factories are places where all employees think. In smart factories, job demarcation is not significant, decision-making hierarchies are flatter, jobs are self-governing, training is ongoing, grievances are rare since cooperation dominates, and there is general knowledge about factory operations and how factories relate to systems of supply and markets. Within smart factories, there is a constant circulation of information about costs, product design, competitive developments, and consumer needs and problems. Smart factories are relatively autonomous in terms of planning, design, engineering know-how, marketing, and procurement. That is, smart factories understand their customers' needs and their supply chain; they are places for product and process design and experimentation, as well as for production. Smart factories can solve many, if not all, of their own production problems. At the same time, they understand the need to focus skills in product-market areas and, within these niches, to fully develop economies of scope and economies of scale.

In BC's forest economy, smart factories set best-practice standards for environmental behaviour, as well as in work organization and technology choice. Thus, they constantly seek ways to ensure that the flow and processing of materials minimize environmental implications. Similarly, smart harvesting operations execute and plan logging and forestry activities by internalizing environmental accountability as part of operational routines. Moreover, smart factories (and harvesting operations) provide opportunities for liaison with R&D professionals – maybe even the beginnings of full-fledged R&D operations (which is where many stand-alone R&D laboratories originated) – and a basis for a collectively stronger, more focused R&D effort within the private sector.

Thus, smart factories seek to become as flexible as possible in creating value and responding to change. As such, they are completely different from their Fordist counterparts which are organized as large-volume producers of simple commodities made by a highly stratified, tightly controlled, Taylorist workforce. Skill was never actually eliminated by Taylorism, but that was always the underlying intent. Smart operations, on the other hand, celebrate learning; similarly the firm of the future is a "learning organization" (Streeck 1989).

Smart Firms

Smart firms are high-performance organizations that seek to realize the intellectual potentials of all their employees in support of continuous policies of innovation – incremental and radical – as the basis for competitive strategy. Smart firms ensure that their operations are best-practice, properly supported and coordinated. Within smart firms, boundaries are often blurred and "permeable," to facilitate knowledge flow, mutual understanding, and feedback (Fruin 1992). In BC's forest economy, smart firms develop long-run strategies that combine market leadership with environmental leadership. Moreover, these goals set standards for the development of supply links beyond the firm, especially with respect to the suppliers of materials and equipment. Indeed, smart firms are not simply exemplars but lobbying forces, emphasizing the need for the BC forest economy as a whole to become an innovation culture.

Smart firms develop explicit technology (defensive and offensive) strategies that promote a continuing commitment to innovation as a competitive strategy. Smart firms in BC's high-cost environment aspire to product leadership based on differentiation and value. In Porter's (1985) terms, smart firms recognize that value maximization is not the same as cost minimization and that in high-cost environments the former rather than the latter is the preferred strategy. From this perspective, priority is given to market focus and value, and then cost minimization is considered, rather than the other way around. As Porter warns, it is hard for most firms to root their competitive advantage in cost as well as value leadership.

Smart firms, at least large and giant firms, typically engage in in-house R&D. Indeed, in-house R&D is perhaps the best single indicator of smartness and its absence limits options by which technology can be used as a competitive weapon in creating market advantage. Aspiring smart firms can still seek technology-based competitive advantages in the absence of in-house R&D, even if these advantages will not be as significant or diverse. These advantages cannot be automatically assumed, however. Thus, the vast amount of scientific and technical information relevant to the forest industries generated in public (for example, universities) and quasi-public (for example, cooperative laboratories) institutions needs to be accessed, evaluated, internally communicated, and acted on in light of each firm's specific operations. Lines of communication have costs, and the nature of communication and evaluation depends on the expertise of the recipients. Smart firms, without in-house R&D, promote the highest levels of engineering expertise and worker skills throughout their operations, encourage (and pay for) multiple lines of communication with outside suppliers of technological know-how, and promote innovation-based flexible operating cultures inside each factory involving teamwork, quality control, and benchmarking. Smart firms are committed to "technological counterpunching skills" and respond as quickly and effectively as possible to new information.

Labour relations agreements, based more on cooperation than on conflict, need to give priority to an innovation culture. As noted (Chapter 8), achieving cooperation is undermined by downsizing, but downsizing should not be a goal itself. Firms need to recognize that the achievement of a smart workforce fundamentally depends on job stability and stable contractual relations, and that smart workers require training and thinking time. Unions need also to establish a similar priority, and to take a proactive rather than passive stance in efforts to promote appropriate forms of job flexibility. The overlapping interests of smart unions and smart firms should further extend to promoting knowledgeable workers whose commitment to quality production meets environmental as well as market obligations.

Smart Production Systems

Smart production systems extol cooperation as well as competition. In smart production systems, firms and factories focus on developing their specialization while contracting out for the services and goods of other smart firms. Smart production systems, therefore, feature high levels of interfirm transactions. Moreover, in smart production systems, these transactions are designed to promote innovative actions. Thus suppliers who offer high-quality inputs and services, unusually innovative products, are rewarded by more business, good prices, and stable relations.

Indeed, smart production systems ensure that indigenous capabilities regarding technology supply are fully developed. Thus, in smart production systems, relations between the users and producers of technology are close; in BC's forest industry this fact implies close relations between forest-product firms and equipment manufacturers. Indeed, a strong, innovative equipment-supply industry would help compensate substantially for a lack of in-house R&D in the forest industry. In smart production systems, transactions between the users and producers of technology are often experimental and uncertain, and not simply about the buying and selling of finished components and machinery. The anticipation is that the short-term costs and uncertainties of such arrangements are more than offset by the gains from innovation.

Even though the forest equipment supply industry has downsized since 1980, opportunities remain to promote its technological capability. As part of the machinery sector, forest equipment suppliers are high tech and because of BC's massive domestic demands, the promotion of this industry makes more sense than some others, such as fast ferries. Forest policy should demand innovative made-in-BC solutions to technological problems as far as possible. Opportunities for BC suppliers may arise to develop leading-edge technologies and related software for solving environmental problems, preferably before the problems occur. Forest policy may also foster links between industry and biotechnology, the latter typically regarded as part of the new industries of the information and communication techno-economic paradigm.

Smart production systems typically have a size distribution of firms in which "giant" firms (and factories), "large" firms (and factories), and SMEs all play distinct but complementary roles. In industries dominated by mature, proven technologies, giant firms tend to dominate because competitive advantage is equated with economies of scale. The BC forest industries during Fordism exemplified this tendency. As the industry restructures to a more dynamic set of technologies and more differentiated set of products, however, specialized large firms become the key to exploiting technological opportunity. In this regard, large firms may be thought of as combining the advantages of SMEs (entrepreneurialism, specialized market focus, and flexible labour relations) with the advantages of the giants (economies of scale and scope in production and marketing). The creation of large firms – firms that are bigger than SMEs but are not giants – reflects the innovativeness of the industry structure. Large firms are created by the growth of SMEs and the restructuring of giants. In the BC forest industry, however, forest policy will have to be flexible enough to cater to these firms, should they emerge.

Forest Policy's Misleading Signals

Forest policy as it has evolved in the 1990s intimates the importance of innovation to the forest industry. Such recognition is to be applauded. From an innovation perspective, however, contemporary forest policy involves a flawed process. The central problem is the failure of government to clearly define overall goals in the allocation of forest tenures. The government has sought to diversify tenure and has sent strong signals of concern about corporate forestry. Indeed, it was the provincial government that initially promoted the idea of timber markets (privatization) because it believed existing tenure arrangements undervalued the resource and protected inefficient corporate behaviour. But the government has never provided overall clarification for diversification or privatization. Moreover, long-awaited decisions regarding community forests were still to be made public in 1999, even though the government was only committed to choosing three community forests, a scale that will scarcely make any difference to forest management in BC. Increased FDI is readily facilitated, despite its negative implications for innovation. At the same time, the government is committed to the treaty process in the absence of a transparent overall tenure reform plan. As a result, the treaty process is adding to the uncertainty in the forest economy, not reducing it.

In the absence of a stable framework of forest allocation allowing firms to assess investments over long time horizons, innovative behaviour, especially involving long-term commitments, is inevitably compromised. Indeed, such uncertainty may have been a contributing factor to MB's decision to close its R&D centre.

From an innovation perspective, another set of important problems has been created by the particular formulation of the high-stumpage regime (see Table 3.10). As a leap of faith, this policy has proven too inflexible to meet the volatile conditions of the 1990s. Indeed, volatility has increased rather than decreased, further eroding the long-term thinking necessary for innovation. Forest Renewal BC was created in part to fund R&D, but this funding was introduced following the stumpage increases and the Forest Practices Code. Because it does not offer a targeted research policy, the outcomes, however beneficial, are likely to be diffuse and in the future. In any case, the fate of Forest Renewal BC's research policy already seems insecure.

Moreover, the provincial government's stumpage policies and related regulations apply indiscriminately to innovators and noninnovators alike. The existing variations are not based on an explicit consideration of innovation. It may even be argued that provincial government stumpage penalizes innovation to the extent that more innovative firms pay higher stumpage by virtue of their greater ability to pay. Certainly, MB has not received any particular favours from the provincial government for its commitment to R&D, and ultimately MB decided it had no obligation to be exceptional. MB's decision to close its R&D centre is a blow to BC and to provincial hopes for an innovation policy. The government should have been far more concerned with the continuance of this research initiative.

The provincial government has sent other puzzling signals regarding innovation and its long-run view of the forest industry. It has stopped funding several long-standing research programs. Just as unfortunate is its decision to subsidize the kraft pulp mill in Prince Rupert, an old commodity mill with long-term problems in attaining profitability. This subsidy, in association with leniency in environmental matters regulations, contrasts sharply with the government's much tougher stance towards MB, the province's leading corporate innovator. If policy rewarded innovators and penalized noninnovators, the government's approach in these cases would have been reversed. The Prince Rupert subsidy sends exactly the wrong kind of signal to the forest industry.

Indeed, a direct implication of the provincial government's intention to promote a value-added sector when fibre supplies are decreasing is that some older mills – specifically those that consume vast quantities of timber to supply a bulk commodity – must close. These closures should be the most unprofitable mills. For communities affected by mill closure, such as Gold River, any continued use of the forest resource should in turn be based on innovative proposals, whether they involve relatively large-scale industrial operations or smaller-scale activities. In the meantime, mill closures reduce pressures on mill supply and provide the provincial government with some flexibility in the allocation of timber rights.

Such misleading signals are disruptive to effective forest policy and contradict the intent of the high-stumpage regime. Instead, forest policy needs to consistently highlight and support the idea of an innovation culture. Moreover, it is worth reiterating that innovation implies institutional, as well as technological, change. Indeed, a central challenge facing an innovation policy for the forest industry is to promote stronger commitments to cooperation (Chapter 12).

Conclusion: Towards a Smart and Green Forest Economy

To survive and prosper, BC's forest economy must be smart and green. If BC is to fully realize economic and environmental values from the forest, a comprehensive commitment to innovation by firms, workers, communities, governments, and environmentalists must be made. Otherwise, BC will continue its trend towards a "technological backwater" (Hamilton 1998f). Support for noninnovative behaviour and penalizing innovative behaviour is counterproductive. A highly unstable business environment may drive out the innovators and leave behind the laggards, with BC losing economically and probably environmentally. Value-added activities depend on innovative firms, and innovative firms are more likely to be environmental leaders, too. Innovation at least offers the hope of a positive-sum game in which BC benefits from industry and its environment. BC, for example, can aspire to be a world leader in eco-certification for its forest firms, environmental technology related to forest producers for its equipment suppliers, environmentally sensitive forest management systems (incorporating ecoforestry and biotechnology potentials), and forest-product and forestry experimentation. More thought must be given to promoting innovation, and forest policy should be developed according to innovation priorities. Only through innovation can the BC forest economy become a global environmental and industrial leader.

12
The BC Forest Economy as a Local Model

> The story is as clear as it is dark.
>
> – Eco 1996: 70

> When you come to a crossroads, take it.
>
> – Advice attributed to Yogi Berra

> However, most of the burden does lie with the provincial government (current and past governments) which must take its responsibility for its mismanagement of the commercial forests. As the forestry ministry itself says, the crisis is unique to B.C.
>
> –*Vancouver Sun* 1999

The story is clear. The BC forest economy is restructuring. In particular, the Fordist model of production that dominated BC's forest economy from the 1940s to the 1970s is transforming into a flexible model of production. More or less in tandem with this industrial transformation, the raw material basis of the forest industries is changing from exploitation of the natural forest to managed second-growth forestry. Mass production is becoming flexible mass production or flexibly specialized as value maximization has replaced cost minimization; giant integrated corporations are becoming more specialized and small firms are rapidly emerging; Taylorized core workers are becoming functionally flexible core workers; commodity trading within the continent is changing to product marketing around the world; the governance of forest towns is shifting from the servility of managerialism to the unruly opportunities of entrepreneurialism; the monopoly of industrial forestry has given way to environmentally enhanced stewardship; innovation is a challenge, but its importance is recognized; and the provincial government has captured the spirit of transformation by replacing a low-stumpage regime with a high-stumpage regime. The prescriptive model of the dynamics of change from Fordism to flexibility is now predictive (see Figure 3.9). Or is it?

The story is dark. The conceptualization of industrial and resource dynamics as dualities, beginning with the overarching transformation from Fordism to flexibility, obscures the richness of reality, encouraging a too linear view of directions of change. Shifts from giant integrated MNCs to tiny SMEs, mass production to flexible production, Taylorized workers to functionally flexible workers, commodity trading to product marketing, or managerial communities to entrepreneurial communities, and so on, define polar cases, not the only cases. The presence of alternative categories – for example, large firms that are neither giants nor SMEs, workers who are skilled and specialized, towns that are neither managerial nor entrepreneurial – as well as the volatility of change, means that the strength of these shifts is hard, if not impossible, to measure. Moreover, industrial and resource dynamics are contested processes, and the very existence of conflicts reveals alternative values and models of the future. If a good case can be made that Fordism is breaking down, the vested interests of Fordism are not giving up.

The story is also dark simply because the BC forest economy is on the geographic margin. The main theories of industrial transformation are rooted in the experience of core countries. As such, they neglect the reality of peripheries as places where cores work out their contradictions by promoting low resource prices and excess capacity, and orchestrating the terms of trade. As Innisians have long known, the diversification and stability of cores typically means specialization and instability in the periphery. The staple trap in which BC finds itself is not an absolute barrier to diversification and stability, but it casts a shadow over attempts to escape it. Moreover, in the forest periphery of BC, industrial dynamics are complicated by resource dynamics in a way that does not happen in cores. Whether cores want extravagant or wiser use of resources, the consequences are simply passed on to the periphery.

The BC forest economy is at a crossroads in industrial and resource dynamics, and a path must be chosen. BC has no choice but to "take the fork in the road" because the path on which it stands is moving. The BC forest economy, however, is not a monolith. Rather, its future is shaped by choices made by firms of varying characteristics, unions, communities, environmentalists, politicians, bureaucrats, and consumers; each group is also internally differentiated. Whether decision makers will collectively make coherent and appropriate choices is not inevitable. The flexibility imperative implies myriad options, not always with clear signposts. Unfortunately, the policies of the provincial government, the central player in the forest economy, are not sending clear signals nor providing a stable framework within which corporations, communities, and individuals can respond coherently in developing cooperative, sustainable forms of local differentiation. From this perspective, there is little practical difference in policies described as experimental from policies that are muddled and ad hoc.

Within the crossroads, the BC forest economy is a troubled landscape (Barnes and Hayter 1997). Environmental protests, Aboriginal claims, reduced forest yields, lower allowable annual cuts, labour-saving technological change, fierce outside competition, trade restrictions, consumer boycotts, and lower wood-product prices reflect on this trouble, providing sharp contrast to the thirty golden years of Fordism. A recent editorial (*Vancouver Sun* 1999), while recognizing the complexity of the forces underlying the forest economy's troubles, ultimately places responsibility on the policies of provincial governments, past and present. This judgment is fair because forest land in BC is overwhelmingly Crown land; provincial governments, past and present, have insisted that they be the owners of the forests. The judgment is also useful because it identifies where responsibility rests for resolving the forest economy's troubles: the provincial government. These troubles are "unique" because of the peculiar role of provincial policy. There is also a deeper sense in which the BC forest economy is unique: it is what Barnes calls (1987, 1996) a "local model," a distinct local economy contributing to the richly differentiated landscapes, or economic geography, of the world as a whole.

Flexible Crossroads is an economic geography of BC's forest economy, and the remainder of this last chapter offers a few thoughts on BC's forest economy as a local model. The first part views the BC forest economy through a global kaleidoscope to reveal the broad range of exogenous and endogenous forces that shape its uniqueness in relation to the rest of the world. This discussion draws particularly on recent conceptualizations of economic geography by Barnes (1987, 1993, 1996) and Patchell (1996) (see also Storper and Salois 1997). The second part makes a plea for greater levels of cooperation in the BC forest economy, assigning central responsibility to the provincial government for a stable framework and vision for forest policy.

BC's Forest Economy: A Local Model in a Global Kaleidoscope

Economic geography seeks to understand how "place makes a difference to the economic process" (Barnes 1987) and represents the global economy as a "regional mosaic" (Scott and Storper 1986), as regional "worlds of production" (Storper and Salois 1997), or as "local models" (Barnes 1987, 1996: 206-28). Local models, to use Johnston's (1984) language, are unique but not singular (Figure 12.1). Thus, local models are formed by the interplay of endogenous and exogenous conditions as wider exogenous forces of globalization in the form of trade, foreign direct investment, financial flows, and migration are interpreted and manipulated by the policies and actions of local populations. In turn, local actions and policies contribute to globalization. Alternatively put, local models exist as places and spaces. Local models are places with specific coordinates, landscapes, and settlement structures within which people live, work, and develop their values and culture and ideas for economic organization. Local models are spaces because patterns

Figure 12.1

The creation of local models

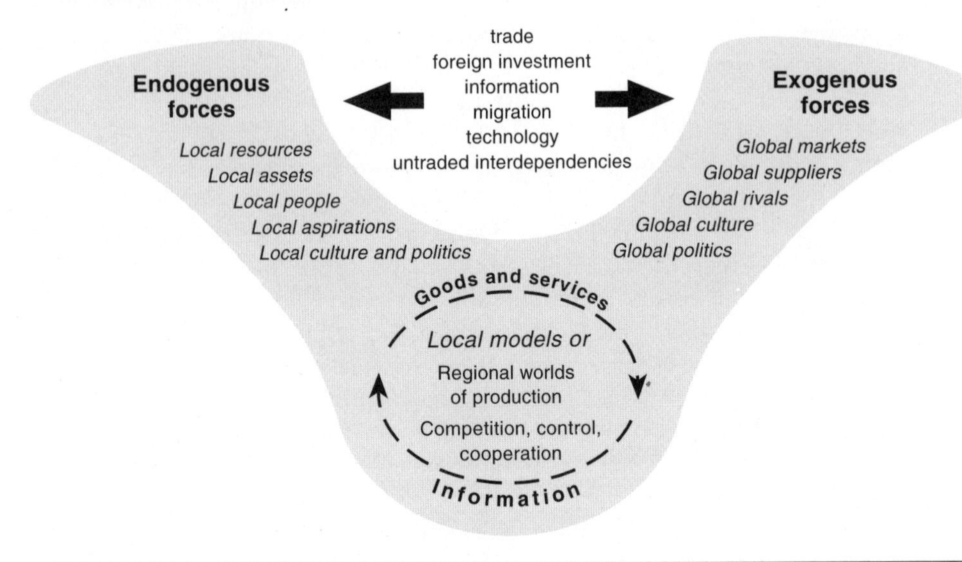

of work and divisions of labour need to fit into global systems of production and exchange.

The wider global "space" economy offers opportunities and threats to "places," with the result that the place-space relation is dynamic and subject to tension. Indeed, over the long run, place-space relations are subject to the full force of the processes of creative destruction. The ability of local populations to cope with these processes is highly varied and defines the substance of the meaning of local models. Patchell (1996) analyzes these variations through the metaphor of the kaleidoscope. Looking down the kaleidoscope, he observes that how people organize themselves economically is based on three fundamental principles of cooperation, competition, and control. But cooperation is the key; in his view, the evolutionary ideal for a local model occurs when cooperation promotes control and competition. In this ideal, competition encourages innovation to penetrate markets or reduce costs, control ensures that complex production systems are coordinated, and cooperation constrains both hyper-competition (which is destructive of long-term thinking and stability) and excessive control (which dampens innovation). But cooperation can degenerate into collusion, competition into conflict, and control into repression.

As we look down the kaleidoscope, the global economy reveals local models that represent different strategies for linking cooperation (and collusion), competition (and conflict), and control (and repression). In the broadest sense, local models are created by the unique interplay of exogenous and endogenous forces. Alternatively put, the kaleidoscope is turned by exogenous and endogenous hands. The relative power of these hands varies over time and space. In core regions, endogenous forces are powerful (but exogenous forces cannot be ignored). On the geographic margin, exogenous forces are inevitably more influential (but endogenous forces cannot be ignored). In the BC forest economy, its very rationale and essence was exogenously created, and exogenous forces have powerfully shaped its evolution. Simultaneously, however, the settlement established by exogenous forces creates endogenous influences and potentials for made-in-BC behaviour and policies.

The Uniqueness of BC's Forest Economy

To a significant degree, BC's distinctiveness as a local model is tied up with the forest industry. For more than 100 years, the forest industry has provided the economic basis for BC's global role, and its technological and organizational structures have defined the character of BC's economy. In turn, the character of BC's forest industry has been profoundly shaped by regional context. The size and richness of the forest resource, geographic marginality, and the lateness of development, beginning in a virtual "empty land" containing a scattered Aboriginal population, have provided defining dimensions for the trajectory of the forest-product industries. In practice, industry structures and regional context have mutually evolved through choices made by many actors, organizations, and institutional arrangements that are hard to disentangle but have contributed towards a unique local model of development in BC. The importance of the forest industry in the province has declined, and the forest economy can no longer be conflated, if it ever could, with BC's economy. But the forest industries are still a powerful presence in BC, and their legacy is considerable.

In the twentieth century, as BC's forest economy has evolved and transformed, it is difficult, perhaps impossible in any precise way, to sort out endogenous (local) from exogenous (global) forces. Yet, these distinctions remain relevant to contemporary debate. Thus, the provincial government interprets the troubled landscape of BC's forest economy in the late 1990s as largely exogenously driven by the Asian economic crisis, American protectionism, and boycotts inspired by international environmentalism. Industry observers, and perhaps the weight of public opinion, interpret these same troubles as endogenously created, specifically by the high-stumpage regime, and related measures, adopted by the provincial government in the 1990s (*Vancouver Sun* 1999; Drushka 1999a, 1999b).

If both perspectives have legitimacy, such a distinction fails to appreciate the extent to which the endogenous and exogenous, the local and the global, mutually shape one another in (politicized) processes operating at various spatial scales recently termed glocalization (Swyngedouw 1997). Thus, American protectionism is a powerful exogenous force constraining BC's forest industries. At the same time, the form American protectionism has taken has been affected by negotiations in which Canada and BC participated. In practice, federal desires for a free trade agreement and provincial excitement over the prospect of higher stumpage shaped Canada's position and helped empower protectionist interests in the American lumber industry, leading to ongoing American interference in BC's forest management policies. To the extent that Canadian governments could have bargained differently – and there are good reasons for believing they could have done so – the Memorandum of Understanding of 1986, the Softwood Lumber Agreement of 1996, and ongoing US interests in BC's stumpage policies cannot simply be interpreted as exogenously imposed constraints on the BC forest economy, beyond any local influence.

It may be thought that the Asian economic crisis is a more "pure" form of an exogenous event beyond the control of private- and public-sector decision makers in BC. Even in this context, however, dependence on Asian markets is not divorced from how American protectionism has been handled, while the development of Asian trade, especially Japanese, has been recent, crisis driven, and largely left under the control of Japanese organizations that now have an important presence in BC. Moreover, the provincial government "used" high Japanese prices to justify its escalation of stumpage. When prices dropped, however, stumpage could not automatically follow. Regarding the environment, the provincial government introduced significant initiatives to meet the demands of global and local environmentalists, but it did so without receiving any quid pro quo.

On the geographic margin, as in BC, it is tempting to give priority to exogenous factors in explanations of local development. Yet, in BC, local institutions and policy makers have influence, some degree of freedom, to contribute to glocalization. Historically, to access global opportunities, provincial forest policy *chose* to strongly support free trade ideology, an open door policy to foreign direct investment, and large-scale industrialization. Such policies, which reached their apogee during Fordism, were explicitly meant to ensure that the BC forest economy would be "open," connected as fully as possible to global opportunities and, by implication, to the full force of exogenous influences, bad and good. These choices did not have to favour large-scale industrialization nor provide foreign direct investment with such easy access. Unintended or not, provincial forest policy during Fordism defined a particular form of glocalization that resulted in a large-scale, but narrowly focused, truncated forest-production system in BC. Indeed, provincial policies literally enforced Fordism by maintaining Crown ownership of forests, leased only for the purposes the provincial government wanted, notably large-scale industry that exported to a few core countries, often organized by MNCs. Provincial forest policy thus underwrote the staple trap.

Restructuring at the Intersection of Industrial and Resource Dynamics
An entirely unintended implication of provincial forest policy during Fordism was to facilitate the conditions for a unique restructuring of BC's forest economy over the last quarter century or so. The uniqueness of this restructuring results from the intersection of industrial and resource dynamics. Because the Fordist production system was so comprehensively adopted in BC's forest economy, the imperatives of flexibility that have so assailed Fordism everywhere have been especially powerful in BC. Moreover, the same commitment to the Fordist production system ensured that BC's forests would be gobbled up at the fastest possible rate. Indeed, during Fordism, conventional wisdom often supposed that the gobbling should be even faster, given the presence of "overmature" timber; that is, timber that is no longer growing and has a high potential to die, decay, and rot, thereby losing value for the purposes of industrial forestry. By 1970, the forest industry was established in all parts of the province, and all the main tree species were utilized (see Figure 2.7).

As a result, the transformation of BC's forest industries stimulated by the shift from Fordism to flexibility coincided with the falldown effect, which in turn inspired environmental opposition to industrial forestry to take on extremely urgent tones. Indeed, environmental opposition to BC's forest industries was supported by growing global concerns for environmental values along with significant improvements in the ability to communicate environmental concerns around the globe.

For some time, neither the provincial government nor industry fully appreciated these transformations, let alone their intersection. Industry was admittedly slow in responding to the strength of environmental influence, local and global, especially to issues related to biodiversity, which places considerable value on mature timber and old-growth forests (Cafferata 1997). Moreover, the spectacular spread of the forest industry throughout the province during Fordism helped reawaken Aboriginal people to rights and claims that had been ignored, not only in practice but also in theory, as a result of the empty land assumption (Willems-Braun 1997a).

The recession and the beginnings of American protectionism combined in the early and mid-1980s to create a crisis in BC's forest economy for stakeholders, upsetting firms, unions, governments, local communities, environmentalists, and Aboriginal peoples. The seeds for fundamental changes in forest policy were also sown. In 1988, the right-wing Socred government revised the stumpage formula, and stumpage increased as a way of appeasing American protectionists and also conveniently increased provincial coffers. Increasing stumpage then became the central plank in the new NDP government's battery of forestry legislation in the 1990s, a policy barrage labelled in this book as the high-stumpage regime (see Figure 3.8; Table 3.10).

The High-Stumpage Regime: A Critique of an Endogenous Policy

The high-stumpage regime, in origin, represents a legitimate, coherent attempt to stimulate the transformation of BC's forest economy from a low-value (commodity) Fordist production system to a flexible, higher-value production system more in tune with contemporary environmental values and the full range of public interests, including those of Aboriginal peoples. The high-stumpage regime generally offered an alternative vision of a new forest economy consistent with flexibility and environmental imperatives. The reality of its unfolding, however, has been problematic; the high expectations of the early 1990s have ended in the crisis of the late 1990s, which is as severe, if not more so, than the recession of the early 1980s. Booming and busting, apparently, is still the modus operandi of BC's forest economy.

The present provincial government may claim, as it has, that its high-stumpage regime is a victim of geographic circumstances beyond its control, notably American protectionism and the Asian economic crisis. But the government has aided and abetted these forces. Its self-description as a victim is a weary argument that is ultimately defeatist. BC is on the geographic margin, but it is a rich margin. Governments on the margin may have to be smart, and a smart policy is exactly what the high-stumpage regime claimed to be.

The present provincial government can claim that it is not only a victim of geography but also a victim of history, in particular a victim of legacy. After all, provincial authorship of Fordism was by other governments from another party that were given several decades to ensure an imprint on BC's forest economy. Restructuring in situ is no doubt a very different challenge from creating new industrial spaces, and once a landscape is created, it shapes the future. But the victim-of-history argument can only be taken so far. History is known, and smart policies cannot ignore this legacy. The high-stumpage regime could have been a superior option to the path chosen in the late 1940s, but that is not what happened.

Unfortunately, the government's high-stumpage regime did not fully appreciate the geographical and historical circumstances of BC's forest economy. Moreover, while decision-making processes regarding forest use are now more transparent, they are also more adversarial, often unrestrained by clear rules of debate. The high-stumpage regime, designed to resolve BC's troubled economy, is in trouble itself. What went wrong? Although the provincial government is the "prime suspect," the government is not the only suspect, but it does control most forest land in BC, and its policies are central to industrial transformation, present and future. Given this caveat, two broad criticisms of the high-stumpage regime address its underlying assumptions and evolution in practice. These criticisms are not rooted in ideology, specifically anti-NDPism.

Higher Stumpage as a Leap of Faith

A central assumption underlying the entire edifice of the high-stumpage regime is that higher stumpage would stimulate higher-value activities, in turn essential to the achievement of both economic and environmental goals (Chapter 3). This assumption constitutes a leap of faith. It may even have backfired. The rationale underlying the rapid ratcheting up of stumpage rates was based on a far too simple cause-and-effect assumption: that if inputs are increasingly costly, outputs will be increasingly more valuable. Not only has this policy underestimated the strength of competition from elsewhere, it has underestimated the impact of historical legacy, including the consequences arising from the nature of resource dynamics. Thus, other jurisdictions, including across Canada, have the same idea of promoting value-added forestry, but their stumpage is lower. Crucially, value-added trends demand innovation so that the key policy stimulus is not higher stumpage but improved performance regarding innovation. BC needs a smart forest economy, along the lines intimated in the last chapter, and stumpage should reward innovation rather than the reverse. Given that the legacy of Fordism imposes significant constraints on innovation, these constraints need to be the policy priority. Furthermore, because the best trees are now gone, resource dynamics further counsel an emphasis on innovation first and stumpage later.

As important related components of the high-stumpage regime, the creation of new parks, support for SMEs, the treaty process, reduced AAC levels, Forest Renewal BC, and the Forest Practices Code are all linked to the theme of higher value. These initiatives relate to one another and validly seek to promote the nonindustrial and the industrial values of the forest, which is now the public expectation. At the same time, none of these initiatives required the government to increase stumpage rates to the extent that it did. Thus, the high-stumpage regime could have focused on enhanced stewardship rather than high stumpage payments. For industry, the Forest Practices Code and the other policies required more environmentally friendly policies, and they added to costs. In effect, the government's high-stumpage regime raised a high level of expectation regarding both environmental goals and economic goals, while increasing its own revenues, a wonderful positive-sum game all depending on increasing the value of forest-industry outputs.

But industry has been unable to increase the value of outputs to maintain economic contributions, and the government has found itself under increasing pressure to relax environmental criteria, and reduce the stumpage formula. Such changes, coming so soon after the policies were introduced, suggest that the high-stumpage regime was of a more experimental nature than the government had intimated. Moreover, the government has sent increasingly mixed signals about its long-term intentions. The large-scale subsidies provided to the kraft pulp mill at Prince Rupert, along with waiving of environmental regulations at this mill and at others threatened by closure, raise the spectre of confusion in government thinking. A direct implication of the shift towards a more flexible, value-oriented forest economy, whose fibre base is declining, is that commodity mills must close. The subsidization of the Prince Rupert mill flies in the face of the government's own intentions to promote value, encourage environmentalism, and reduce the AAC. Meanwhile, the government did nothing to stop the closure of in-house R&D programs vital to long-run innovation; it may also have contributed to their demise by raising costs for the firms in question. That is, government policy rewarded noninnovation and penalized innovation.

Confusion over the direction of government policy has not been helped by long delays in the timber supply reviews, associated AAC adjustments, and the award of community forests. The delay in 1998-9 in discussing the Nisga'a treaty is another example. The delays also send mixed messages. Thus, the general expectation that the AAC will be reduced has been offset by announcements that managed forests in some areas are growing faster than anticipated and by informed opinion that, with appropriate management, harvest levels may be increased significantly. Such delays contribute to uncertainty over forest use and raise questions about investing in long-term forestry. Delays in policy formulation also reflect a shift from a secretive, industry-dominated model of decision making to a democratic model in which all interest groups participate. This welcome shift necessarily implies a more unruly process, but debate has tended to be adversarial rather than cooperative, a tendency that government has not resolved and has made worse. Indeed, the provincial government has itself not always been transparent in executing forest policy, as a recent scathing decision by a BC Supreme Court judge regarding Carrier Lumber of Prince George underlined (Hamilton 1999d). In this decision, the judge determined the government had breached the terms of a timber harvesting agreement with Carrier Lumber "to play politics" in secretly making a deal with Chilcoten-area Natives. If this decision directly reflects (badly) on the integrity of the government, it also intimates the government's failure to clarify how it wanted to restructure tenure among competing stakeholders in its high-stumpage regime. In fact, as another consequence of the high-stumpage regime, provincial autonomy to make such a clarification is now much reduced.

The Loss of Provincial Autonomy: BC Forests as Contested Space

A second set of criticisms of the high-stumpage regime is that its formulation and implementation have served to undermine provincial government autonomy over the forestry resource and, unintended or not, to contribute towards BC forests as contested space. Several points support this unusual accusation.

First, the government accepted both the American view that low stumpage constituted a subsidy and the various ill-conceived Canadian government attempts to resolve this issue. In so doing, it has legitimized American interference in BC's forest resource policy, which before 1986 was not the norm.

Second, the Forest Practices Code, Clayoquot Sound Compromise, AAC revisions, new parks, and community forest proposals represent a comprehensive and costly commitment to meeting environmental goals. In return, however, the government received no assurances of cooperation from environmentalists. Indeed, as the Clayoquot Sound Compromise revealed, environmentalists did better by refusing to cooperate. There is only mixed evidence that environmentalists agree to the norms of democratic practice within BC.

Third, the treaty process, designed to empower natives, was restarted by the provincial government, although this issue is a federal responsibility, in the belief that Aboriginal peoples have inherent rights. But the scope of this empowerment is uncertain and contentious. Indeed, it is not clear that the treaty process, contrary to the claims of the provincial government, will eliminate uncertainties over Aboriginal claims. Any treaties will shift power (and stumpage rights) over forests to Aboriginal peoples. In the meantime, the federal government, given the Indian Act and the Supreme Court of Canada, has a direct and increasing say over how provincial forests can be used.

Fourth, privatization of the forests has recently been effectively lobbied by industry as a way to compensate for loss of land acquired for land claims and parks, to provide an incentive for better forestry, and to reduce countervail threats from the US. The government may not be able to privatize forests, however, even if it should wish to do so, because of opposition from environmentalists and Aboriginal peoples who would probably be able to stall any such plans in the courts.

Fifth, while the high-stumpage regime was a barrage of social experiments regarding forest policy, the government failed to define an overall plan for tenure reform within which these polices could be worked through and constrained. In the absence of such a plan, forces have now been unleashed that are undermining provincial autonomy and making overall tenure reform both more urgent and less likely. BC has literally become a contested space (see Figure 10.4). Industry, environmentalists, and Aboriginal groups are all pursuing their interests, if necessary to courts of appeal beyond BC's borders, including the federal government, the Supreme Court of Canada, various trade adjudication bodies, and the court of public opinion. From this perspective, the high-stumpage regime has opened a Pandora's box. Moreover, in addition to Aboriginal land claims that cover the entire province and environmental "land claims" that cover much of the province, the US federal government is strongly supporting demands for BC's forest policy to conform to American priorities, namely, privatization with the explicit intention of allowing log exports. Indeed, recent provincial government willingness to permit higher levels of log exports to help companies become profitable is itself an indictment of the high-stumpage regime which pinned so much hope on creating value-added jobs.

A special note might be added regarding provincial government attitudes towards American protectionism. The previous provincial government, not just the present one, along with the federal government, kowtowed to American toughness and the view that stumpage constituted a subsidy. In so doing, Canadian governments legitimized American interference in BC's forest policy. But, the world's biggest superpower needs no invitation to interfere. Yet, according to an informed, if not entirely disinterested source, the present CEO of MacMillan Bloedel and former participant on the American side, even the Americans were surprised at Canadian passivity to their demands in the negotiation of the Softwood Lumber Agreement of 1996, the most recent attempt to appease the US. The problem with this agreement is not only that it limits BC's exports to the US, but that it institutionalizes political bickering within BC (and Canada) and American interference in what should be an area of provincial autonomy.

Amazingly, Canadian governments keep missing the point that a legitimate aspect of their sovereignty has been compromised. The present provincial government is not alone in this failure, but as an NDP party – the party of economic nationalism – it should know better. Moreover, the implications of the failure to defend legitimate rights of sovereignty are still being felt as the provincial government is under substantial pressure to privatize Crown forests, at least in part to further appease American interests. Privatization may offer substantial benefits, but the issues are not straightforward, and any such policy should be implemented to enhance innovation within BC. If the US imposes tariffs, so be it. That is its right. Meanwhile, so-called solutions to this threat have made the formulation of forest policy in BC more difficult.

These harsh criticisms of the high-stumpage regime require context. The provincial government inherited almost dinosaurian attitudes in the corporate and union sectors, resource towns passive to issues of development, Aboriginal peoples with unclear aspirations in relation to mainstream society, and environmentalists with single-issue intransigence that stretches democratic practice to the limit and beyond. A recent demonstration of the difficulty facing the government is environmentalist objections to privatization as a threat to the Forest Practices Code and proper environmental management. Yet, the environmental problems of current forest practices have largely evolved within a framework of public ownership. The fact is that private forests do not have to be unregulated forests.

The provincial government has also had to live with an unfortunate legacy of federal government disinterest in BC and its forest economy. For 100 years the federal government failed to meet its obligations regarding Aboriginal treaties; in recent decades, greatly aided by transfer payments from BC, it has subsidized BC's forest product competitors in central and eastern Canada; its handling of the softwood lumber dispute has been inept; and it is increasingly relying on the Supreme Court, empowered by the 1982 Charter of Rights, to resolve political dilemmas, most notably Aboriginal land claims. Since that time, Supreme Court decisions have widened Aboriginal powers, but the basis of these decisions, as exemplified in the *Delgamuukw* decision, is vague and contentious. For BC, the lack of federal leadership on land claims and reliance on a remote court to resolve what are fundamentally political and social problems have served to add to uncertainty and conflict over forest policy.

Moreover, effective dialogue between the federal and BC provincial governments appears to have broken down, a rift underlined by the extraordinarily high-handed (and possibly unconstitutional) federal expropriation of BC territory, namely the Nanoose Bay test range. In 1999, the federal government also refused to consider the BC government's plea for financial help for the forest industry, although it has initiated some talks directly with industry itself (O'Neil and Beatty 1999). At a time when the problems of BC's forest economy require a much stronger commitment to cooperation and compromise among various stakeholders, both senior levels of government are setting the worst possible example. Regardless of who is most to blame, the deep-seated antagonism between the provincial and federal governments is not helping the former develop a coherent forest policy.

The extent to which the provincial government has authored its own loss of autonomy over the forest resource may be debated. There is little doubt, however, that there has been a profound change in this autonomy. Provincial control over resources is a foundation stone of the Canadian Federation, and in 1980 BC's provincial government was deemed to be in full control of Crown land and it was able to allocate forest resources as it deemed fit, consistent with laws for which it was also responsible. In 1999, provincial authority over the forest resource is challenged from all sides, in and out of court. To some extent at least, one of the most activist provincial governments in BC has seemingly undermined itself.

The provincial government can also claim that it attempted the most comprehensive forest reregulation in BC since the Sloan Commission of the 1940s to address extremely complex problems. If this reregulation is not evolving as anticipated, the experience can be used to construct a more effective alternative or at least modification. In light of this experience, the reregulation of the reregulation should emphasize the themes of cooperation, stability, and enhanced stewardship.

Cooperation in Support of Enhanced Stewardship

Looking down Patchell's (1996) kaleidoscope, BC's forest economy during

Fordism featured strong competition in global product markets, and competition for timber supplies that was closely controlled by provincial policy favouring large corporations, including MNCs. Dominated by hierarchically organized corporations (that is, by internal division of labour) in the highly competitive pursuit of commodity markets, BC's forest economy has evolved only weak forms of cooperation among firms (and the social division of labour). This weakness is especially apparent in the development of technology. While BC has developed important technological capabilities, little sustained collaboration exists between equipment suppliers and forest firms in BC of the kind required to establish BC as a globally important supplier of forest technology. Instead, reliance on imported technology is high. In addition, in higher-value segments, subcontracting among small firms does exist, but it is not particularly sophisticated and geared to sustaining innovative behaviour. In general, truncated internal divisions of labour have characterized BC's forest industry, and the development of complex social divisions of labour that exploit rival expertise through cooperative networking remains a challenge.

In the shift from Fordism to flexibility, competition for global product markets is perhaps even stronger, while provincial authority over the timber resource has been increasingly questioned. If corporate control of timber licences remains in evidence, environmentalism, aboriginalism, and local communities represent vocal, often dissenting voices in favour of alternative forms of control and of nonindustrial users of the forest. New forms of cooperation have emerged, for example, with respect to flexible specialization, eco-certification, silvicultural subcontracting, and joint ventures with Aboriginal peoples, but these trends need to be strengthened and focused on the theme of innovation. In the terms of Patchell's kaleidoscope, it is the benefits that can be derived from cooperation that are underrepresented.

The long-run problem for the BC forest economy is not lack of competitiveness but lack of cooperation, not too much technology but too little. Smart production systems are both competitive and cooperative, and they maintain control through a locally coherent social division of labour that, if the scale and complexity of production allows, combines the talents of giant, large, and small firms. A chorus of pleas has urged the BC forest economy to become more innovative and to develop smart production systems – smart factories, workers, firms, and interfirm relations – to remain viable. These pleas, however, do not automatically translate into imperatives.

In an age of flexibility and globalization, much debate arises about what governments can and cannot do to shape economic development. If policy discretion is less than it once was, according to Hirst and Zeitlin (1991), an important priority for governments is to help develop appropriate business climates by facilitating cooperative attitudes and developing coherent, long-run agendas that ensure various interest groups, business, labour, local communities, and environmentalists work together. Such a task should be a priority for BC, especially for an innovation-driven forest policy. Admittedly, engendering a stronger spirit of cooperation will not be easy, given BC's tradition of economic liberalism, with its emphasis on competition, independence, and individualism, the poor relations between federal and provincial governments, and the deep cleavages between the forest industry and other communities. Nevertheless, if the full range of forest values is to be realized, and if the forest industry is to become a more complex, innovative production system, greater levels of cooperation and trust are essential. The provincial government faces a tremendous task in this regard, but greater clarity about long-run intentions regarding tenure reform would be a start. The reestablishment of professional relations with the federal government is another priority, and the federal government must show leadership in resolving Aboriginal land claims via negotiation rather than via the courts. With a stable framework for tenure and control, BC's forests can then be used to achieve economic and environmental global leadership.

From the perspective of the forest industry, the basic thrust of forest policy should centre on innovation. If industrial forest policy is considered as a form of innovation policy, then diversification and privatization of tenure can play important roles in helping stimulate stronger commitments to innovation (Chapter 11). A key assumption of the development of an innovation-centred forest policy is the establishment of a stable and transparent framework for forest tenure and use throughout the province. Stability is absolutely essential if firms, workers, communities, and households are to commit themselves to training, education, research, and innovation, all of which require investment and time. Given stability and transparency, an important priority is to ensure wood is allocated to innovators. Moreover, innovative behaviour should be rewarded rather than penalized, as is often the case now, when higher values are immediately absorbed by the government in the form of higher stumpage. Charging higher stumpage to innovators and reducing stumpage to old, marginal fibre-guzzling pulp mills make no sense. Rather, the government needs to consistently apply the principle of "contribution to innovation" in all its forest-policy decision making. The notion of sustainability cannot be divorced from the impact of technological change.

The reallocation of a portion of forest tenures to local communities potentially makes sense from an innovation perspective to the extent it widens decision-making influences, brings forests under closer and more careful management, and creates stable supplies of small wood-fibre supplies. The realization of innovation potentials in community forests may be thwarted, however, if the local management base is too complicated and unruly. The government also needs to speed its decision making on a long-delayed designation of three new community forests. In any case, three new forests will scarcely affect overall forest policy in BC. In the diversification of forest tenure, community forests can be assigned a much bigger role.

In contemplating forest policy, including the central issues of stumpage and tenure, concern over what the US will think and do has become an important, complicating issue. From this perspective, the government faces a real sovereignty problem, for which it is partially responsible. If sovereignty is important, the frank response would be to reject further American interference in BC's forest policies. If such a rejection leads to the imposition of American tariffs, BC at least would retain legitimacy by looking after its own resource; a tariff, which can be eliminated, may also be better than a quota. A tariff can be seen for what it is. In the meantime, it needs to be recognized that whenever BC (and Canada) legitimizes American interference, such interference is guaranteed. In the long run, then, innovation potentials can best be realized if BC reestablishes sovereignty over its own resource.

In the late 1990s, in a state of crisis, it is not easy to adopt a long-term view. However, it is important that long-term goals are not compromised by short-term expediency or the need to solve immediate crisis. The development of an innovation-based forest policy and a smart forest-production system can only be achieved in the long run.

Flexible Conclusions

The BC forest economy is in a state of crisis, and its problems are long term and structural. The forest industries have been restructuring since at least the early 1980s, and the present crisis reflects the severity of the underlying problem of transformation. The BC forest economy is at a crossroads, defined by industrial and resource dynamics, and the crossroads is flexible. For forestry policy, there is a basic division between the ecological-utopian path (M'Gonigle and Parfitt 1994) and the modified industrial-forestry path based on multiple values within zones or among a set of zones (Binkley 1997a). Neither option is mutually exclusive. Another environmentally sponsored option claims priority for the environment, even if much of the coastal industry has to be shut down (Marchak et al. 1999).

A more intriguing challenge, one attempted by the high-stumpage regime, is for BC to aspire to the full realization of industrial and environmental benefits from the forest. In theory, such a positive-sum game is possible, and BC should be smart enough to translate theory into practice. The policy demands of a smart forest policy are inevitably more demanding, but the social benefits are greater. In this way, geographically marginal BC can define a new global role in which it is a leader, not a follower. Such a path is possible at the crossroads.

References

Aitken, H.G.H. 1961. *American Capital and Canadian Resources*. Cambridge, MA: Harvard University Press.

Allen Hopgood Enterprises Ltd. 1986. *The Potential for New Technologies in Canada's Forest Sector*. Ottawa: Ministry of State for Science and Technology.

Amin, A. 1993. The globalization of the economy: An erosion of regional networks? In *The Embedded Firm: On the Socioeconomics of Industrial Networks,* ed. B. Hogut, W. Shan, and G. Walker, 278-95. London: Routledge.

Anderson, F.J. and N.C. Bonsar. 1985. *The Ontario Pulp and Paper Industry: A Regional Profitability Analysis*. Toronto: Ontario Economic Council.

Anderson, M., and J. Holmes. 1995. High-skill, low wage manufacturing in North America: A case study from the automotive parts industry. *Regional Studies* 29: 655-71.

Anderson, R.S., and W. Huber. 1988. *The Hour of the Fox: Tropical Forests, the World Bank, and Indigenous People in Central India*. Seattle: University of Washington Press.

Atkinson, J. 1985. The changed corporation. In *New Patterns of Work,* ed. D. Clutterback, 13-35. Aldershot: Gower.

–. 1987. Flexibility or fragmentation? The United Kingdom labour market in the 1980s. *Labour and Society* 12: 87-105.

Austin, I. 1996. Greenpeace: It all started here: Environmental colossus is 25 today. *Province* September 15: A26.

Auty, R.M. 1990. *Resource-Based Industrialization: Sowing the Oil in Eight Developing Countries*. Oxford: Clarendon Press.

–. 1993. *Sustaining Development in Mineral Economies: Too Few Producers*. London: Routledge.

–. 1995. Industrial policy, sectoral maturation, and postwar economic growth in Brazil: The resource curse thesis. *Economic Geography* 71: 257-72.

Averitt, R.T. 1968. *The Dual Economy: The Dynamics of American Industry*. New York: Norton.

Baldwin, R.E. 1956. Patterns of development in newly settled regions. *Manchester School of Economics and Social Studies* 24: 161-79.

Barker, M., and D. Soyez. 1994. Think locally – act globally? The transnationalisation of Canadian resource-use conflicts. *Environment* 36: 12-20, 32-6.

Barnes, T.J. 1987. Homo economicus, physical metaphors, and universal models in economic geography. *Canadian Geographer* 32: 347-50.

–. 1993. Innis and the geography of communications and empire. *Canadian Geographer* 37: 357-9.

–. 1996. *Logics of Dislocation: Models, Metaphors, and Meanings of Economic Space*. New York: Guilford Press.

–, and R. Hayter. 1992. The little town that did: Flexible accumulation and the community response in Chemainus, British Columbia. *Regional Studies* 26: 647-63.

–, and R. Hayter. 1994. Economic restructuring, local development and resource towns: Forest communities in coastal British Columbia. *Canadian Journal of Regional Science* 17: 289-310.

–, and R. Hayter, eds. 1997. *Troubles in the Rainforest: British Columbia's Forest Economy in Transition*. Canadian Western Geographical Series No. 33. Victoria: Western Geographical Press.

–, R. Hayter, and E. Grass. 1990. Corporate restructuring and employment change: A case study of MacMillan Bloedel. In *Corporate Firm in a Changing World Economy*, ed. M. de Smidt and E. Wever, 145-65. London: Routledge.

Barr, B.M., and K.J. Fairbairn. 1974. Some observations on the environment of the firm: Locational behaviour of kraft pulp mills in the interior of British Columbia. *Professional Geographer* 26: 19-26.

Baskerville, G.L. 1990. Canadian sustained yield management – Expectations and realities. *Forestry Chronicle* February: 25-8.

BC Wild. 1998. Overcut: British Columbia forest policy and the liquidation of old growth forests. Vancouver: BC Wild.

Beatty, J. 1999. B.C. wants $500 million from Ottawa to save 18,000 forest industry jobs. *Vancouver Sun* May 12: A1, A2.

Bedford, N.S. N.d. *The Market for British Columbia Logging and Sawmilling Equipment in Southeast Asia*. Victoria: Government of British Columbia.

Behrisch, T. 1995. *Preparing for Work: A Case Study of Secondary School Students in Powell River, B.C.* MA thesis, Simon Fraser University.

Bell, C. 1998. New directions in the law of aboriginal rights. *The Canadian Bar Review* 77: 36-72.

Bell, D. 1974. *The Coming of the Post-Industrial Society: A Venture in Social Forecasting*. London: Heinemann.

Best, M.H. 1990. *The New Competition*. Cambridge, MA: Harvard University Press.

Binkley, C.S. 1993. Creating a knowledge-based forest sector. *Forestry Chronicle* 69: 294-9.

–. 1995. Designing an effective forest sector research strategy for Canada. *Forestry Chronicle* 71: 589-95.

–. 1997a. A crossroad in the forest: The path to a sustainable forest sector in British Columbia. In *Troubles in the Rainforest: British Columbia's Forest Economy in Transition*, ed. T. Barnes and R. Hayter, 15-35. Canadian Western Geographical Series No. 33. Victoria: Western Geographical Press.

–. 1997b. Preserving nature through plantation forestry: The case for forestland allocation with illustrations from British Columbia. *Forestry Chronicle* 73: 553-9.

–, and O.L. Forgacs. 1997. Status of forest sector research and development in Canada. Unpublished paper.

–, C.F. Raper, and C.L. Washburn. 1996. Institutional ownership of US timberland. *Journal of Forestry* 94: 21-8.

–, and S.B. Watts. 1992. The status of forestry research in British Columbia. *Forestry Chronicle* 68: 730-5.

–, and S.B. Watts. 1999. The status of and recent trends in forest sector research in British Columbia. *Forestry Chronicle* 75: 607-13.

Blahna, D.J. 1990. Social bases for resource conflict in areas of reverse migration. In *Community and Forestry*, ed. R.G. Lee, D.R. Field, and W.R. Burch Jr., 159-78. Boulder, CO: Westview Press.

Blomley, N.K. 1996. Shut the province down: First nations' blockades in British Columbia. *B.C. Studies* 3: 5-35.

Bradbury, J.H. 1978. The instant towns of British Columbia: A settlement response to the metropolitan call on the productive base. In *Vancouver: Western Metropolis*, ed. L.J. Evenden, 117-33. Western Geographical Series No. 16. Victoria: University of Victoria.

Bramham, D. 1993. Our jobs are leaving town, so are people. *Vancouver Sun* November 6: A1.

British Columbia, Government of. 1963. *Financial and Economic Review*. Victoria: Ministry of Finance.

–. 1967. *Financial and Economic Review*. Victoria: Ministry of Finance.

–. 1980. *Forest and Range Resource Analysis Technical Report*. Victoria: Ministry of Forests.

–. 1988. *Financial and Economic Review*. Victoria: Ministry of Finance.

–. 1995. *Financial and Economic Review*. Victoria: Ministry of Finance.

–. 1996. *Summary of Timber Supply Results 1992-96*. Victoria: Ministry of Forests.

Britton, J.N.H. 1991. Reconsidering innovation policy for small and medium sized enterprises: The Canadian case. *Environment and Planning C* 9: 189-206.
–. 1996. High-tech Canada. In *Canada and the Global Economy,* ed. J.N.H. Britton, 255-72. Montreal-Kingston: McGill-Queen's University Press.
–, and J. Gilmour. 1978. *The Weakest Link: A Technological Perspective on Canadian Industrial Underdevelopment.* Background Study No. 43. Ottawa: Science Council of Canada.
Bromley, D. 1992. Property rights as authority systems: The role of rules in resource management. In *Emerging Issues in Forest Policy,* ed. P. Nemetz, 453-70. Vancouver: UBC Press.
Brunelle, A. 1990. The changing structure of the forest industry in the Pacific Northwest. In *Community and Forestry,* ed. R.G. Lee, D.R. Field, and W.R. Burch Jr., 107-24. Boulder, CO: Westview Press.
Burda, C. 1999. Leave forests in public hands. *Vancouver Sun* May 25: A13.
–, F. Gale, and M. M'Gonigle. 1998. Eco-forestry versus the state(us) quo: Or why innovative forestry is neither contemplated nor permitted within the state structure of British Columbia. *BC Studies* 119: 45-72.
Cabarle, M., R.J. Hrubes, C. Elliot, and T. Synnott. 1995. Certification-Accreditation. *Journal of Forestry* 93: 12-16.
Cafferata, W. 1997. Changing forest practices on coastal British Columbia. In *Troubles in the Rainforest: British Columbia's Forest Economy in Transition,* ed. T. Barnes and R. Hayter, 53-63. Canadian Western Geographical Series No. 33. Victoria: Western Geographical Press.
Cail, R.E. 1974. *Man, Land and the Law: The Disposal of Crown Lands in British Columbia, 1871-1913.* Vancouver: UBC Press.
Canada, Government of. 1972. *Foreign Direct Investment.* Ottawa: Queen's Printer.
–. 1974. *Foreign Owned Subsidiaries in Canada 1964-71.* Ottawa: Department of Industry, Trade and Commerce.
–. 1979. *Single-Sector Communities.* Ottawa: Ministry of Supply and Services.
–. 1996. *Nisga'a Treaty Negotiations: Agreement in Principle in Brief.* Issued jointly by the Government of Canada, the Province of British Columbia, and the Nisga'a Tribal Council.
Canadian Council of Forest Ministers. 1997. *Compendium of Canadian Forestry Statistics.* Ottawa: Canadian Council of Forest Ministers.
Canadian Forest Industries Council. 1986. *The Management of Canadian Forests.* Ottawa: Canadian Forest Industries Council.
Canadian Press. 1990. US resists attempts to reopen lumber deal. *Vancouver Sun* November 2: C5.
Cannon, J.B., and P. McGloughlin. 1990. Trends in regional policy in Australia and Canada. In *Industrial Transformation in Canada and Australia,* ed. R. Hayter and P. Wilde, 259-73. Ottawa: Carleton University Press.
Carrothers, W.A. 1938. Forest industries of British Columbia. In *The North American Assault on the Canadian Forest,* ed. A.R.M. Lower, 225-344. Toronto: Ryerson.
Castells, M. 1996. *The Information Age: Economy, Society and Culture. Volume 1: The Rise of the Network Society.* Oxford: Blackwell.
Chancellor Partners. 1997. *The Economic Impact of the Forest Industry on British Columbia and Metropolitan Vancouver.* Commissioned by Forest Alliance of B.C. and Vancouver Board of Trade, Vancouver.
–. 1998. *The Potential Economic Impact of the Y2Y Initiative on the Forest Industries and on the Economy of British Columbia.* Vancouver: Forestry Alliance of B.C.
Clague, M., and L. Flavell. 1989. *Final Evaluation of Year III of the Alberni Enterprise Project: An Evaluation of a Multi-Element Approach to Community Economic Development.* Ottawa: Ministry of Supply and Services.
Clapp, R.A. 1995. Creating competitive advantage: Forest policy as industrial in Chile. *Economic Geography* 71: 273-96.
–. 1998a. Regions of refuge and the agrarian question: Peasant agriculture and plantation forestry in Chilean Araucanía. *World Development* 26: 571-89.
–. 1998b. The resource cycle in forestry and fishing. *Canadian Geographer* 42: 129-44.
–. 1998c. Waiting for the forest law: Resource-led development and environmental politics in Chile. *Latin American Research Review* 33: 3-36.
Clark, G.L. 1981. The employment relation and the spatial division of labour. *Annals of the Association of American Geographers* 71: 412-24.

–. 1986. The crisis of the midwest auto industry. In *Production, Work and Territory,* ed. A.J. Scott and M. Storper, 127-48. London: Allen Unwin.

Clark-Jones, M. 1987. *A Staple State: Canadian Industrial Resources in Cold War.* Toronto: University of Toronto Press.

Clawson, M. 1979. Forests in the long sweep of American history. *Science* 204: 1168-74.

–, and R. Sedjo. 1984. History of sustained yield concept and its application to developing countries. In *History of Sustained Yield Forestry: A Symposium,* ed. H.K. Steen, 3-15. Portland, OR: Forest History Society.

Clement, W. 1989. Debates and directions: A political economy of resources. In *New Canadian Political Economy,* ed. W. Clement and G. Williams, 36-53. Kingston-Montreal: McGill-Queen's University Press.

–, and G. Williams, eds. 1989. *New Canadian Political Economy.* Montreal: McGill-Queen's University Press.

Coffey, W., and M. Polèse. 1985. Local development: Conceptual bases and policy implications. *Regional Studies* 19: 85-93.

COFI. [Council of Forest Industries.] 1992. *British Columbia Forest Industry Fact Book.* Vancouver: Council of Forest Industries.

–. 1994. *A Proposed Aboriginal Forestry Strategy for the B.C. Forest Industry.* Developed by E.B. Experts and Clayton Resources. Vancouver: Council of Forest Industries.

–. 1995. *British Columbia Forest Industry Statistical Tables.* Vancouver: Council of Forest Industries.

–. 1997. *British Columbia Fact Book – 1997.* Vancouver: Council of Forest Industries.

Cohen, D.H. 1992. Adding value incrementally: A strategy to enhance solid wood exports to Japan. *Forest Products Journal* 42(4): 40-4.

–. 1993. Preliminary assessment of market potential for finger-jointed lumber in Japanese residential construction. *Forest Products Journal* 43(5): 21-7.

–, and D.C. Mowery. 1984. Firm heterogeneity and R&D: An agenda for research. In *Strategic Management for R&D,* ed. B. Bozeman and A. Link, 197-232. Lexington, MA: D.C. Heath.

–, and P.M. Smith. 1992. Global marketing strategies for forest product industries. *Canadian Journal of Forestry Research* 22: 124-31.

Cooke, P., and K. Morgan. 1993. The network paradigm: New departures in corporate and regional development. *Environment and Planning D* 11: 543-64.

Copithorne, L. 1979. *Natural Resources and Regional Disparities.* Economic Council of Canada. Ottawa: Supply and Services Canada.

Cox, K.R., and A. Mair. 1988. Locality and the community in the politics of local development. *Annals of the Association of American Geographers* 78: 307-25.

Cox, T.R. 1974. *Mills and Markets: A History of the Pacific Coast Lumber Industry to 1900.* Seattle: University of Washington Press.

Croon, I. 1988. The Scandinavian approach to the future of pulp and paper. In *Global Issues and Outlook in Pulp and Paper,* ed. G.F. Schreuder, 268-75. Seattle: University of Washington Press.

Daly, D.J. 1979. Weak links in the weak link. *Canadian Public Policy* 3: 307-17.

Davis, H.C. 1993. Is the metropolitan Vancouver economy uncoupling from the rest of the province? *B.C. Studies* 98: 3-19.

–, and T.A. Hutton. 1989. The two economies of British Columbia. *B.C. Studies* 82: 3-15.

Delgamuukw v. *British Columbia,* [1997] 3 S.C.R. 1010.

District of North Cowichan. N.d. *Our Communities, Our Forests, Our Future* (brochure). Duncan: District of North Cowichan.

Dobie, J. 1971. *Economies of Scale in Sawmilling in B.C.* PhD dissertation, University of Oregon.

Doeringer, P., and M. Piore. 1971. *Internal Labour Markets and Manpower Analysis.* Lexington, MA: D.C. Heath.

Drache, D. 1982. Harold Innis and Canadian capitalist development. *Canadian Journal of Political and Social Theory* 6: 25-49.

–, ed. 1995. *Staples, Markets and Cultural Change: Harold A. Innis.* Montreal and Kingston: McGill-Queen's University Press.

Drengson, A.R., and D.M. Taylor, eds. 1997. *Ecoforestry: The Art and Science of Sustainable Forest Use*. Gabriola Island, BC: New Society Publishers.
Drushka, K. 1993. Forest tenure: Forest ownership and the case for diversification. In *Touch Wood: BC Forests at the Crossroads,* ed. K. Drushka, B. Nixon, and R. Travers, 1-22. Madeira Park, BC: Harbour Publishing.
–. 1999a. A case of stunted growth. *Vancouver Sun* March 31: D1-D2, D19.
–. 1999b. No quick, easy fixes for the forest industry. *Vancouver Sun* May 19: H2.
–, B. Nixon, and R. Travers, eds. 1993. *Touch Wood: BC Forests at the Crossroads*. Madeira Park, BC: Harbour Publishing.
Durrant, P. 1990a. Lumber deals nags B.C. *Province* January 14: 33.
–. 1990b. U.S. takes tough stand. *Province* November 18: 48.
Eco, U. 1996. *The Island of the Day Before*. New York: Penguin.
Edgell, M.C.R. 1987. Forestry. In *British Columbia: Its Resources and People,* ed. C.N. Forward, 109-38. Western Geographical Series No. 22. Victoria: University of Victoria.
Edgington, D.W. 1992. *Japanese Direct Investment in Canada: Recent Trends and Prospects*. B.C. Geographical Series No. 49. Department of Geography, University of British Columbia.
–, and R. Hayter. 1997. International trade, production chains and corporate strategies: Japan's timber trade with British Columbia. *Regional Studies* 31: 149-64.
Egan, B., and S. Klausen. 1998/9. Female in a forest town: The marginalization of women in Port Alberni's economy. *B.C. Studies* 118: 5-40.
Ekins, P. 1992. *A New World Order: Grassroots Movements for Global Change*. London: Routledge.
Ernst and Young. 1998. *Technology and the B.C. Forest Products Sector*. Vancouver: Science Council of British Columbia.
Farley, A.L. 1972. The forest resource. In *British Columbia: Studies in Canadian Geography,* ed. J. L. Robinson, 87-118. Toronto: University of Toronto Press.
–. 1979. *Atlas of British Columbia*. Vancouver: UBC Press.
Federal Treaty Negotiations Office. N.d. Treaty Negotiations in British Columbia (map). Issued jointly by the Federal Treaty Negotiations Office, Indian and Northern Affairs Canada, and the Information Management Branch, British Columbia Ministry of Aboriginal Affairs. Np.
Financial Post Company. 1998. *MacMillan Bloedel Ltd*. Toronto: The Financial Post Data Group.
Fisher, R. 1977. *Contact and Conflict: Indo-European Relations in British Columbia, 1774-1890*. Vancouver: UBC Press.
Flanagan, W.F. 1998. Piercing the veil of real property law: *Delgamuukw v. British Columbia*. *Queen's Law Journal* 24: 279-326.
Florida, R. 1996. Lean and green: The move to environmentally conscious manufacturing. *California Management Review* 39: 80-105.
Forest Industries. 1982. New Zealand firm to buy Crown Zellerbach assets. *Forest Industries* December: 11.
Forest Research Council of British Columbia. 1986. *Fifth Annual Report*. Richmond: Secretariat on Forestry Research and Development.
Forestry Canada. 1996. *Selected Forestry Statistics 1995*. Ottawa: Forestry Canada, Economics Branch.
Forgacs, O. 1997. The British Columbia forest industry: Transition or decline? In *Troubles in the Rainforest: British Columbia's Forest Economy in Transition,* ed. T. Barnes and R. Hayter, 167-78. Canadian Western Geographical Series No. 33. Victoria: Western Geographical Press.
Forintek Canada, and J. McWilliams. 1993. *Structure and Significance of the Value-Added Wood Product Industry in British Columbia*. Victoria: Government of Canada.
Freeman, C. 1974. *The Economics of Industrial Innovation*. London: Penguin.
–. 1987. *Technology Policy and Economic Performance: Lessons from Japan*. London: Pinter Publishers.
–. 1988. Japan: A new national system of innovation? In *Technical Change and Economic Theory,* ed. G. Dosi, C. Freeman, R. Nelson, G. Silverberg, and L. Soete, 330-48. London: Pinter Publishers.
–. 1995. The national system of innovation in historical perspective. *Cambridge Journal of Economics* 19: 5-24.

–, and C. Perez. 1988. Structural crises of adjustment, business cycles and investment behaviour. In *Technical Change and Economic Theory,* ed. G. Dosi, C. Freeman, R. Nelson, G. Silverberg, and L. Soete, 38-66. London: Pinter Publishers.

Freudenburg, W.R. 1992. Addictive economies: Extractive industries and vulnerable localities in a changing world economy. *Rural Sociology* 57: 305-32.

Friedmann, J., and C. Weaver. 1978. *Territory and Function: The Evolution of Regional Planning*. London: Edward Arnold.

Fruin, M. 1992. *The Japanese Enterprise System*. Oxford: Clarendon Press.

Fulton, J. 1999. Trees and jobs: We have a plan. *Vancouver Sun* January 20: A11.

Galbraith, J.K. 1952. *American Capitalism*. Boston: Houghton Mifflin.

–. 1966. *The New Industrial State*. Boston: Houghton Mifflin.

Galois, R., and R. Hayter. 1991. The wheel of fortune: British Columbia lumber and the global economy. In *Essays in Honour of Archie MacPherson,* ed. P.M. Koroscil, 169-202. Mission, BC: R.S. Graphics.

Garner, R. 1996. *Environmental Politics*. New York: Prentice Hall/Harvester Wheatsheaf.

Gertler, M. 1988. The limits to flexibility: Comments on the post-Fordist vision of production and its geography. *Transactions of the Institute of British Geographers* 17: 259-78.

Gibbon, A. 1999a. U.S. lumber move could cost B.C. 1,000 jobs. *Globe and Mail* June 9: B11.

–. 1999b. BC approves MacMillan Bloedel sale. *Globe and Mail* October 6: B3.

Gould, E. 1975. *Logging: British Columbia Logging History*. Saanichton, BC: Hancock House Publishers.

Graham, J., and K. St. Martin. 1990. Resources and restructuring in the international solid wood products industry. *Geoforum* 21: 289-302.

Grass, E. 1987. *Employment Changes During Recession: The Case of the British Columbia Forest Products Manufacturing Industries*. MA thesis, Simon Fraser University.

–. 1990. Employment and production: The mature stage in the life-cycle of a sawmill: Youbou, British Columbia, 1929-89. PhD thesis, Simon Fraser University.

–, and R. Hayter. 1989. Employment change during recession: The experience of forest product manufacturing plants in British Columbia, 1981-1985. *Canadian Geographer* 33: 240-52.

Guest, S., J.K. Wright, and E.M. Teclaffy, eds. 1956. *A World Geography of Forest Resources*. American Geographical Society, Special Publication No. 33, 155-69. New York: Ronald Press.

Gunton, T. 1997. Forestry land use and public policy in British Columbia: The dynamics of change. In *Troubles in the Rainforest: British Columbia's Forest Economy in Transition,* ed. T. Barnes and R. Hayter, 65-74. Canadian Western Geographical Series No. 33. Victoria: Western Geographical Press.

–, and J. Richards, eds. 1987. *Resource Rents and Public Policy in Western Canada*. Halifax: Institute for Research on Public Policy.

Guthrie, J.A., and G.A. Armstrong. 1961. *Western Forest Industries*. Baltimore: Johns Hopkins.

Haden-Guest, S., J.K. Wright, and E.M. Teclaffy, eds. 1956. *A World Geography of Forest Resources*. American Geographical Society Special Publication No. 33. New York: Ronald Press.

Haglund, D.G., ed. 1989. *The New Geopolitics of Minerals*. Vancouver: UBC Press.

Haley, D. 1980. A regional comparison of stumpage values in British Columbia and the United States Pacific Northwest. *Forestry Chronicle* 56: 225-30.

–. 1985. The forest tenure system as a constraint on efficient timber management – Problems and solutions. *Canadian Public Policy* 11 Supplement: 315-20.

Halseth, G. 1999. Resource town employment: Perceptions in small town British Columbia. *Tijdschrift voor Economische en Sociale Geografie* 90: 196-210.

–. Forthcoming. "We came for the work": Situating employment in BC's small resource-based communities. *Canadian Geographer.*

Hamilton, G. 1994. New lath turns things around at old Fletcher panel plant. *Vancouver Sun* June 9: D10, D12.

–. 1996. Forest firms seek fortunes outside B.C. *Vancouver Sun* March 30: B1, B7.

–. 1997. Victoria forms forest tenure committee. *Vancouver Sun* December 4: D4.

–. 1998a. 2,700 jobs disappearing as MB returns to basics. *Vancouver Sun* January 22: D1, D8.

–. 1998b. Restructuring cost gives MacBlo biggest forest-industry loss. *Vancouver Sun* February 12: D1, D8.
–. 1998c. Forest assistance too late, report says. *Vancouver Sun* March 5: D1.
–. 1998d. Forest industry settles stumpage rate cut split. *Vancouver Sun* March 23: D1, D2.
–. 1998e. B.C. chops red tape to save forestry firms $300 million. *Vancouver Sun* April 3: A1-A2.
–. 1998f. Departing dean calls for a change in forestry. *Vancouver Sun* August 4: F1, F3.
–. 1998g. Trash NAFTA to save B.C. forest industry, CEO says. *Vancouver Sun* October 9: F1, F2.
–. 1998h. Flush Fletcher buys Asian firm. *Vancouver Sun* October 29: F1, F2.
–. 1998i. Smaller sawmillers feel pinch but stay alive. *Vancouver Sun* November 23: C1, C3.
–. 1998j. Old-growth campaign targets business leaders. *Vancouver Sun* December 8: D1-D2.
–. 1998k. Greenpeace, MB teaming to market Clayoquot timber. *Vancouver Sun* December 10: F1, F13.
–. 1998l. IWA threatens Hallmark boycott. *Vancouver Sun* December 9: D1-D2.
–. 1998m. Hallmark remains B.C. buyer. *Vancouver Sun* December 12: E1, E20.
–. 1999a. New CEO wins kudos for MacBlo turnaround. *Vancouver Sun* February 12: F1, F3.
–. 1999b. TimberWest phases out clear cut logging in B.C. *Vancouver Sun* May 11: D1, D11.
–. 1999c. U.S. wood-products ruling threatens jobs Zirnhelt says. *Vancouver Sun* June 8: C1, C11.
–. 1999d. Forest firm wins huge case in Indian land deal. *Vancouver Sun* July 30: F1.
–. 1999e. MacBlo-Weyerhaeuser deal losing some allure. *Vancouver Sun* September 25: D1-D2.
–. 1999f. US seeks end to lumber pact, open market for log exports. *Vancouver Sun* October 1: A1-A2.
Hammond, H. 1991. *Seeing the Forest among the Trees*. Winlaw, BC: Polestar Press.
–. 1993. Forest practices: Putting wholistic forest use into practice. In *Touch Wood: BC Forests at the Crossroads,* ed. K. Drushka, B. Nixon, and R. Travers, 96-136. Madeira Park, BC: Harbour Publishing.
Hanel, P. 1985. *La Technologie et les Exportations Canadiennes du Materiel pour la Filière Bois-papier*. Montreal: L'Institut de Recherches Politiques.
Hardwick, W.G. 1963. *Geography of the Forest Industry of British Columbia*. Occasional Papers in Geography No. 5. Vancouver: Canadian Association of Geographers, British Columbia Division.
–. 1964. Port Alberni, British Columbia, Canada: An integrated forest industry in the Pacific Northwest. In *Focus on Geographical Activity: A Collection of Original Essays,* eds. R.S. Thoman and D.J. Patton, 60-66. New York: McGraw Hill.
–. 1974. *Vancouver*. Toronto: Collier-Macmillan.
Harris, R.C. 1997. On distance, power and (indirectly) the forest industry. In *Troubles in the Rainforest: British Columbia's Forest Economy in Transition,* ed. T. Barnes and R. Hayter, 207-32. Canadian Western Geographical Series No. 33. Victoria: Western Geographical Press.
Harvey, D. 1988. The geographical and geopolitical consequences of the transition from Fordist to flexible accumulation. In *America's New Market Geography: Nation, Region and Metropolis,* eds. G. Sternlieb and J.W. Hughes, 101-34. Rutgers: Centre for Urban Policy Research.
–. 1989. From managerialism to entrepreneurialism: The transformation in urban governance in late capitalism. *Geografiska Annaler* 71B: 3-17.
–. 1990. *The Condition of Post-Modernity: An Inquiry into the Nature of Cultural Change*. Oxford: Blackwell.
Hay, E. 1993. *Recession and Restructuring in Port Alberni: Corporate, Household and Community Coping Strategies*. MA thesis, Simon Fraser University.
Hayter, R. 1973. *An Examination of Growth Patterns and Locational Behaviour of Multi-Plant Forest Product Corporations in British Columbia*. PhD dissertation, University of Washington.
–. 1976. Corporate strategies and industrial change in the Canadian forest products industries. *Geographical Review* 66: 209-28.
–. 1978a. Forestry in British Columbia: A resource basis of Vancouver's dominance. In *Vancouver: Western Metropolis,* ed. L.J. Evenden, 95-115. Western Geographical Series No. 16. Victoria: University of Victoria.

–. 1978b. Identifying "quits" and "stays" from employee records: A note. *Industrial Management* 20: 11-3.
–. 1978c. Locational decision-making in a resource based manufacturing sector: Case studies from the pulp-and-paper industry of British Columbia. *Professional Geographer* 30: 240-9.
–. 1979. Labour supply and resource based manufacturing in isolated communities: The experience of pulp-and-paper mills in north central British Columbia. *Geoforum* 10: 163-77.
–. 1980. Technological capability in the forest-product sector of British Columbia: An exploratory inquiry. Discussion Paper No. 13. Department of Geography, Simon Fraser University.
–. 1981. Patterns of entry and the role of foreign-controlled investments in the forest-product sector of British Columbia. *Tijdschrift voor Economische en Sociale Geografie* 78: 99-113.
–. 1982a. Research and development in the Canadian forest-product sector – Another weak link? *Canadian Geographer* 26: 256-63.
–. 1982b. Truncation, the international firm and regional policy. *Area* 14: 277-82.
–. 1985. The evolution and structure of the Canadian forest-product sector: An assessment of the role of foreign ownership and control. *Fennia* 163: 439-50.
–. 1986a. The export dynamics of firms in traditional industries during recession. *Environment and Planning A* 18: 729-50.
–. 1986b. Export performance and export potentials: Western Canadian exports of manufactured end products. *Canadian Geographer* 30: 26-39.
–. 1987. Technology and jobs: Innovation policy in British Columbia and the forest-product sector. In *Technical Change, Employment and Spatial Policy,* ed. K. Chapman and G. Humphreys, 215-32. Oxford: Basil Blackwell.
–. 1988. *Technology and the Canadian Forest-Product Industries: A Policy Perspective.* Background Study 54. Ottawa: Science Council of Canada.
–. 1990. Canada's trade and investment links in the Pacific region. In *Industrial Transformation in Australia and Canada,* ed. R. Hayter and P.D. Wilde, 226-38. Ottawa: Carleton University Press.
–. 1992. International trade relations and regional industrial adjustment: The implications of the 1982-86 Canadian-US softwood lumber dispute for British Columbia. *Environment and Planning A* 24: 153-70.
–. 1993. International trade and the Canadian forest industries: The paradox of the North American free trade agreements. *Zeitschrift für Kanada-Studien* 23: 81-94.
–. 1996. Technological imperatives in resource sectors: Forest products. In *Canada and the Global Economy,* ed. J.N.H. Britton, 101-22. Montreal-Kingston: McGill-Queen's University Press.
–. 1997a. *The Dynamics of Industrial Location: The Factory, The Firm and the Production System.* London: Wiley.
–. 1997b. High performance organizations and employment flexibility: A case study of in-situ change at the Powell River paper mill, 1980-94. *Canadian Geographer* 41: 26-40.
–, and T. Barnes. 1990. Innis' staple theory, exports and recession: British Columbia, 1981-86. *Economic Geography* 66: 156-73.
–, and T. Barnes. 1992. Labour market segmentation, flexibility and recession: A British Columbian case study. *Environment and Planning C* 10: 333-5.
–, and T. Barnes. 1997. The restructuring of British Columbia's coastal forest sector: Flexibility perspectives. In *Troubles in the Rainforest: British Columbia's Forest Economy in Transition,* ed. T. Barnes and R. Hayter, 181-202. Victoria: Western Geographical Series.
–, and D.W. Edgington. 1997. Cutting against the grain: A case study of MacMillan Bloedel's Japan strategy. *Economic Geography* 73: 187-213.
–, and D.W. Edgington. 1999. Getting tough and getting smart: Politics of the North American-Japan wood products trade. *Environment and Planning C: Government and Policy* 17: 319-44.
–, E. Grass, and T. Barnes. 1994. Labour flexibility: A tale of two mills. *Tijdschrift voor Economische en Sociale Geografie* 85: 25-38.

–, and J. Holmes. 1993. *Booms and Busts in the Canadian Paper Industry: The Case of the Powell River Paper Mill.* Discussion Paper No. 27. Department of Geography, Simon Fraser University.

–, and J. Holmes. 1994. *Recession and Restructuring at Powell River, 1980-94: Employment and Employment Relations in Transition.* Discussion Paper No. 28. Department of Geography, Simon Fraser University.

–, and J. Patchell. 1993. Different trajectories in the social division of labour: The cutlery industry in Sheffield, England and Tsubame, Japan. *Urban Studies* 30: 1427-45.

–, J. Patchell, and K. Rees. 1999. Business segmentation and location revisited: Innovation and the terra incognita of large firms. *Regional Studies* 33: 425-42.

–, and D. Soyez. 1996. Clearcut issues: German environmental pressure and the British Columbia forest sector. *Geographische Zeitschrift* 84: 143-56.

Hecht, S.B., and A. Cockburn. 1989. *The Fate of the Forest: Developers, Destroyers and Defenders of the Amazon.* New York: Verso.

Hiebert, D. 1990. Discontinuity and the emergence of flexible production: Garment production in Toronto 1901-1931. *Economic Geography* 66: 229-53.

Hirst, P., and G. Thompson. 1992. The problem of globalization: International economic relations, national economic management and the formation of trading blocs. *Economy and Society* 21: 357-95.

–, and J. Zeitlin. 1991. Flexible specialization versus post-Fordism: Theory, evidence and policy implications. *Economy and Society* 20: 1-56.

Hogben, D. 1998. Environmentalists toast MB's clear-cut decision. *Vancouver Sun* June 11: D1, D8.

–, and S. Hunter. 1998. MacBlo president talks of halting clear-cuts in old-growth forests. *Vancouver Sun* April 24: A1-A2.

Holm, B. 1998. Value-added wood is a growth industry. *Vancouver Sun* June 23: A14.

Hopgood, A. (Enterprises Ltd.). 1986. *The Potential for New Technologies in Canada's Forest Sector.* Ottawa: Ministry of State for Science and Technology.

Hull, J.P. 1985. *Science and the Canadian Pulp and Paper Industry 1903-33.* PhD dissertation, York University.

Humphrey, C.R. 1990. Timber-dependent communities. In *American Rural Communities,* ed. A.E. Luloff and L.E. Swanson, 34-60. Boulder, CO: Westview Press.

Hunter, J. 1998. MacBlo to end clearcutting in old-growth coast forests. *Vancouver Sun* June 10: A1-A2.

–. 1999. End feud, new COFI boss urges executives. *Vancouver Sun* April 24: H1.

–, and C. McInnes. 1999. First Nations vow to kill MB land-settlement deal. *Vancouver Sun* April 29: D1, D12.

Hutton, T., and D. Ley. 1987. Location, linkages and labour: The downtown complex of corporate activities in a medium sized city. *Economic Geography* 63: 126-41.

Hutton, T.A. 1997. Vancouver as a control centre for British Columbia's resource hinterland: Aspects of linkage and divergence in a provincial staple economy. In *Troubles in the Rainforest: British Columbia's Forest Economy in Transition,* ed. T. Barnes and R. Hayter, 233-61. Canadian Western Geographical Series No. 33. Victoria: Western Geographical Press.

Ide, S., and A. Takeuchi. 1980. Jiba Sangyo: Localized industry. In *Geography of Japan,* ed. T. Shoin, 299-319. Tokyo: Association of American Geographers.

Industry, Science and Technology Canada. 1989. *British Columbia's Producers of Manufactured and Speciality Solid Wood Products.* Vancouver: Industry, Science and Technology Canada.

Innis, H.A. 1930. *The Fur Trade in Canada: An Introduction to Canadian Economic History.* Toronto: University of Toronto Press.

–. 1933. *Problems of Staple Production in Canada.* Toronto: Ryerson.

–. 1946. *Political Economy in the Modern State.* Toronto: Ryerson.

–. 1956. The teaching of economic history in Canada. In *Essays in Canadian Economic History,* ed. M.Q. Innis, 3-16. Toronto: University of Toronto Press.

–. 1967. The importance of staple products. In *Approaches to Canadian Economic History,* ed. W.T. Easterbrook and M.H. Watkins, 16-9. Toronto: McClelland and Stewart.

Jacques, R., and G.R. Fraser. 1989. The forest sector's contribution to the Canadian economy. *Forestry Chronicle* April: 93-6.
Johnston, R.J. 1984. The world is our oyster. *Transactions of the Institute of British Geographers* 9: 443-59.
Jordan, D. 1998. Value-added isn't adding much value. *Business in Vancouver* October 20-6: 1, 7-8.
Jurasek, L., and M.G. Paice. 1984. *Biotechnology in the Pulp-and-paper industry*. Manuscript report. Ottawa: Science Council of Canada.
Kato, T. 1992. Structural changes in Japanese forest product imports during the 1980s. In *The Current State of Japanese Forestry (VIII): Its Problems and Future,* 87-102. Japanese Forest Economic Society, Tokyo.
Keohane, R.O., and J.S. Nye. 1977. *Power and Interdependence: World Politics in Transition.* Boston, MA: Little Brown.
Kerr, D.P. 1966. Metropolitan dominance in Canada. In *Canada: A Geographical Interpretation,* ed. J. Warkentin, 531-55. Toronto: Methuen.
Kimmins, H. 1992. *Balancing Act: Environmental Issues in Forestry*. Vancouver: UBC Press.
Knight, R. 1978. *Indians at Work: An Informal History of Native Indian Labour in British Columbia 1858-1930*. Vancouver: New Star Books.
Lee, R.G. 1984. Sustained yield and social order. In *History of Sustained-Yield Forestry: A Symposium,* ed. H.K. Steen, 90-100. Portland: Forest History Society.
LeHeron, R.B. 1990. Resource developments and the evolution of New Zealand forestry companies. In *Industrial Transformation and Challenge in Australia and Canada,* ed. R. Hayter and P.D. Wilde, 195-212. Ottawa: Carleton University Press.
–, and M. Roche. 1985. Expanding exotic forestry and the extension of a competing use for rural land in New Zealand. *Journal of Rural Studies* 1: 211-29.
Levitt, K. 1970. *Silent Surrender: The Multinational Corporation in Canada*. Toronto: Macmillan.
Ley, D., and T.A. Hutton. 1987. Vancouver's corporate complex and producer services sector: Linkages and divergence within a provincial staples economy. *Regional Studies* 21: 413-24.
Lipietz, A. 1986. New tendencies in the international division of labour: Regimes of accumulation and modes of regulation. In *Production, Work and Territory: The Geographical Anatomy of Industrial Capitalism,* ed. A.J. Scott and M. Storper, 16-40. London: Routledge.
Lorenz, E.H. 1992. Trust, community and cooperation: Toward a theory of industrial districts. In *Pathways to Industrialization and Regional Development,* ed. M. Storper and A.J. Scott, 195-204. London: Routledge.
Lower, A. 1973. *Great Britain's Woodyard: Britain, America and the Timber Trade 1763-1867.* Montreal: McGill-Queen's University Press.
Lucas, R.A. 1971. *Minetown, Milltown, Railtown: Life in Canadian Communities of Single Industry*. Toronto: University of Toronto.
Lundie, J.A. 1960. An outline of the history and development of the Powell River Company. In *Fifty Years of Paper Making*. Powell River: Powell River News.
Lundvall, B.A., ed. 1992. *National Systems of Innovation: Toward a Theory of Innovation and Interactive Learning*. London: Pinter.
Lush, P. 1992. Primex: A cut above at pleasing the Japanese. *Globe and Mail* February 17: B1, B4.
Lyke, J. 1996. Forest product certification revisited. *Journal of Forestry* 94: 16-20.
Mackay, D. 1982. *Empire of Wood: The MacMillan Bloedel Story*. Vancouver: Douglas and McIntyre.
Mackintosh, W.A. 1967. Economic factors in Canadian history. In *Approaches to Canadian Economic History,* ed. W.T. Easterbrook and M.H. Watkins, 1-15. Toronto: McClelland and Stewart.
Maclean's. 1991a. A powerful screen attack. May 13: 36.
Maclean's. 1991b. A clear cut fight. B.C. logging becomes an international issue. June 10: 50.
MacMillan Bloedel. 1965. *Annual Report*. Vancouver.
–. 1979. *Annual Report*. Vancouver.
–. 1990. *Annual Report*. Vancouver.
–. 1996a. *Annual Report*. Vancouver.

–. 1996b. *Annual Environmental Report.* Vancouver.
–. 1997. *Annual Report.* Vancouver.
Mandel, E. 1980. *Long Waves of Capitalist Development.* Cambridge: Cambridge University Press.
Marchak, M.P. 1983. *Green Gold: The Forest Industry in British Columbia.* Vancouver: UBC Press.
–. 1990. Forestry industry towns in British Columbia. In *Community and Forestry,* ed. R.G. Lee, D.R. Field, and W.R. Burch Jr., 95-106. Boulder, CO: Westview Press.
–. 1991. For whom the bell tolls: Restructuring in the global forest industry. *B.C. Studies* 90: 3-24.
–. 1995. *Logging the Globe.* Montreal: McGill-Queen's University Press.
–. 1997. A changing global context for British Columbia's forest industry. In *Troubles in the Rainforest: British Columbia's Forest Economy in Transition,* ed. T. Barnes and R. Hayter, 149-64. Canadian Western Geographical Series No. 33. Victoria: Western Geographical Press.
–, S.L. Aycock, and D.M. Herbert. 1999. *Falldown: Forestry Policy in British Columbia.* Vancouver: David Suzuki Foundation.
Marshall, H., F.A. Southard, and K.W. Taylor. 1936. *Canadian-American Industry: A Study in International Investment.* Toronto: Ryerson.
Marshall, R., and M. Tucker. 1992. *Thinking for a Living: Work Skills and the Future of the American Economy.* New York: Basic Books.
Maser, C. 1990. *The Redesigned Forest.* Toronto: Stoddart.
Mather, A.S. 1990. *Global Forest Resources.* London: Bellhaven.
–. 1992. The forest transition. *Area* 24: 367-79.
–, and C.L. Needle. 1998. The forest transition: A theoretical base. *Area* 30: 117-224.
–, C.L. Needle, and J. Fairbairn. 1999. Environmental kuznets curves and forest trends. *Geography* 84: 55-65.
McAllister, I., and K. McAllister. 1997. *The Great Bear Rain Forest.* Madeira Park, BC: Harbour Publishing.
McWilliams, J. 1991. *Structure and Significance of the Value-Added Wood Products Industry in British Columbia.* Victoria: Forest Resource Development.
Meiggs, R. 1982. *Trees and Timber in the Ancient Mediterranean World.* Oxford: Clarendon Press.
Melody, W.H., L. Salter, and P. Heyer, eds. 1981. *Culture, Communication, and Dependency. The Tradition of H.A. Innis.* Norwood, NJ: Ablex.
Mensch, A. 1979. *Stalemate in Technology: Innovation Overcame the Depression.* Cambridge, MA: Balinger.
Mertyl, S. 1999. Fletcher considers paper-arm revamp. *Vancouver Sun* May 8: D1, D12.
M'Gonigle, R.M. 1997. Reinventing British Columbia: The path to a sustainable forest sector in British Columbia. In *Troubles in the Rainforest: British Columbia's Forest Economy in Transition,* ed. T. Barnes and R. Hayter, 15-35. Canadian Western Geographical Series No. 33. Victoria: Western Geographical Press.
–, and B. Parfitt. 1994. *Forestopia.* Madeira Park, BC: Harbour Publishing.
Ministry of Forests. 1971-1992. *Annual Reports.* Victoria: Ministry of Forests.
–. 1984. *Forest and Range Resource Analysis.* Victoria: Ministry of Forests.
–. 1990. *Timber Supply Area Summary Report.* Victoria: Ministry of Forests.
–. 1993. *Annual Report 1992/3.* Victoria: Ministry of Forests.
–. 1996a. *Annual Report 1995/6.* Victoria: Ministry of Forests.
–. 1996b. *Summary of Timber Supply Review Results 1992-96.* Victoria: Ministry of Forests.
–. 1997. Forest Jobs for BC. News release. Government of British Columbia.
–. n.d. *Forest Jobs for BC: Jobs and Timber Accord.* Victoria: Government of British Columbia.
Moore, W.E. 1963. *Social Change.* Englewood Cliffs, NJ: Prentice Hall.
Munro, M. 1997. Forest firms reap benefits for cuts to gas emissions. *Vancouver Sun* December 12: A1, A10.
Nakamura, M., and I. Vertinsky. 1994. *Japanese Economic Policies and Growth: Implications for Business in Canada and North America.* Edmonton: University of Alberta.
Nathan, H. 1993. Aboriginal forestry: The role of First Nations. In *Touch Wood: BC Forests at the Crossroads,* ed. K. Drushka, B. Nixon, and R. Travers, 137-70. Madeira Park, BC: Harbour Publishing.

Neill, R. 1972. *A New Theory of Value: The Canadian Economics of H.A. Innis*. Toronto: University of Toronto Press.
Nelson, R. 1988. Institutions supporting technical change in the United States. In *Technical Change and Economic Theory,* ed. G. Dosi, C. Freeman, R. Nelson, G. Silverberg, and L. Soete, 312-39. New York: Pinter.
Nemetz, P., ed. 1992. *Emerging Issues in Forest Policy*. Vancouver: UBC Press.
Nixon, R. 1993. Public participation: Changing the way we make forest decisions. In *Touch Wood: BC Forests at the Crossroads,* ed. K. Drushka, B. Nixon, and R. Travers, 23-66. Madeira Park, BC: Harbour Publishing.
NLK and Associates. 1992. *The Pulp and Paper Sector in British Columbia: The Next Twenty Years*. NLK and Associates, no place of publication.
North, D.C. 1955. Location theory and regional economic growth. *Journal of Political Economy* 63: 243-58.
Ofori-Amoah, B. 1990. *Technology Choice in a Global Industry: The Case of the Twin-Wire in Canada*. PhD dissertation, Simon Fraser University.
–, and R. Hayter. 1989. Labour turnover characteristics at the Kitimat pulp and paper mill: A log-linear analysis. *Environment and Planning A* 21: 1491-1510.
Olsen, W.H. 1963. *Water over the Wheel*. Chemainus: Chemainus Valley History Society.
O'Neil, P. 1999. Ottawa shuts out Victoria in talks with B.C. forest industry. *Vancouver Sun* May 22: A1-A2.
–, and J. Beatty. 1999. Ottawa rejects plea to bail out B.C.'s troubled forest industry. *Vancouver Sun* May 13: A1-A2.
O'Riordan, T. 1976. *Environmentalism*. London: Pion.
Owen, S. 1995. *British Columbia's Strategy for Sustainability*. Victoria: Commission on Resources and Environment.
Paehlke, R. 1996. Environmental challenges in democratic practice. In *Democracy and the Environment,* ed. W.M. Lafferty and J. Meadowcroft, 18-38. Cheltenham: Edward Edgar.
Palmer, V. 1998. The premier misses a point in his pulp mill musings. *Vancouver Sun* July 24: A18.
Parker, I. 1988. Harold Innis as a Canadian geographer. *Canadian Geographer* 32: 63-9.
Patchell, J. 1993. From production systems to learning systems: Lessons from Japan. *Environment and Planning A* 25: 797-815.
–. 1996. Kaleidoscope economies: The processes of cooperation, competition, and control in regional economic development. *Annals of the Association of American Geographers* 86: 481-506.
–, and R. Hayter. 1992. Dynamics of adjustment and the social division of labour in the Tsubame cutlery industry. *Growth and Change* 23: 199-216.
Pearse, P.H. 1976. *Timber Rights and Forests Policy in British Columbia: Report of the Royal Commission on Forest Resources*. Victoria: Queen's Printer.
Peck, J. 1996. *Work Place: The Social Regulation of Labour Markets*. New York: Guilford.
Peel Commission. 1991. *Forest Resources Commission: The Future of Our Forests*. Victoria: Forest Resources Commission.
Percy, M.B. 1986. *Forest Management and Economic Growth in British Columbia*. A study prepared for the Economic Council of Canada. Ottawa: Supply and Services.
–, and C. Yoder. 1987. *The Softwood Lumber Dispute and Canada-US Trade in Natural Resources*. Halifax: Institute for Public Research.
Persson, R. 1974. World forest resources: Review of the world's forest resources in the early 1970s. *Research Note 17*. Stockholm: Royal College of Forestry.
Pfister, R.L. 1963. External trade and regional growth: A case study of the Pacific Northwest. *Economic Development and Cultural Change* 11: 134-51.
Pinfield, L., C.G. Hoyt, R.J. Clifford, and R.D. Algar. 1974. *Manpower Planning in Northern British Columbia*. Discussion Paper No. 7401-2. Department of Economics and Commerce, Simon Fraser University.
Piore, M., and C. Sabel. 1984. *The Second Industrial Divide: Possibilities for Prosperity*. New York: Basic Books.
Pollard, S. 1981. *Peaceful Conquest: The Industrialization of Europe 1760-1970*. Oxford: Oxford University Press.

Porteous, J.D. 1987. Single enterprise communities. In *British Columbia: Its Resources and People,* ed. C.N. Forward, 383-400. Western Geographical Series No. 22. Victoria: University of Victoria.

Porter, M.E. 1985. *Competitive Advantage.* New York: Free Press.

–. 1993. *Canada at the Crossroads: The Reality of a New Competitive Environment.* Ottawa: Monitor Company Canada.

Powles, J.M. 1993. Cooperative program behind Western Canada's wood products exports to Japan. *Canada-Japan Trade Council Newsletter* May-June: 5-9.

Price Waterhouse. 1992a. *Canada-British Columbia Forest Resource Development Agreement: Performance of the Value-Added Wood Products Industry in British Columbia.* Vancouver: Price Waterhouse.

–. 1992b. *The Forest Industry in British Columbia 1991.* Vancouver: Price Waterhouse.

–. 1992c. *Performance of the Value-Added Wood Products Industry in British Columbia.* Vancouver: Price Waterhouse.

–. 1995. *Forest Alliance of British Columbia: Analysis of Recent British Columbia Government Forest Policy and Land Use Initiatives.* Vancouver: Price Waterhouse.

–. 1997. *The Forest Industry in British Columbia 1997.* Vancouver: Price Waterhouse.

–. 1998. *The Forest Industry in British Columbia 1998.* Vancouver: Price Waterhouse.

Princen, T. 1994. NGOs: Creating a niche in environmental diplomacy. In *Environmental NGOs in World Politics,* ed. T. Princen and M. Finger, 29-47. London: Routledge.

Pynn, L. 1999. MB, environmentalists agree to pact on Clayoquot logging. *Vancouver Sun* June 16: A3.

Randall, J.E., and R.G. Ironside. 1996. Communities on the edge: An economic geography of resource-dependent communities in Canada. *Canadian Geographer* 40: 17-35.

Raumolin, J. 1984. Formation of sustained-yield forestry system in Finland. In *History of Sustained Yield Forestry: A Symposium,* ed. H.K. Steen, 155-69. Portland, OR: Forest History Society.

–. 1992. The diffusion of technology in the forest and mining sector in Finland. In *Mastering Technology Diffusion – The Finnish Experience,* ed. S. Vuori and P. Ylä-Antilla, 321-77. Helsinki: Research Institute of the Finnish Economy, Series B.

Reed, F.L.C., and Associates. 1978. *Forest Management in Canada.* Ottawa: Environment Canada, Canadian Forestry Service.

–. N.d. (c. 1996). Why B.C.'s annual timber harvest is falling (it's not what you think). Editorial. Vancouver: Forestry Alliance.

Reed, M., and A.M. Gill. 1997. Community economic development in a rapid growth setting: A case study of Squamish. In *Troubles in the Rainforest: British Columbia's Forest Economy in Transition,* ed. T. Barnes and R. Hayter, 263-85. Canadian Western Geographical Series No. 33. Victoria: Western Geographical Press.

Rees, J.A. 1985. *Natural Resources: Allocation, Economics and Policy.* London: Methuen.

Rees, K.G. 1993. Flexible specialization and the case of the remanufacturing industry in the Lower Mainland of British Columbia. MA thesis, Simon Fraser University.

–, and R. Hayter. 1996. Flexible specialization, uncertainty and the firm: Enterprise strategies in the wood-remanufacturing industry of the Vancouver metropolitan area, British Columbia. *Canadian Geographer* 40: 203-19.

Reiffenstein, T. 1999. *A Production Chain in Development: The Export of Premanufactured Homes from BC to Japan.* MA thesis, Simon Fraser University.

Resnick, P. 1985. B.C. capitalism and the empire of the Pacific. *B.C. Studies* 67: 29-46.

Rinehart, D. 1998. Historic deal with Nisga'a a wonderful day for B.C. *Vancouver Sun* July 16: A1, A14.

Rosenau, J.N. 1990. *Turbulence in world politics.* Princeton: Princeton University Press.

Rosenbluth, G. 1977. Canadian policy on foreign ownership and control of business. In *Canadian Economy: Problems and Policies,* ed. G.C. Ruggeri, 73-8. Toronto: Gage Educational Press.

Ross, D.A. 1982. Canadian foreign policy and the Pacific Rim: From national security anxiety to creative economic cooperation. In *Politics and the Pacific Rim: Perspectives on the 1980s,* ed. F.Q. Quo, 27-58. Burnaby, BC: Simon Fraser University.

Ross, P.S. 1973. *A Study of Manpower in the Logging and Sawmilling Industry of B.C.* Vancouver.

Rotstein, A. 1977. Innis: The alchemy of fur and wheat. *Journal of Canadian Studies* 12: 6-31.

Rumelt, R.P. 1974. *Strategy, Structure and Economic Performance*. Cambridge, MA: Harvard University Press.

Runyon, K.L. 1991. *Canada's Timber Supply: Current Status and Outlook*. Ottawa: Forestry Canada, Information Report E-X-45.

Sabel, C.F., and J. Zeitlin. 1985. Historical alternatives to mass production: Politics, markets and technology in nineteenth century industrialization. *Past and Present* 108: 133-76.

Safarian, A.E. 1979. Foreign ownership and industrial behaviour: A comment on the "weakest link." *Canadian Public Policy* 3: 318-35.

Saurin, J. 1996. International relations, social ecology and the globalisation of environmental change. In *Environment and International Relations*, ed. J. Vogler and M.F. Imber, 77-98. London: Routledge.

Schoenberger, E. 1987. Technical and organisational changes in automobile production: Spatial implications. *Regional Studies* 21: 199-214.

Schumpeter, J.A. 1943. *Capitalism, Socialism and Democracy*. New York: Harper.

Schwindt, R. 1977. *The Existence and Exercise of Corporate Power – A Case Study of MacMillan Bloedel Ltd*. Ottawa: Supply and Services Canada.

–. 1979. Pearse Commission and the Industrial Organization of the British Columbia Forest Industry. *B.C. Studies* 41: 3-35.

–. 1987. British Columbia forest sector: The pros and cons of the stumpage system. In *Resource Rents and Public Policy in Western Canada*, ed. T. Gunton and J. Richards, 181-214. Halifax: Institute for Research on Public Policy.

–, and T. Heaps. 1996. *Chopping up the Money Tree: Distributing the Wealth from British Columbia's Forests*. Vancouver: David Suzuki Foundation.

Science Council of British Columbia. 1983a. *Creating the Climate for Innovation*. Vancouver: Spark Report (Planning reports of the forests and forest products sector committee).

–. 1983b. *Increased Value from a Changing Resource*. Vancouver: Spark Report (Planning reports of the forests and forest products sector committee).

–. 1989. *Forestry Research and Development in British Columbia: A Vision for the Future*. Vancouver: Forest Planning Committee.

Science Council of Canada. 1992. *Canadian Forest Products*. Technology Sector Strategy Series No. 9. Ottawa: Ministry of Supply and Services.

Scott, A.J. 1988. *New Industrial Spaces*. London: Pion.

–, and M. Storper. 1986. *Production, Work, Territory: The Geographical Implications of Industrial Capitalism*. London: Allen Unwin.

Seager, A. 1988. Workers, class and industrial conflict in New Westminster, 1900-1930. In *Workers, Capital and the State in British Columbia: Selected Papers*, ed. P. Warburton and D. Coburn, 117-40. Vancouver: UBC Press.

Sector Task Force. 1978. *Canadian Forest Products Industry*. Ottawa: Department of Industry, Trade and Commerce.

Servan-Schreiber, J.J. 1968. *The American Challenge*. London: Hamish Hamilton.

Shearer, R.A., J.H. Young, and G.R. Munro. 1973. *Trade Liberalization and a Regional Economy: Studies of the Impact of Free Trade on British Columbia*. Toronto: University of Toronto Press.

Sloan, G. 1945. *Report of the Commissioner Relating to the Forest Resources of British Columbia*. Victoria: King's Printer.

–. 1956. *Forest Resources of British Columbia: Report of the Commissioner*. Vols. 1 and 2. Victoria: Queen's Printer.

Silversides, C.R. 1984. Mechanized forestry: World War II to the present. *Forestry Chronicle* August: 231-5.

Simpson, S. 1997. Mission leads municipal logging trend. *Vancouver Sun* November 22: B1.

Sinclair, W.F. 1974. *The Socioeconomic Importance of Maintaining Quality of Recreational Resources in Northern British Columbia: The Case of Lakelse Lake*. Vancouver: Evergreen Press.

Sjoholt, S. 1987. New trends in promoting regional development in local communities in Norway. In *International Economic Restructuring and the Regional Community*, ed. H. Muegge and W. Stohr, 277-93. Aldershot: Avebury Press.

Smith, J.G., and G. Lessard. 1970. *Forest Resources Research in Canada*. Background Study No. 14. Science Council of Canada. Ottawa: Information Canada.

Solandt, O.M. 1979. *Forest Research in Canada*. Ottawa: Canadian Forestry Advisory Council.
Sopow, E. 1984. Running out of trees. *Province* March 11: C22.
Sorensen, J. 1990. Konnichi wa, Interfor. *BC Business* September: 38-44.
Soyez, D. 1988. Scandinavian silviculture in Canada: Entry and performance barriers. *Canadian Geographer* 32: 133-40.
Spry, I.M. 1981. Overhead costs, rigidities of productive capacity and the price system. In *Culture, Communication and Dependency,* ed. W.H. Melody, L. Salter, and P. Heyer, 155-66. Norwood, NJ: Ablex Publishing Corporation.
Starks, R. 1972. BCFP complex opens up inaccessible north. *Canadian Pulp-and-Paper Industry* August: 28.
Statistics Canada, cat. 31-203. 1973. 1979. 1989. 1996. *Manufacturing Industries of Canada: National and Provincial Areas*. Ottawa: Statistics Canada.
–, cat. 01-517. 1977. *Foreign Ownership of Canadian Industry.* Ottawa: Statistics Canada.
–, cat. 61-205 and 61-206. 1990. 1993. 1996. *Public and Private Investment in Canada*. Ottawa: Statistics Canada.
–, cat. 25-202-XPB. 1996. *Canadian Forestry Statistics*. Ottawa: Statistics Canada.
Steen, H.K., ed. 1984. *History of Sustained-Yield Forestry: A Symposium*. Portland: Forest History Society.
Stephens, T. 1999. Finding your way in the woods. *Vancouver Sun* April 2: A15.
Stoffman, O. 1990. A cut above. *Globe and Mail Report on Business*, November: 98-106.
Storper, M., and R. Salois. 1997. *Worlds of Production*. Cambridge, MA: Harvard University Press.
Streeck, W. 1989. Skills and the limits to neoliberalism: The enterprise of the future as a place of learning. *Work, Employment and Society* 3: 89-104.
–. 1992. Training and the new industrial relations: A strategic role for unions? In *The Future of Labour Movements*, ed. M. Regini, 250-69. London: Sage.
Swift, J. 1983. *Cut and Run*. Toronto: Between the Lines.
Swyngedouw, E. 1997. Neither global nor local: "Glocalization" and politics of scale. In *Spaces of Globalization,* ed. K.R. Cox, 137-66. New York: Guilford.
Taylor, O.W. 1975. *Timber: History of the Forest Industry in British Columbia*. Vancouver: J.J. Douglas.
Taylor, R.P. 1996. Democracy and environmental ethics. In *Democracy and the Environment,* ed. W.M. Lafferty and J. Meadowcroft, 86-107. Cheltenham: Edward Elgar.
Thirgood, J.V. 1981. *Man and the Mediterranean Forest.* London: Academic Press.
Tiebout, C.M. 1956. Exports and regional economic growth. *Journal of Political Economy* 64: 160-69.
Tillman, D.A. 1985. *Forest Products: Advanced Technologies and Economic Analyses*. New York: Academic Press.
Times-Colonist. 1999. Lumber fight all about power. *Times-Colonist* May 14: A12.
Totman, C. 1989. *Green Archipelago*. Berkeley: University of California Press.
Toyama, W.T. 1993. *A Study of the Japanese Market for Manufactured Building Products*. Canadian Embassy, Tokyo.
Travers, O.R. 1993. Forest policy: Rhetoric and reality. In *Touch Wood: BC Forests at the Crossroads,* ed. K. Drushka, B. Nixon, and R. Travers, 171-224. Madeira Park, BC: Harbour Publishing.
Trist, E. 1979. New directions of hope: Recent innovations interconnecting organizational, industrial, community and personal development. *Regional Studies* 13: 439-51.
Uhler, R.S., G.M. Townsend, and L. Constantino. 1991. Canada-US trade and the product-mix of the Canadian pulp-and-paper industry. In *Canada-United States Trade in Forest Products,* ed. R.S. Uhler, 106-22. Vancouver: UBC Press.
Vancouver Sun. 1998. Forestry needs more than a door stopper. *Vancouver Sun* November 10: A12.
–. 1999. Forestry needs more than federal money. *Vancouver Sun* May 14: A14.
Varty, Alex. 1997. High tech marries low tech. *The Georgia Straight* November 20-27: 15, 17.
Wallis, A., D. Stokes, G. Westcott, and T. McGee. 1997. Certification and labelling as a new tool for sustainable forest management. *Australian Journal of Environmental Management* 4: 224-38.

Watkins, M.H. 1963. A staple theory of economic growth. *Canadian Journal of Economics and Political Science* 29: 141-58.
–. 1981. The Innis tradition in Canadian political economy. *Canadian Journal of Political and Social Theory* 6: 12-34.
Westoby, J. 1989. *Introduction to World Forestry*. Oxford: Basil Blackwell.
Weyerhaeuser Canada Ltd. 1975. *Submission to the Royal Commission on Forest Resources*. Vancouver.
Wilkinson, B.W. 1997. Globalization of Canada's resource sector: An Innisian perspective. In *Troubles in the Rainforest: British Columbia's Forest Economy in Transition,* ed. T. Barnes and R. Hayter, 131-47. Canadian Western Geographical Series No. 33. Victoria: Western Geographical Press.
Willems-Braun, B. 1997a. Colonial vestiges: Representing forest landscapes on Canada's west coast. In *Troubles in the Rainforest: British Columbia's Forest Economy in Transition,* ed. T. Barnes and R. Hayter, 99-127. Canadian Western Geographical Series No. 33. Victoria: Western Geographical Press.
–. 1997b. Buried epistemologies: The politics of nature in "post" colonial British Columbia. *Annals of the Association of American Geographers* 87: 3-31.
Wilson, F. 1997. B.C. strike pits workers against foreign capital. *Vancouver Sun* July 24: A13.
Wilson, J. 1987/88. Forest conservation in British Columbia, 1935-85: Reflections on a barren political debate. *B.C. Studies* Winter: 3-32.
–. 1997. Implementing forest policy change in British Columbia: Comparing the experiences of the NDP governments of 1972-75 and 1991-?. In *Troubles in the Rainforest: British Columbia's Forest Economy in Transition,* ed. T. Barnes and R. Hayter, 75-97. Canadian Western Geographical Series No. 33. Victoria: Western Geographical Press.
–. 1998. *Talk and Log: Wilderness Politics in British Columbia, 1965-96*. Vancouver: UBC Press.
Woodbridge, P. 1988. Marketing and production issues in the pulp-and-paper industry of the future. In *Global Issues and Outlook in Pulp and Paper,* ed. G.F. Schreuder, 276-84. Seattle: University of Washington Press.
Woodbridge, Reed and Associates. 1984. *British Columbia's Forest Products Industry: Constraints to Growth*. Report prepared for the Ministry of State for Economic and Regional Development. Vancouver.
Yaffe, B. 1999. Clark and company too unruly for Ottawa to handle. *Vancouver Sun* May 26: A15.
Yamakawa, M. 1993. New trends in home building. *Life in Contemporary Japan* 3: 45-52.
Young, A. 1979. *The Sogo Shosha: Japan's Multinational Trading Companies*. Boulder, CO: Westview Press.

Index